T0073504

Tauseef Gulrez and Aboul Ella Hassanien (Eds.)

Advances in Robotics and Virtual Reality

Intelligent Systems Reference Library, Volume 26

Editors-in-Chief

Prof. Janusz Kacprzyk
Systems Research Institute
Polish Academy of Sciences
ul. Newelska 6
01-447 Warsaw
Poland
E-mail: kacprzyk@ibspan.waw.pl

Prof. Lakhmi C. Jain
University of South Australia
Adelaide
Mawson Lakes Campus
South Australia 5095
Australia
E-mail: Lakhmi.jain@unisa.edu.au

Further volumes of this series can be found on our homepage: springer.com

Tauseef Gulrez and Aboul Ella Hassanien (Eds.)

Advances in Robotics and Virtual Reality

 Springer

Tauseef Gulrez
COMSATS Institute of Information
Technology
Department of Computer Science
M.A. Jinnah Campus
Defence Road
Lahore, Pakistan
E-mail: gtauseef@ieee.org

Aboul Ella Hassanien
Cairo University
Faculty of Computer and Information
Information Technology Department
5 Ahmed Zewal St.
Orman, Giza
E-mail: Aboitcairo@gmail.com

ISBN 978-3-642-23362-3   e-ISBN 978-3-642-23363-0

DOI 10.1007/978-3-642-23363-0

Intelligent Systems Reference Library   ISSN 1868-4394

Library of Congress Control Number: 2011937018

© 2012 Springer-Verlag Berlin Heidelberg

Typeset & Cover Design: Scientific Publishing Services Pvt. Ltd., Chennai, India.

Printed on acid-free paper

9 8 7 6 5 4 3 2 1

springer.com

Foreword

Perhaps you've noticed. Our society has allowed something once feared to take a prominent role in our daily lives. This transition was gradual at first, but the proliferation of robotic systems now common in our everyday lives cannot be ignored. And rather than fear this encroachment, we have embraced it. Why? Because robotic devices, once depicted in books and movies as capable of taking over the world, then relegated to the drudgery of the assembly line, have shown to be quite beneficial in many domains.

In our homes, robots sweep our floors and mow our lawns. At the local hospital, robots facilitate minimally invasive surgical procedures, which in turn reduce hospital stays and speed recovery. Physical therapists employ robotic systems to rehabilitate walking and arm movements after stroke and spinal cord injury. And military organizations use robotic systems extensively to augment troops, providing additional surveillance, search and rescue, and even advanced artificial limbs when life-changing injuries occur.

A similar shift is seen in the field of virtual reality (VR). Once the fodder of science fiction authors and playwrights, virtual reality technology now enables advanced simulation-based training of doctors and military personnel, not to mention the entertainment of millions of teenagers! Virtual reality has enabled research advances, and has been conceived as a tool to liberate consciousness. With its capacity to allow the creation and control of dynamic three-dimensional, ecologically-valid stimulus environments, within which behavioral responses can be recorded and measured, VR offers clinical assessment and rehabilitation options that are not available using traditional methods.

These shifts in acceptance, coupled with significant advances in enabling technologies, have facilitated amazing achievements in the fields of robotics and virtual reality, a sample of which are described in this book. Advancements in sensor and actuator technologies, materials, imaging, computational engineering, and algorithms have taken robots out of the controlled, programmable assembly line environments, and thrust them into the mainstream. Increased investments in research by governments around the globe have pushed technological advancements in the areas of medicine and health care; autonomous navigation by robotic vehicles on land, sea, and air; natural interactions between robots, computers, and humans; and augmented virtual reality systems that operate on mobile, real-time platforms.

In this book, Professors Gulrez and Hassnein have identified cutting edge research in the fields of robotics and virtual reality, and give an excellent overview of the state-of-the-art. Their goal in compiling this text was to exchange ideas and stimulate research at the intersection of robotics and virtual reality. They succeed in presenting technologies, algorithms, system architectures, and applications at this intersection, and whetting the reader's appetite for what may be commonplace in the future.

As a robotics and virtual reality researcher who has had the privilege to collaborate with roboticists and engineers at NASA and a number of esteemed surgeons of the Texas Medical Center, I remember a time when the topics I was studying were difficult for family members to believe. Humanoid robots, haptic feedback technology, surgical robots -- this was the stuff of science fiction. Now, it is popular science. A few years ago my father benefited from minimally invasive robotic surgery. A few months ago my sons watched in amazement as R2, a humanoid robot developed by NASA and General Motors, launched into space to become an astronaut assistant on the International Space Station. What is next? Read on...

Marcia K. O'Malley, PhD
Rice University
Director, Mechatronics and Haptics Interfaces Laboratory
Chair, IEEE Technical Committee on Haptics
Past Chair, ASME Dynamic Systems and
Controls Division Robotics Technical Committee

Preface

This book brings together topics from two widely discussed subject areas of the 21st century, i.e. Robotics and Virtual Reality (VR). The coupling of Robotics and VR with recent advances in both fields has made a paradigm shift for a number of application e.g. rehabilitation robotics (including brain machine interfaces), training simulations and artificial environments etc.

The development of advanced robotic and VR systems has set forth an exponentially increasing number of experiments aimed at the creation of user friendly computer systems and restoration of quality of human life. These developments have also revealed the lack of presentation of these advances in the form of a comprehensive tool. This book purposely presents the most recent advances in robotics and VR known to date from the perspective of some of the leading experts in the field and presents an interesting array of developments put into fifteen comprehensive chapters. The chapters are presented in a way that the reader will get a seamless impression of the current concepts of optimal modes of both experimental and applicable roles of robotic devices.

The uniqueness of the book is a considerable emphasis on practical applicability in solving real-time problems. There is also an in depth look at the role of robotics from a mechatronics and virtual reality standpoint. The book is divided into three parts, where it begins by exploring the inherent and unique challenges of rehabilitation and presents the robotic platforms upon which promising preliminary results were noted. It then explores the key elements of interaction between robot and human control, an area of great concern in the computer field especially in digital robotic media at present. The third section involves the interaction-control between artificial environments and humans. VR promises to be a valuable supplementary tool to all those involved in non-destructive testing. It also acts as a platform upon which researchers can gain a solid and evidence based approach towards the initiation of future projects.

We assured our best that the book would provide a comprehensive knowledge of the field of robotics and VR and would satisfy a large group of readers, including researchers in the field, graduate and postgraduate students, and designers that use the robotics and 3D VR technology. We believe that the book will become a representative selection of the latest trends and achievements on the robotics area. We would be extremely happy if such an important goal would be achieved.

This Volume comprises of 15 chapters including an overview chapter providing an up-to-date and state-of-the research on the application of robotics, human-robotic interaction using virtual tools, and advances in virtual reality.

The book is divided into 3 parts

Part-I: Advanced Robotic Systems in Practice
Part-II: Enabling Human Computer Interaction
Part-III: 3D Virtual Reality Environments

Part-I on Advanced Robotic Systems in Practice contains five chapters that describe several robotics applications.

In Chapter (1), "High-field MRI-Compatible Needle Placement Robots for Prostate Interventions: Pneumatic and Piezoelectric Approaches" Hao Su et.al reviews two distinct MRI-compatible approaches for image-guided transperineal prostate needle placement. It articulates the robotic mechanism, actuator and sensor design, controller design and system integration for a pneumatically actuated robotic needle guide and a piezoelectrically actuated needle placement system. The two degree-of-freedom (DOF) pneumatic robot with manual needle insertion has a signal to noise ratio (SNR) loss limited to 5% with alignment accuracy under servo pneumatic control better than 0:94mm per axis. While the 6-DOF piezoelectrically actuated robot is the first demonstration of a novel multi piezoelectric actuator drive with less than 2% SNR loss for high-field MRI operating at full speed during imaging. Preliminary experiments in phantom studies evaluate system MRI compatibility, workflow, visualization and targeting accuracy.

Chapter (2), "A Robot for Surgery: Design, Control and Testing" by Basem Yousef and Rajni Pate designed a dexterous robot arm with a sophisticated configuration and joint-structure to perform as a gross positioner that can carry, accurately position and orient different types of surgical tools and end effectors for image-guided robot-assisted therapy.

In Chapter (3), "Force Sensing and Control In Robot-Assisted Suspended Cell Injection System" Haibo Huang et.al. presents a robotic cell-injection system for automatic injection of batch-suspended cells. To facilitate the process, these suspended cells are held and fixed to a cell array by a specially designed cell holding device, and injected one by one through an "out-of-plane" cell injection process. Starting from image identifying the embryos and injector pipette, a proper batch cell injection process, including the injection trajectory of the pipette, is designed for this automatic suspended cell injection system. A micropipette equipped with a PVDF micro force sensor to measure real time injection force, is integrated in the proposed system.

In Chapter (4), "Advanced Hybrid Technology for Neurorehabilitation: the HYPER Project" by Alessandro De Maur et.al present a development of a new rehabilitation therapy based on an integrated ER-MNP hybrid systems combined with virtual reality and brain neuro-machine interface (BNMI). This solution, based on improved cognitive and physical human-machine interaction, aims to overcome the major limitations regarding the current available robotic-based therapies.

In Chapter (5), "Toward Bezier Curve Bank-Turn Trajectory Generation for Flying Robot" by Affiani Machmudah et. al. presents a UAV maneuver planning with an objective is to minimize a total maneuvering time and a load factor when the UAV follows an agile maneuvering path. The Genetic algorithm (GA) combined by fuzzy logic approach will lead this challenging work to meet the best maneuvering trajectories in an UAV operational area. It considers UAV operational limits such as a minimum speed, a maximum speed, a maximum acceleration, a maximum roll angle, a maximum turn rate, as well as a maximum roll rate. The feasible maneuvering path will be searched first, and then it will be processed to generate the feasible flight trajectories. The intelligent GA-fuzzy approach has succeeded to discover the best trajectory which satisfies all UAV constraints when they follow the agile maneuvering path. The changing speed strategy is very challenging to reduce the load factor during the UAV maneuver. To achieve the target mission, the fuzzy membership function can be constructed as the guidance to keep the GA in the right way

Part-II on **Enabling Human Computer Interaction** contains five chapters discussing many approaches in human computer interaction including audio-visual speech processing; pervasive interaction based assistive technology, and Gesture recognition in conceptual design.

Chapter (6), "Audio-Visual Speech Processing for Human Computer Interaction" by Siew Wen Chin et.al. by presents an audio-visual speech recognition (AVSR) for Human Computer Interaction (HCI) that mainly focuses on 3 modules: (i) the radial basis function neural network (RBF-NN) voice activity detection (VAD) (ii) the watershed lips detection and H∞ lips tracking and (iii) the multi-stream audio-visual back-end processing. The importance of the AVSR as the pipeline for the HCI and the background studies of the respective modules are first discussed follow by the design details of the overall proposed AVSR system. Compared to the conventional lips detection approach which needs a prerequisite skin/non-skin detection and face localization, the proposed watershed lips detection with the aid of H∞ lips tracking approach provides a potentially time saving direct lips detection technique, rendering the preliminary criterion obsolete. Alternatively, with a better noise compensation and a more precise speech localization offered by the proposed RBF-NN VAD compared to the conventional zero-crossing rate and short-term signal energy, it has yield to a higher performance capability for the recognition process through the audio modality. Lastly, the developed AVSR system which integrates the audio and visual information, as well the temporal synchrony audio-visual data stream has proved to obtain a significant improvement compared to the unimodal speech recognition, also the decision and feature integration approaches.

Chapter (7), "Solving Deceptive Tasks in Robot Body-Brain Co-Evolution By Searching For Behavioral Novelty", by Peter Krcah applies novelty search to the problem of body-brain co-evolution and demonstrate that novelty search significantly outperforms fitness-based search in a deceiving barrier avoidance task but does not provide an advantage in the swimming task where a large unconstrained behavior space inhibits its efficiency. The authors show that the advantages of novelty search previously demonstrated in other domains can also be utilized in the more complex domain of body-brain co-evolution, provided that the task is deceiving and behavior space is constrained.

In Chapter (8), "From Dialogue Management to Pervasive Interaction based Assistive Technology" by Yong Lin and Fillia Makedon introduces a pervasive interaction based planning and reasoning system for individuals with cognitive impairment, for their activities of daily living. The introduced system is a fusion of speech prompt, speech recognition as well as events from sensor networks. The system utilizes Markov decision processes for activity planning, and partially observable Markov decision processes for action planning and executing. Multimodal and multi-observation is the characteristics of a pervasive interaction system. Experimental results demonstrate the flexible effect the reminder system works for activity planning.

Chapter (9), "De-SIGN: Robust Gesture Recognition in Conceptual Design, Sensor Analysis and Synthesis" by Manolya Kavakli1 and Ali Boyali1 developed a Gesture Recognition System (De-SIGN) in various iterations. De-SIGN decodes design gestures. In this chapter, we present the system architecture for De-SIGN, its sensor analysis and synthesis method (SenSe) and the Sparse Representation-based Classification (SRC) algorithm have developed for gesture signals, and discussed the system's performance providing the recognition rates. The gesture recognition algorithm presented here is highly accurate regardless of the signal acquisition method used and gives excellent results even for high dimensional signals and large gesture dictionaries. They findings that gestures can be recognized with over 99% accuracy rate using the Sparse Representation-based Classification (SRC) algorithm for user-independent gesture dictionaries and 100% for user-dependent.

Chapter (10), "Image segmentation of cross-country scenes captured in IR spectrum" by Artem Lenskiy elaborates on the problem of segmenting cross-country scene images in IR spectrum using salient features. Salient features are robust to variations in scale, brightness and angle of view. As for salient features the author chooses Speeded-Up Robust Features (SURF). The author provides a comparison of two SURF implementations. SURF features are extracted from input image and for each of features terrain class membership values are calculated. The values are obtained by means of multi-layer perception. The features class membership values and their spatial positions are then applied to estimate class membership values for all pixels in the image. The values are used to assign a terrain class to each pixel in the image. To decrease the effect of segmentation blinking and speed up segmentation, the system is tracking camera position and predicts positions of features.

Part-III on **Part-III 3D Virtual Reality Environments contains five chapters** discussing many virtual reality applications including articulated robots, GPU-based real time virtual reality modeling, virtual reality technology for blind and virtual impaired people

In Chapter (11), "Virtual Reality as a Surrogate Sensory Environment" Theodore Hall et.al. examine certain aspects of virtual reality systems that contribute to their utility as surrogate sensory environments. These systems aim to provide users with sensory stimuli that simulate other worlds. The fidelity of the simulation depends on the input data, the software, the display hardware, and the physical environment that houses it all. Robust high-fidelity general-purpose simulation requires a collaborative effort of modelers, artists, programmers, and system administrators. Such collaboration depends on standards for modeling and data representation, but these standards lag behind the leading-edge capabilities of processors and algorithms. We illustrate this through a review of the evolution of a few of the leading standards and case studies of projects that adhered to them to a greater or lesser extent. Multi-modal simulation often requires multiple representations of elements to accommodate the various algorithms that apply to each mode.

In Chapter (12), "Operating High-DoF Articulated Robots Using Virtual Links and Joints" by Marsette Vona presents the theory, implementation, and application of a novel operations system for articulated robots with large numbers (10s to 100s) of degrees-of-freedom (DoF), based on virtual articulations and kinematic abstractions. Such robots are attractive in some applications, including space exploration, due to their application exibility. But operating them can be challenging: they are capable of many different kinds of motion, but often this requires coordination of many joints. Prior methods exist for specifying motions at both low and high-levels of detail; the new methods fill a gap in the middle by allowing the operator to be as detailed as desired. The presentation is fully general and can be directly applied across a broad class of 3D articulated robots.

In Chapter (13), "GPU-Based Real-Time Virtual Reality Modeling and Simulation of Seashore" Minzhi Luo et. Al., devoted to efficient algorithms for real-time rendering of seashore which take advantage of both CPU calculation and programmable Graphics Processing Unit (GPU). The scene of seashore is a usual component of virtual environment in simulators or games and should be realistic and real-time. With regard to the modeling of seashore, concept models based on Unified Modeling Language (UML) as well as precise mathematical models of ocean wave, coastline and breaking wave are presented. While regarding the simulation of seashore, optics effects imitation of the ocean surface and the simulation of foam and spray are realized.

Digitization of real objects into 3D models is a rapidly expanding field, with ever increasing range of applications. The most interesting area of its application is in the creation of realistic 3D scenes, i.e. virtual reality. On the other hand, virtual reality systems are typical examples of highly modular systems of data processing. These modules are considered as subsystems dealing with selected types of inputs from the modeled world, or producing this world including all providing effects respectively. **Chapter (14)** "Data Processing for Virtual Reality" by Csaba Szabó, et.al., focuses on data processing tasks within a virtual reality system. Data processing tasks considered are as follows: 3D modeling and 3D model acquisition; spoken term recognition as part of the acoustic subsystem; and visualization, rendering and stereoscopy as processes of output generation from the visual subsystem.

Virtual reality technology enables people to become immersed in a computer-simulated and three-dimensional environment. In **Chapter (15) on** "Virtual reality technology for blind and visual impaired people: recent advances and future directions" Neveen Ghali et.al., investigate the effects of the virtual reality technology on disabled people such as blind and visually impaired people (VIP) in order to enhance their computer skills and prepare them to make use of recent technology in their daily life. As well as, they need to advance their information technology skills beyond the basic computer training and skills. This chapter describes what best tools and practices in information technology to support disabled people such as deaf-blind and visual impaired people in their activities such as mobility systems, computer games, accessibility of e-learning, web-based information system, and wearable finger-braille interface for navigation of deaf-blind. Moreover, the authors show how physical disabled people can benefits from the innovative virtual reality techniques and discuss some representative examples to illustrate how virtual reality technology can be utilized to address the information technology problem of blind and visual impaired people. Challenges to be addressed and an extensive bibliography are included.

We are very much grateful to the authors of this volume and to the reviewers for their great efforts by reviewing and providing interesting feedback to authors of the chapter. The editors would like to thank Dr. Thomas Ditzinger (Springer Engineering Inhouse Editor, Intelligent Systems Reference Library Series), Professor Janusz Kacprzyk (Editor-in-Chief, Springer Intelligent Systems Reference Library Series) for the editorial assistance and excellent cooperative collaboration to produce this important scientific work. We hope that the reader will share our joy and will find it useful!

June 2011   Editors

Tauseef Gulrez
COMSATS Institute of Information Technology,
Lahore, Pakistan

Aboul Ella Hassanien
Cairo University, Egypt

Contents

Part III: 3D Virtual Reality Environments

Part I

Advanced Robotic Systems in Practice

High-Field MRI-Compatible Needle Placement Robots for Prostate Interventions: Pneumatic and Piezoelectric Approaches

Hao Su, Gregory A. Cole, and Gregory S. Fischer

Automation and Interventional Medicine (AIM) Laboratory,
Department of Mechanical Engineering, Worcester Polytechnic Institute,
Worcester, MA, 01609, USA
Tel.: (001)508-831-5261
{haosu,gfischer}@wpi.edu
http://aimlab.wpi.edu

Abstract. Magnetic resonance imaging (MRI) can be a very effective imaging modality for live guidance during surgical procedures. The rationale of MRI-guided surgery with robot-assistance is to perform surgical interventions utilizing "real-time" image feedback while minimize operation time and improves the surgical outcomes. However, challenges arise from electromagnetic compatibility within the high-field (1.5T or greater) MRI environment and mechanical constraints due to the confined close-bore space. This chapter reviews two distinct MRI-compatible approaches for image-guided transperineal prostate needle placement. It articulates the robotic mechanism, actuator and sensor design, controller design and system integration for a pneumatically actuated robotic needle guide and a piezoelectrically actuated needle placement system. The two degree-of-freedom (DOF) pneumatic robot with manual needle insertion has a signal to noise ratio (SNR) loss limited to 5% with alignment accuracy under servo pneumatic control better than $0.94mm$ per axis. While the 6-DOF piezoelectrically actuated robot is the first demonstration of a novel multi piezoelectric actuator drive with less than 2% SNR loss for high-field MRI operating at full speed during imaging. Preliminary experiments in phantom studies evaluates system MRI compatibility, workflow, visualization and targeting accuracy.

1 Introduction

Prostate cancer is the most common cancer in males and the second most commonly found cancer in human. The estimated new prostate cancer cases (192,280) in 2009 account for 25% incident cases in men in the United States [24]. Each year approximately 1.5 million core needle biopsies are performed, yielding about 220,000 new prostate cancer cases. Over 40,000 brachytherapy radioactive seed implantation procedures are performed in the United States each year, and the number is steadily rising. Prostate specific antigen blood

T. Gulrez, A.E. Hassanien (Eds.): Advances in Robotics & Virtual Reality, ISRL 26, pp. 3–32.
springerlink.com   © Springer-Verlag Berlin Heidelberg 2012

tests and digital rectal exams are the two major preliminary prostate cancer diagnosis methods. However, prostate biopsy is the conclusive approach to confirm cancer diagnosis. Prostate brachytherapy and cryotherapy are often used for early treatment.

Magnetic resonance imaging can provide high resolution multi-parametric imaging, large soft tissue contrast, and interactive image updates making it an ideal guidance modality for needle-based prostate interventions. In this chapter, two MRI-compatible robotic assistants utilizing pneumatic and piezoelectric actuation for needle placement are designed for image-guided prostate interventions. Both systems are intended to provide increased positioning accuracy through the use of precision motion and image feedback. The actuated systems enable acquisition of interactively updated images during robot motion in situ. These images enable real-time guidance and of needle tracking in MRI which may be used for image-guided closed loop control of needle placement.

The motivation of deploying robotic system in prostate interventions comes from three major aspects. First, robot-assisted surgery guarantees good geometric accuracy as it relies upon rigid mechanical structures and motion/force control of the robot, thus overcoming the accuracy limits of traditional mechanical templates ($5mm$) and allowing needle angulation. Second, robots can be designed with appropriate scale to fit into open scanner bores or cylindrical closed-bores. Third, since robots are stable and untiring, they can stay inside the scanner bore and perform interventional procedures via teleoperated control more easily than human surgeons. This eliminates or reduces the necessity of the iterative and time-consuming procedure that usually takes several cycles of imaging for registration inside bore, interventions out of bore (due to space limit) and confirmation inside bore. In general, MRI-guided surgery with robot-assistance can ultimately minimize operation time and maximize placement accuracy thus improves the surgical outcomes. Moreover, it can greatly reduce the equipment cost and overhead.

Contents of the Chapter: This chapter is organized as follows: Section 3 analyzes and articulates the design requirement. Section 4 describes the 2-DOF pneumatic needle placement robot design with manual insertion and Section 5 describes the 6-DOF fully actuated needle placement robot design with piezoelectric actuation. Section 6 presents the system architecture and workflow utilizing these two robots. A number of experiments in 3 Tesla closed-bore scanner and analysis are elaborated in Section 7. Finally, a discussion of the system is presented in Section 8 that summarizes the experimental insight and future work.

2   Background and Literature Review

Transrectal ultrasound (TRUS) is the current "gold standard" for guiding both biopsy and brachytherapy due to its real-time nature, low cost and ease of use. Fig. 1 shows the traditional template-based TRUS-guided approach to brachytherapy seed placement. However, TRUS-guided biopsy has a detection rate of only $20 - 30\%$ [47]. The low success rate is largely due to the inherent limitation of ultrasound imaging itself and the mechanical template used in the procedure to guide the needles. Ultrasound imaging is inferior to MRI for prostate cancer treatment due to its limited resolution and inability to display implanted radiation seeds. Furthermore, the ultrasound probe deforms the prostate and can induce seed migration. Since the dosimetry plan is usually performed on the manual segmentation of pre-operative ultrasound images, the seed placement accuracy is deteriorated because of the low image quality and probe induced tissue deformation. On the other hand, the template used to guide needles in TRUS is a $6cm \times 6cm$ mechanical grid with $5mm \times 5mm$ space distance which limits the entrance location at the perineum and the positioning resolution of needle insertion. In addition, this mechanically constrained template limits the ability for needle angulation to adjust needle orientation for better targeting or to avoid pubic arch interference (PAI), thus restricting the eligible patient population.

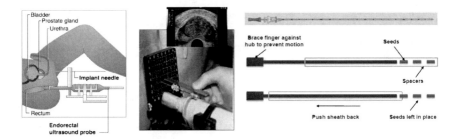

Fig. 1. Traditional template-based TRUS-guided approach to prostate brachytherapy (left and middle). Brachytherapy needle from CP Medical and the schematic procedure of preloaded needles: after insertion, the sheath is withdrawn over the stylet, leaving the seeds in the place (modified from [51]) (right).

Conversely, the MRI-based medical paradigm offers several advantages over other imaging counterparts. First, MRI has multiple mechanisms to provide high-fidelity soft tissue contrast and spatial resolution. Second, MRI is a three-dimensional imaging modality that permits arbitrary imaging plane selection, even in a dynamic manner. Third, MRI does not produce ionizing radiation thus imposes no radiation safety hazard to the patient or practitioner. The clinical efficacy of MRI-guided prostate brachytherapy and biopsy was demonstrated by D'Amico *et al.* at the Brigham and Women's Hospital using a 0.5T open-MRI scanner [10]. MR images were used to plan and monitor transperineal needle placement. The needles were inserted manually using a guide comprising a grid of holes, with the patient in the lithotomy position, similarly to the TRUS-guided approach.

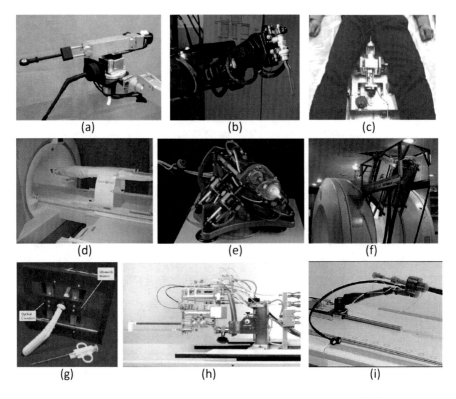

Fig. 2. MRI guided robots and assistance device: Masamune *et al.* [30] (a), Suther-land *et al.* [45](b), Goldenberg *et al.* [20] (c), INNOMOTION [31] (d), Stoianovici *et al.* (e), Chinzei *et. al* [7] (f), Elhawary *et. al* [13] (g), van den Bosch *et. al* [5] (h) and Krieger *et al.* [26] (i).

The challenges, however, arise from electromagnetic compatibility in the high-field (1.5T or greater) MRI environment and mechanical constraints due to the confined closed-bore space. Manifestation of bidirectional MRI compatibility requires that both the device not disturb the scanner function and not create image artifacts and the scanner should not disturb the device functionality. Thus, in the actuation level, conventional DC/AC motors that reply on electromagnetism for conventional robotic actuation is not feasible in MRI. In the material level, conductive materials such as metallic components can induce heating thus special means should be taken to avoid the side-effects of these materials or even avoid using them. Moreover, the confined physical space in closed-bore high-field MRI presents stringent challenges for mechanical design.

Thorough reviews of MRI-compatible systems to date for image-guided interventions are presented by Tsekos, *et al.* [50] and by Elhawary, *et al.* [12]. The potential role of robotics in MRI-guided prostate interventions was thoroughly discussed in [17] and a clinical overview of MRI-guided biopsy of the prostate is presented in [33]. Fig. 2 shows a collection of the MRI-compatible assistants for prostate interventions developed to date. Robotic assistance has been investigated for guiding instrument placement in MRI, beginning with neurosurgery [30]. A most recent MRI-guided neurosurgical robot [45] was developed by Sutherland *et. al.* Goldenberg *et al.* [20] proposed an MRI-compatible robotic system for image-guided prostatic interventions. INNOMOTION [31] is the first commercially available MRI-guided robot that utilizes pneumatic actuation and was recently acquired by Synthes©. Stoianovici *et al.* [37] described a MRI-compatible pneumatic stepper motor PneuStep, which has a very low level of image interference. Chinzei *et. al* [7] developed a general-purpose robotic assistant for open MRI that was subsequently adapted for transperineal intraprostatic needle placement [11]. Elhawary *et. al* [13] performed transrectal prostate biopsy. van den Bosch *et. al* [5] performed MRI-guided needle placement and seed delivery in the prostate. Krieger *et al.* [26] presented a 2-DOF passive, un-encoded, and manually manipulated mechanical linkage to aim a needle guide for transrectal prostate biopsy with MRI guidance. Song *et. al* [36] presented a pneumatic robot for MRI-guided transperineal prostate biopsy and brachytherapy. Two recent developments of pneumatically actuated robots [34,5] further investigated the actuation feasibility in terms of compatibility and accuracy. The last decade also witnessed the development of MRI-guided robots for other surgical applications, including cardiac surgery [56,29,28] and breast cancer diagnosis [55], etc.

Generally, there are four actuation principles utilized in MRI applications, namely remote actuation, hydraulic, pneumatic and ultrasonic/piezoelectric actuators [19]. A substantial armamentarium of actuation methods has been studied in a number of image-guided interventions and rehabilitation scenarios. Early MRI-compatible robotic systems focus on manual driven or ultrasonic motor driven and the latter cannot run during imaging due to significant signal loss. Remote actuation suffers from bulky structure, low bandwidth and lower resolution and is not preferable for robotic applications. Hydraulic and pneumatic actuation are considered as the silver bullet for MRI applications in the sense that these system can completely avoid electrical and magnetic noise inside the scanner by using dielectric materials, thus providing high signal noise ratio. Hydraulic systems renders very large output force, but usually suffer from cavitation and fluid leakage. Pneumatic actuation is easier to maintain but is back-drivable due to the compressibility of air and is more favorable for high bandwidth applications, especially ideal for force control. The disadvantages of pneumatic actuation is also because of compressibility of air and the induced time delay, that makes it difficult to control at the millimeter level in which medical robotics usually tries to maintain. Ultrasonic/piezoelectric actuation provides high positioning resolution [23,22] and does not rely upon magnetism, but since there is still high frequency electrical signal and the piezoelectric elements are often embedded inside ferromagnetic or paramagnetic materials, piezoceramic motors using commercially available motor controllers have been evaluated by [13] which negatively impacted image quality. The difficulty arises from the actuator driving controller that usually induces significant image artifact when utilizing off-the-shelf control circuits. However, the scalability, simplicity, size and inherent robustness of electromechanical systems present a clear advantage over pneumatically actuated systems.

Objective and Contributions: *Generally, no prior work has considered the exploitation of modular robotic systems for in-situ needle placement in high-field closed-bore MRI, with its many merits, in a transperineal manner which alleviates the requirement to perform the implant procedures in a different pose than used for preoperative imaging.* This paper reviews the design of two approaches that comply with the compatibility while address *in-situ* needle placement with reduced cost and system complexity. By leveraging the pneumatic and piezoelectric actuation approaches, the underlying philosophy of this research effort aims to explore the two distinct methods to demonstrate two design examples that provides semi-automatic and teleoperated needle placement robots for prostate interventions. Even the focus of the design targets prostate cancer, this interventional system is possible to be adopted for other similar percutaneous applications.

3 Design Requirement

3.1 Surgical Procedure Analysis and Design Implication

Robotic design starts from the surgical procedure analysis which provides the guideline and "optimal" solution of designing clinically practical robots. This includes the analysis of prostatic intervention classification based on interventional approach (transrectal, transperineal and transurinal) and the motion analysis of needle insertion during biopsy, brachytherapy and ablation.

The transperineal approach is preferable for the following reasons [3]. First, transrectal approach samples the peripheral zone (the most common location of the cancer) across its smallest dimension and has a higher possibility of missing the cancer than the transperineal approach. Second, the transrectal method does not allow accessing some regions of the prostate that are blocked by the urethra. Third, the transrectal technique is not extendible to therapy (brachytherapy or cryotherapy) which requires multiple needle placements because of increased rectal injury that has a high risk of infection.

As shown in Fig. 1 right, the clinical 18Gauge (1.27mm) needles for prostate brachytherapy have an inner stylet and hollow sheath. Radioactive seeds are pre-loaded with 5.5mm spacers between them before staring the surgery. During the insertion, one hand holds the cannula and the other hand brace against stylet hub to prevent relative motion. After insertion, the sheath is withdrawn over the stylet while leaving the seeds in place. To mimic the physician preload needle type brachytherapy procedure, the needle driver provides 1-DOF cannula rotation about its axis with 1-DOF translational insertion. Another 1-DOF of translational stylet motion is implemented to coordinate the motion with respect to the cannula. The rotation motion of the cannula may be used for bevel-based steering to limit deflection or may be used for active cannula operation [54].

3.2   System Requirement

The system's principal function is accurate transperineal needle placement in the prostate for diagnosis and treatment, primarily in the form of biopsy and brachytherapy seed placement, respectively. The patient is positioned in the supine position with the legs spread and raised as shown in Fig. 3 (left). The patient is in a similar configuration to that of TRUS-guided brachytherapy, but the typical MRI bore's constraint ($\leq$ 60cm diameter) necessitates reducing the spread of the legs and lowering the knees into a semi-lithotomy position. The robot operates in the confined space between the patient's legs without interference with the patient, MRI scanner components, anesthesia equipment, and auxiliary equipment present as shown in the cross-section shown in Fig. 3 (right).

The size of an average prostate is $50mm$ in the lateral direction by $35mm$ in the anterior-posterior direction by $40mm$ in length. The average prostate volume is about 35cc; by the end of a procedure, this volume enlarges by as much as 25% due to swelling [51]. For our system, the standard $60mm$ $\times$ $60mm$ perineal window of TRUS-guided brachytherapy was increased to $100mm$ $\times$ $100mm$, in order to accommodate patient variability and lateral asymmetries in patient setup. In depth, the workspace extends to $150mm$ superior of the perineal surface. Direct access to all clinically relevant locations in the prostate is not always possible with a needle inserted purely along apex-base direction due to PAI. Needle angulation in the sagittal and coronal planes will enable procedure to be performed on many of these patients where brachytherapy is typically contraindicated due to PAI as described by Fu, *et al.* [16].

The robot is mounted to a manual linear slide that positions the robot in the access tunnel and allows fast removal for reloading brachytherapy needles or collecting harvested biopsy tissue. The primary actuated motions of the robot include two prismatic motions which replicate the DOF of a traditional template-base approach, and two rotational motions for aligning the needle axis to help avoid PAI and critical structures. In addition to these base motions, application-specific motions are also required; these include needle insertion, canula retraction or biopsy gun actuation, and needle rotation. The accuracy of the individual servo-controlled joints is targeted to be 0.1mm, and the needle placement accuracy of the robotic system itself is targeted to be better than 1.0mm. This target accuracy approximates the voxel size of the MR images used which represents the finest possible targeting precision. The overall system accuracy, however, is expected to be somewhat less when effects such as imaging resolution, needle deflection, and tissue deformation are taken into account. The MR image resolution is typically 1mm and the clinically significant target is typically $5mm$ in size (the precision of a traditional TRUS template).

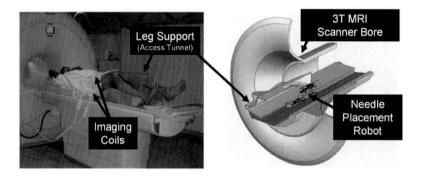

Fig. 3. Positioning of the patient in the semilithotomy position on the leg support (left). The robot accesses the prostate through the perineal wall, which rests against the superior surface of the tunnel within the leg rest (right). [15] ©2008 IEEE

3.3 MRI Compatibility Requirements

Significant complexity is introduced when designing a system operating inside the bore of high-field 1.5-3T MRI scanners since traditional mechatronics materials, sensors and actuators cannot be employed. This section addresses the additional system requirements arising from compatibility with the MRI scanner. A thorough description of the issues relating to MR Safety is described by Shellock [35]. The MR-Safe definitions are according to the ASTM Standard F2052 [1] while MR-compatible is the commonly used term:

MR-Safe: The device, when used in the MR environment, has been demonstrated to present no additional risk to the patient or other individual, but may affect the quality of the diagnostic information. The MR conditions in which the device was tested should be specified in conjunction with the term MR safe since a device that is safe under one set of conditions may not be found to be so under more extreme MR conditions.

MR-Compatible: A device is considered MR-compatible if it is MR safe and if it, when used in the MR environment, has been demonstrated to neither significantly affect the quality of the diagnostic information nor have its operations affected by the MR device. The MR conditions in which the device was tested should be specified in conjunction with the term MR-compatible since a device that is safe under one set of conditions may not be found to be so under more extreme MR conditions.

4   Needle Placement Robot: Pneumatic Approach

The first iteration of a pneumatic system [15] provides the two prismatic motions in the axial plane over the perineum and an encoded manual needle guide, as shown in shown in Fig. 4. This represents an automated high-resolution needle guide, functionally similar to the template used in conventional brachytherapy. The primary base DOFs are broken into two decoupled planar motions. Motion in the vertical plane is based on a modified version of a scissor lift mechanism that is traditionally used for plane parallel motion. Motion in the horizontal plane is accomplished with a second planar bar mechanism. The base of the manipulator has a modular platform that allows for different end effectors to be mounted on it. The two initial end effectors will accommodate biopsy guns and brachytherapy needles. Both require an insertion phase; the former requires activating a single-acting button to engage the device and a safety lock. The latter requires an additional controlled linear motion to accommodate the cannula retraction to release brachytherapy seeds. Rotation of the needle about its axis may be implemented to either "drill" the needle in to limit deflection, or to steer the needle using bevel steering techniques such as those described in [54]. Sterility has been taken into consideration for the design of the end effectors. In particular, the portions of the manipulator and leg rest that come in direct contact with the patient or needle will be removable and made of materials that are suitable for sterilization. The remainder of the robot will be draped. An alternative solution is to enclose the entire leg rest with the robot in a sterile drape, thus completely isolating the robot from the patient except for the needle.

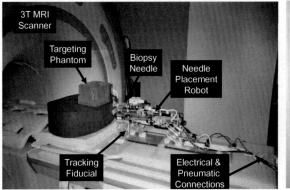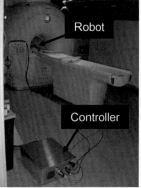

Fig. 4. Pneumatic robotic needle placement mechanism with two active DOF and one passive, encoded needle insertion. Dynamic global registration is achieved with the attached tracking fiducial (left). The configuration of the robot with the controller in the scanner room (right).

4.1   Mechanical Design

Mechanism design is particulary important since the robot is non-metallic and must operate in a confined space. The design was developed such that the kinematics can be simplified, control can be made less complex, motions may be decoupled, actuators can be aligned appropriately, and system rigidity can be increased. Based upon analysis of the workspace and the application, the following additional design requirements have been adopted: 1) prismatic base motions should be able to be decoupled from angulation since the majority of procedures will not require the two rotational DOFs; 2) actuator motion should be in the axial direction (aligned with the scanner's axis) to maintain a compact profile; and 3) extension in both the vertical and horizontal planes should be telescopic to minimize the working envelope.

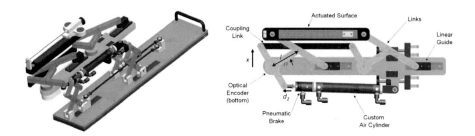

Fig. 5. This mechanism provides for motion in the vertical plane. Coupling the forward and rear motion provides for vertical travel, independently moving the rear provides for elevation angle adjustment (left). This mechanism provides for motion in the horizontal plane. The design shown provides prismatic motion only; rotation can be enabled by actuating rear motion independently by replacing coupling link (L_c) with a second actuator. The modular, encoded needle guide senses the depth during manual needle insertion and can be replaced with different end effectors for other procedures (right).

The primary base DOFs are broken into two decoupled planar motions. Motion in the vertical plane includes 100 mm of vertical travel, and optionally up to 15° of elevation angle. This is achieved using a modified version of a scissor lift mechanism that is traditionally used for planar parallel motion. By coupling two such mechanisms, as shown in Fig. 5 (left), 2-DOF motion can be achieved. Stability is increased by using a pair of such mechanisms in the rear. For purely prismatic motion, both slides move in unison; angulation is generated by relative motions. To aid in decoupling, the actuator for the rear slide can be fixed to the carriage of the primary motion linear drive, thus allowing one actuator to be locked when angulation is unnecessary. As shown in Fig. 5 (left), the push rods for the front and rear motions are coupled together to maintain only translational motion in the current prototype.

Motion in the horizontal plane is accomplished with a second planar bar mechanism. This motion is achieved by coupling two straight line motion mechanisms, as shown in Fig. 5 (right), generally referred to as Scott-Russell mechanisms. By combining two such straight-line motions, both linear and rotational motions can be realized in the horizontal plane. The choice of this design over the use of the previously described scissor-type mechanism is that this allows for bilateral motion with respect to the nominal center position. Fig. 5 (right) shows the mechanism where only translation is available; this is accomplished by linking the front and rear mechanisms with a connecting bar. A benefit of this design is that it is straightforward to add the rotational motion for future designs by replacing the rigid connecting bar (L_c) with another actuator. Due to the relative ease of manufacturing, the current iteration of the system is made primarily out of acrylic. In future design iterations, the links will be made out of high strength, dimensionally stable, highly electrically insulating, and sterilizable plastic (e.g., Ultem or polyetheretherketone (PEEK)).

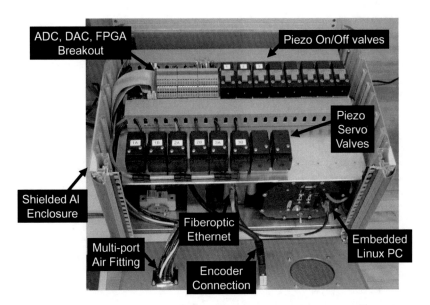

Fig. 6. Controller contains the embedded Linux PC providing low-level servo control, the piezoelectric valves, and the fiber-optic Ethernet converter. The EMI shielded enclosure is placed inside the scanner room near the foot of the bed. Connections to the robot include the multi-tube air hose and the encoder cable; connection to the planning workstation is via fiber-optic Ethernet.

4.2   Pneumatic Actuator Design

The MRI environment places severe restrictions on the choice of sensors and actuators. Many mechatronic systems use electrodynamic actuation, however, the very nature of an electric motor precludes its use in high-field magnetic environments. Therefore, it is necessary to either use actuators that are compatible with the MR environment, or to use a transmission to mechanically couple the manipulator in close proximity to the scanner to standard actuators situated outside the high field. MR-compatible actuators such as piezoceramic motors have been evaluated in [13, 46]; however, these are prone to introducing noise into MR imaging, and therefore, negatively impacting image quality. Mechanical coupling can take the form of flexible drive shafts [26], push-pull cables, or hydraulic (or pneumatic) couplings [18].

Pneumatic cylinders are the actuators of choice for this robot. Accurate servo control of pneumatic actuators using sliding mode control (SMC) with sub-millimeter tracking accuracy and 0.01mm steady-state error (SSE) have been demonstrated in [4]. Although pneumatic actuation seems ideal for MRI, most standard pneumatic cylinders are not suitable for use in MRI. Custom MR compatible pneumatic cylinders have been developed for use with this robot. The design of these cylinders is based upon Airpel 9.3 mm bore cylinders. These cylinders were chosen because the cylinder bore is made of glass and the piston and seals are made of graphite, giving them inherently low friction and MR compatibility. In collaboration with the manufacturer, we developed the cylinders shown in Fig. 5 (right) bottom that are entirely nonmetallic except for the brass shaft. The cylinders can handle up to 100 psi (6.9 bar), and therefore, can apply forces up to 46.8 N. Our newly developed SMC controller has shown sub-millimeter tracking accuracy using this pneumatic actuator [53].

In addition to moving the robot, it is important to be able to lock it in position to provide a stable needle insertion platform. Pneumatically operated, MR-compatible brakes have been developed for this purpose. The brakes are compact units that attach to the ends of the previously described cylinders, as shown in Fig. 5 (right) top, and clamp down on the rod. The design is such that the fail-safe state is locked and applied air pressure releases a spring-loaded collet to enable motion. The brakes are disabled when the axis is being aligned and applied when the needle is to be inserted or an emergency situation arises.

Proportional pressure regulators are implemented for this robot because they allow for direct control of air pressure, thus the force applied by the pneumatic cylinder. This is an advantage because it aids in controller design and also has the inherent safety of being able to limit applied pressure to a prescribed amount. Most pneumatic valves are operated by a solenoid coil; unfortunately, as with electric motors, the very nature of a solenoid coil is a contraindication for its use in an MR environment.With pneumatic control, it is essential to limit the distance from the valve to the cylinder on the robot; thus, it is important to use valves that are safe and effective in the

MR environment. By placing the controller in the scanner room near the foot of the bed, air tubing lengths are reduced to 5 m. The robot controller uses piezoelectrically actuated proportional pressure valves, thus permitting their use near MRI. A pair of these valves provide a differential pressure of ± 100 psi on the cylinder piston for each actuated axis. A further benefit of piezoelectrically actuated valves is the rapid response time (4 ms). Thus, by using piezoelectric valves, the robot's bandwidth can be increased significantly by limiting tubing lengths and increasing controller update rate.

4.3   Robot Controller Hardware

MRI is very sensitive to electrical signals passing in and out of the scanner room. Electrical cables passing through the patch panel or wave guide can act as antennas, bringing stray RF noise into the scanner room. For that reason, and to minimize the distance between the valves and the robot, the robot controller is placed inside the scanner room. The controller comprises an electromagnetic interference (EMI) shielded enclosure that sits at the foot of the scanner bed, as shown in Fig. 4 right; the controller has proved to be able to operate $3m$ from the edge of the scanner bore. Inside of the enclosure is an embedded computer with analog I/O for interfacing with valves and pressure sensors and a field programmable gate array (FPGA) module for interfacing with joint encoders (see Fig. 6). Also, in the enclosure are the piezoelectric servo valves, piezoelectric brake valves, and pressure sensors. The distance between the servo valves and the robot is minimized to less than $5m$, thus maximizing the bandwidth of the pneumatic actuators. Control software on the embedded PC provides for low-level joint control and an interface to interactive scripting and higher level trajectory planning. Communication between the low-level control PC and the planning and control workstation sitting in the MR console room is through a 100-FX fiber-optic Ethernet connection. In the prototype system, power is supplied to the controller from a filtered dc power supply that passes through the patch panel; a commercially available MR-compatible power supply will be used in future design iterations. No other electrical connections pass out of the scanner room, thus significantly limiting the MR imaging interference.

5   Needle Placement Robot: Piezoelectric Approach

The newly developed piezoelectric robot [42,43,39,41], shown in Fig. 7, offers the capability of real-time *in-situ* needle steering in high-field MRI. It consists of a modular 3-DOF needle driver with optical tracking frame coupled similar to that of the pneumatic robot. The modular needle driver simultaneously provides needle cannula rotation and independent cannula and stylet prismatic motion. For experimental purposes, it is shown with a generic MRI-compatible 3-DOF $x - y - z$ stage; however, application specific designs will be used for the appropriate clinical applications. The piezoelectric motors

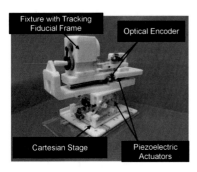
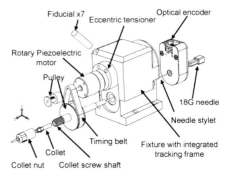

Fig. 7. Piezoelectric robotic needle placement mechanism with three DOF needle rotation and collinear translation for prostate brachytherapy on a three DOF Cartesian stage(left). Detailed design of the needle driver platform (right). [43] ©2011 IEEE

are actuated using custom motor control drive circuits integrated into a controller architecture adapted from that of the pneumatic robot. To overcome the loss of needle tip proprioception information, a multi-DOF fiber optic force sensor is integrated with the piezoelectric robot.

The needle placement robot consists of a needle driver module (3-DOF) and Cartesian positioning module (3-DOF). The material is rapid prototyped with ABS and laser cut acrylic. Considering the supine configuration and the robot workspace, the width of the robot is limited to 6cm. As shown in Fig. 7 (left), the lower layer of the needle driver module is driven with linear piezoelectric motor and the upper layer provides cannula rotation motion and stylet prismatic motion.

5.1 Mechanical Design

To design a needle driver that allows a large variety of standard needles, a new clamping device shown in Fig. 7 (right) rigidly connects the needle shaft to the driving motor mechanism. This structure is a collet mechanism and a hollow screw made of stereolithography ABS is twisted to fasten the collet thus rigidly locking the needle shaft on the clamping device. The clamping device is connected to the rotary motor through a timing belt that can be fastened by a eccentric belt tensioner. The clamping device is generic in the sense that we have designed 3 sets of collets and each collet can accommodate a width range of needle diameters. The overall needle diameter range is from 25 Gauge to 7 Gauge. By this token, it can not only fasten brachytherapy needle, but also biopsy needles and most other standard needles instead of designing some specific structure to hold the needle handle.

Once a preloaded needle or biopsy gun is inserted, the collet can rigidly clamp the cannula shaft. Since the linear motor is collinear with the collet and

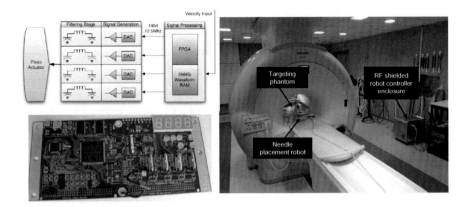

Fig. 8. Piezoelectric actuator driver architecture and prototype using FPGA generated waveform (left), and the configuration of the robot with the controller in the scanner room(right).

shaft, we need to offset the shaft to manually load the needle. We designed a brass spring preloaded mechanism that provides lateral passive motion freedom. The operator can squeeze the mechanism and offset the top motor fixture then insert the loaded needle through plain bearing housing and finally lock with the needle clamping. This structure allows for easy, reliable and rapid loading and unloading of standard needles.

5.2   Piezoelectric Actuator Driver

The piezoelectric actuators (PiezoMotor, Uppsala, Sweden) chosen are non-harmonic piezoelectric motors which have two advantages over a harmonic drive: the noise caused by the driving wave is much easier to suppress, and the motion produced by the motors is generally at a more desirable speed and torque relative to harmonic piezoelectric motors. Even though piezoelectric motors do not generate magnetic fields, commercial motor driver boards usually induce significant image artifacts due to electrical noise according to the most recent result [25]. A new low noise driver [8,9] was developed and its architecture is shown in Fig. 8(left) and Fig. 8 (right) shows the configuration of the robot with the controller in the scanner room. Waveform tables are stored in RAM and utilized by a synthesizer running on the FPGA to generate four independent control waveforms of arbitrary phase and frequency. These control waveforms are then streamed out to the analog amplification stage at 25 mega samples per second.

6   Surgical Visualization and Workflow

6.1   Surgical Visualization

The user interface for the robot is based on 3D Slicer open-source surgical navigation software (3D Slicer Software, http://www.slicer.org). The navigation software runs on a Linux-based workstation in the scanner's console room, and communicates to the robot controller over a fiberoptic connection. A customized graphical user interface (GUI) specially designed for the prostate intervention with the robot is described in [32]. The interface allows smooth operation of the system throughout the clinical workflow including registration, planning, targeting, monitoring and verification (Fig. 9). The workstation is connected to the robot and the scanner's console via Ethernet. OpenIGT Link [48], is used to exchange various types of data including control commands, position data, and images among the components. Fig. 10 shows the configuration used in the system as described in [49].

In the planning phase, pre-operative images are retrieved from a DICOM server and loaded into the navigation software. Registration is performed between the pre-operative planning images and intra-operative imaging using techniques such as those described by Haker, *et al.* [21]. Target points for the needle insertion are selected according to the pre-operative imaging, and the coordinates of the determined target points are selected in the planning GUI. Once the patient and the robot are placed in the MRI scanner, a 2D image of the fiducial frame is acquired and passed to the navigation software to calculate the 6-DOF pose of the robot base for the robot-image registration. The position and orientation of the robot base is sent through the network from the navigation software to the robot controller. After the registration phase, the robot can accept target coordinates represented in the image (patient) coordinate system in standard Right-Anterior-Superior (RAS) coordinates.

During the procedure, a target and an entry point are chosen on the navigation software, and the robot sends the coordinates then aligns the needle guide appropriately. In the current system, the needle is inserted manually while the needle position is monitored by an encoded needle guide and displayed in real-time on the display. Needle advancement in the tissue is visualized on the navigation software in two complementary ways: 1) a 3D view of needle model combined with pre-operative 3D image re-sliced in planes intersecting the needle axis, and 2) 2D real-time MR images acquired from the planes along or perpendicular to the needle path and continuously transferred from the scanner through the network. The former provides a high refresh rate, allowing a clinician to manipulate the needle interactively. The latter provides the changing shape or position of the target lesion with relatively slower rate depending on the imaging speed (typically 0.5Hz).

The interface software enables "closed-loop" needle guidance, where the action made by the robot is captured by the MR imaging, and immediately fed back to a physician to aid their decision for the next action. The reason for keeping a human in the loop is to increase the safety of the needle

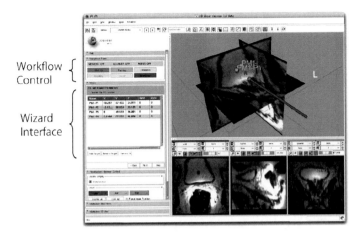

Fig. 9. 3D Slicer planning workstation showing a selected target and the real-time readout of the robot's needle position. The line represents a projection along the needle axis and the sphere represents the location of the needle tip. [15] ©2008 IEEE

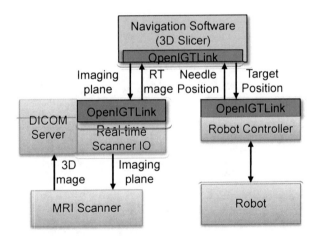

Fig. 10. Diagram shows the connection and the data flow among the robot components. OpenIGTLink, an open-source device communication tool, is used to exchange control, position, and image data.

insertion, and to allow for monitoring progress via live MR images. Once the needle is fully aligned before patient contact, the placement is adjusted in response to the MR images. The physician performs the insertion under real-time imaging. Fig. 9 shows the planning software with an MR image of the phantom loaded and real-time feedback of the robot position is used to generate the overlaid needle axis model.

6.2   Needle Placement Robot Tracking and Localization

Dynamic global registration between the robot and scanner is achieved by passive tracking the fiducial frame in front of the robot as shown in Fig. 7 (right). The rigid structure of the fiducial frame is made of ABS and seven MR Spot fiducials (Beekley, Bristol, CT) are embedded in the frame to form a Z shape passive fiducial. Any arbitrary MR image slicing through all of the fiducial rods provides the full 6-DOF pose of the frame, and thus the robot, with respect to the scanner. Thus, by locating the fiducial attached to the robot, the transformation between patient coordinates (where planning is performed) and the robot is known. To enhance the system reliability and robustness, multiple slices of fiducial images are used to register robot position using principal component analysis method. The end effector location is then calculated from the kinematics based on the encoder positions.

6.3   System Workflow

The workflow of the system mimics that of traditional TRUS-guided prostate needle insertions, and are as follows (shown in Fig. 11):

1. Acquire pre-procedural MRI volume of the patient's prostate and surrounding anatomy.
2. Select specific needle tip targets as shown in Fig. 9.
3. Define corresponding needle trajectories.
4. Acquire an MR image of the robot's tracking fiducial.
5. Register the robotic system to patient/scanner coordinates.
6. Load the biopsy needle or pre-loaded brachytherapy needle into the robot's needle driver.
7. Send coordinates in patient coordinates to the robot.
8. Automatically align the needle guide and lock in place.
9. Manually insert the needle along prescribed axis as virtual needle guide is displayed on real-time MR images intersecting the needle axis.
10. Confirm correct placement.
11. Harvest tissue or deliver therapy.
12. Retract the needle guide and remove biopsy or brachytherapy needle from robot.
13. Update surgical plan as necessary based on volumetric imaging of the prostate and knowledge of the intervention performed.
14. Repeat for as many needles as necessary.

6.4   Teleoperated Needle Insertion

Manual insertion was the preferred technique due to the need for tactile feedback during the insertion phase. However, it was found that the ergonomics of manual insertion along this guide proved very difficult in the confines of the scanner bore. The limited space in closed-bore high-field MRI scanners

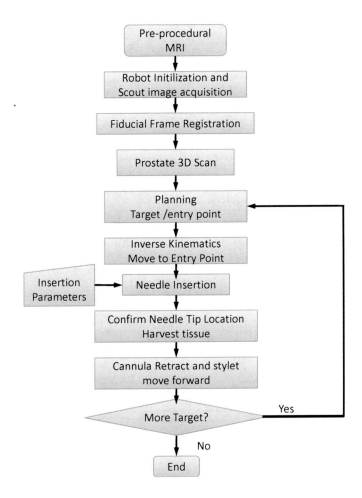

Fig. 11. Workfow of MRI-guided needle placement that mimics that of traditional TRUS-guided prostate needle intervention.

requires a physical separation between the surgeon and the imaged region of the patient. To overcomes the loss of needle tip proprioception information, we are developing a teleoperated haptic system with optical force-torque sensor, to be integrated with a 3-DOF robotic needle driver for MR-guided prostate needle placement. The piezoelectrically actuated robot incorporates tactile feedback and teleoperation. As shown in Fig. 12 , an MRI compatible robot controller sits inside the scanner room and communicates to the navigation software and scanner interface running on a laptop in the console room through a fiber optic connection. The optical force sensor interface is incorporated into the in-room robot controller and the needle interaction forces are transmitted back to the navigation software console along with the

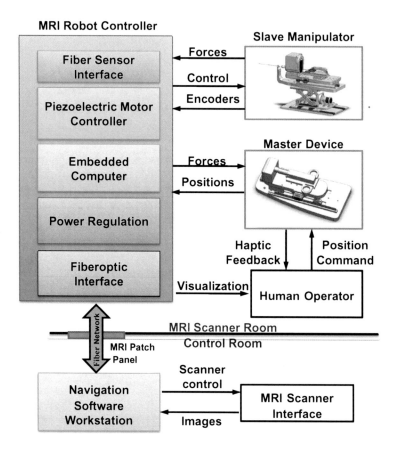

Fig. 12. System architecture for the master - slave haptic interface of the piezo-electric robot. The fiber optic force sensor and robot are placed near the isocenter of the MRI scanner, the master manipulator is connected to the navigation software interface, and the two are couple through the robot controller in the scanner room using a fiber optic network connection.

robot position. The haptic feedback device is integrated into the navigation software framework to provide forces to the operator and control back to the robot.

A direct force feedback architecture [6] is used to control the teleoperated needle placement system. In this iteration of the system, a commercially available Novint Falcon (Novint Technologies, Inc., Albuquerque, NM) haptic device is used as the master robot. It has 3 position DOF and can be used to position the needle in the Cartesian space. Other high quality haptic devices [27] can be utilized to enhance the teleoperation fidelity. In the future version, the haptic device would be designed with piezoelectric actuation and optical encoding as shown in Fig. 12. It would be operated inside the scanner room for better observation of patient's physiological condition and other surgical procedures. The human operator positions obtained from the haptic interface are used for trajectory generation and control of the motion of the slave manipulator. The slave robot in this design is the second generation of the 2-DOF pneumatically actuated robotic assistant system which overcomes many design difficulties and promises safe and reliable intraprostatic needle placement inside closed high-field MRI scanners. However, the addition of force feedback allows incorporation of an actuated needle driver and firing mechanism with needle rotation. The contact forces between needle and the tissue can be measured by the force/toque sensor [40, 44] and further fed back to the haptic device through its interface.

7   Experiments

The first iteration of the needle placement robots has been constructed and is operational. All mechanical, electrical, communications, and software issues have been evaluated. The current state of the pneumatic manipulator is two actuated DOFs (vertical and horizontal) with an encoded passive needle insertion stage. Evaluation of the robot is in two distinct phases: 1) evaluation of the MR compatibility of the robot; and 2) evaluation of the workspace and accuracy. The piezoelectric actuated robot is 6-DOF and the MRI compatibility has been evaluated using similar protocols. This section reports the experimental setup and results of the two robots.

7.1   Imaging Protocols

Four imaging protocols as shown in Table 1, were selected for evaluation of compatibility of the system: 1) diagnostic imaging T1-weighted fast gradient echo (T1 FGE/FFE), 2) diagnostic imaging T2-weighted fast spin echo (T2 FSE/TSE), 3) high-speed real-time imaging fast gradient echo (FGRE), and 4) functional imaging spin echo-planar imaging (SE EPI). Details of the scan protocols are shown in Table 1. All sequences were acquired with a slice thickness of 5mm and a number of excitations (NEX) of one. Three configurations were evaluated and used in the comparison: 1) baseline of the phantom only, 2) motor powered with controllers DC power supply turned on and 3) system servoing inside MRI board. Three slices were acquired per imaging protocol for each configuration.

Compatibility was evaluated on a 3T Philips Achieva scanner. A 110 mm, fluid-filled spherical MR phantom was placed in the isocenter and the robot placed such that the tip was at a distance of 120 mm from the center of the phantom (a representative depth from perineum to prostate), as shown in Fig. 4(right). For the pneumatic robot, the phantom was imaged using four standard prostate imaging protocols: 1) T1W FFE: T1 weighted fast field

Table 1. Scan Parameters for Compatibility Evaluation – 3T Philips Achieva

Protocol	FOV	TE	TR	Flip Ang	Bandwidth
T1W	240 mm	2.3 ms	225 ms	75°	751 Hz/pixel
T2W	240 mm	90 ms	3000 ms	90°	158 Hz/pixel
FGRE	240 mm	2.1 ms	6.4 ms	50°	217 Hz/pixel
EPI	240 mm	45 ms	188 ms	90°	745 Hz/pixel

gradient echo; 2) T2W TSE: T2 weighted turbo spin echo; 3) TFE (FGRE): "real-time" turbo field gradient echo.

A baseline scan with each sequence was taken of the phantom with no robot components using round flex coils similar to those often used in prostate imaging. The following imaging series were taken in each of the following configurations: 1) phantom only; 2) controller in room and powered; 3) robot placed in scanner bore; 4) robot electrically connected to controller; and 5) robot moving during imaging (only with T1W imaging). For each step, all three imaging sequences were performed and both magnitude and phase images were collected.

The amount of degradation of SNR was used as the measure of negative effects on image quality. SNR of the MR images was defined as the mean signal in a 25 mm square at the center of the homogeneous sphere divided by the standard deviation of the signal in that same region. The SNR was normalized by the value for the baseline image, thus limiting any bias in choice of calculation technique or location. SNR was evaluated at seven 3 mm thick slices (representing a 25 mm cube) at the center of the sphere for each of the three imaging sequences. When the robot was operational, the reduction in SNR of the cube at the phantom's center for these pulse sequences was 5.5% for T1W FFE, 4.2% for T2W TSE, and 1.1% for TFE (FGRE). Further qualitative means of evaluating the effect of the robot on image quality are obtained by examining prostate images taken both with and without the presence of the robot. Fig. 13 shows images of the prostate of a volunteer placed in the scanner bore on the leg rest. With the robot operational, there is no visually identifiable loss in image quality of the prostate.

7.2   MR Compatibility of Pneumatic Actuated Needle Placement Robot

As can be seen in Fig. 13, the motors and encoders provide very small visually identifiable interference with the operation of the scanner. Fig. 15 depicts one slice of the tracking fiducial frame which provides the full position information of the robot. We utilize SNR as the metric for evaluating MR compatibility with baseline phantom image comparison. For comparison, the SNR of each configuration was normalized by the average SNR of the 3 baseline images for each imaging protocol. SNR of the MR images is defined as the mean

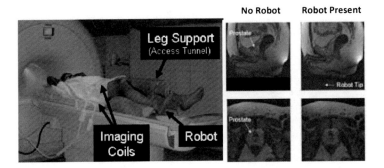

Fig. 13. Qualitative analysis of prostate image quality for pneumatic robot. Patient is placed on the leg support and the pneumatic robot sits inside of the support tunnel inside the scanner bore. T2 weighted sagittal and transverse images of the prostate taken when no robot components were present and when the robot was active in the scanner. [15] ©IEEE 2008

signal in a 25mm square at the center of the homogeneous sphere divided by the standard deviation of the signal in that same region [52]. The SNR was normalized by the value for the baseline image. The technique used for measuring SNR is equivalent to that described by the NEMA standard [2]. The SNR is calculated as described in Method 4 of the cited standard:

$$SNR = \frac{S}{imagenoise} \qquad (1)$$

where, S is the mean pixel value within the Measurement Region of Interest (MROI) and *imagenoise* is the standard deviation (SD) of the selected Noise Measurement Region of Interest (NMROI). The location of the NMROI has a significant impact on calculated SNR; the selected NMROI in the top-left corner minimizes variance between image slices and is outside of ghosting artifacts as described by Firbank, *et al* [14]. Statistical analysis with a Tukey Multiple Comparison confirms that no pair shows significant signal degradation with a 95% confidence interval.

7.3 Accuracy of Pneumatic Actuated Needle Placement Robot

Accuracy assessment is broken into two parts: localization and placement. These two must be distinguished, especially in many medical applications. In prostate biopsy, it is essential to know exactly where a biopsy comes from in order to be able to form a treatment plan if cancer is located. In brachytherapy treatment, radioactive seed placement plans must be made to avoid cold spots where irradiation is insufficient; by knowing where seeds are placed, the dosimetry and treatment plan can be interactively updated. Based on encoder resolution, localization accuracy of the robot in free space is better than 0.1mm in all directions.

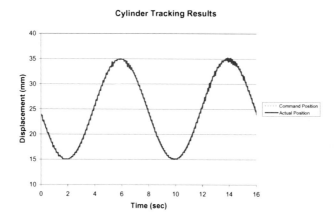

Fig. 14. Dynamic tracking results for the pneumatic cylinder for an 8sec period, 20mm amplitude sine wave.

Positioning accuracy is dependent on the servo pneumatic control system. The current control algorithm for pneumatic servo control is based upon sliding mode control techniques. The pneumatic cylinder was commanded to move back and forth between two points and the steady state error for each move was recorded. The average positioning accuracy was 0.26mm RMS error. In addition to point-to-point moves, the positioning accuracy of the cylinder alone in free space has been evaluated for dynamic trajectory tracking. The cylinder was commanded to follow a 0.125Hz, 20mm amplitude sine wave. Fig. 14 shows the tracking results for the cylinder. The RMS error for this task was 0.23mm which approximates the set deadband of 0.30mm.

7.4   MR Compatibility of Piezoelectric Actuated Needle Placement Robot

The four imaging protocols shown in Table 1 were evaluated in the same setup and the same metric (SNR). As can be seen in Fig. 15 (left), the motors and encoders provide very small visually identifiable interference with the operation of the scanner. Fig. 15 (right) depicts one slice of the tracking fiducial frame which provides the full position information of the robot. Statistical analysis with a Tukey Multiple Comparison confirms that no pair shows significant signal degradation with a 95% confidence interval. This result presents significant improvement over recent research result in [25]. It implies that piezoelectric actuation is also feasible inside scanner bore during imaging, and thus makes it a good candidate in robotic design.

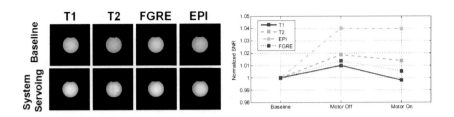

Fig. 15. Representative results showing the images obtained of baseline and system servoing inside scanner bore conditions (left), Results of the normalized SNR under each protocol. No statistically significant variations exist between baseline and motor running conditions.(right).

8 Conclusions

This chapter presents two MRI-compatible approaches to image-guided transperineal prostate needle placement. Described is the mechanism, actuator and sensor design, controller design and system integration for a pneumatically actuated robotic needle guide and a piezoelectrically actuated needle steering platform. The 2-DOF pneumatic robot with manual needle insertion has an SNR loss limited to 5% with alignment accuracy under servo pneumatic control better than $0.94mm$ per axis. While the 6-DOF piezoelectrically actuated robot is the first demonstration of a novel multi piezoelectric actuator drive with less than 2% SNR loss for high field MRI operating at full speed during imaging. The preliminary experiments in phantom studies evaluates system MRI compatibility, workflow, visualization and targeting accuracy.

Attaining an acceptable level of MR compatibility required significant experimental evaluation. Several types of actuators, including piezoelectric motors and pneumatic cylinder/valve pairs, were evaluated. Pneumatic actuators have great potential for MRI-compatible mechatronic systems. Since no electronics are required, they are fundamentally compatible with the MR environment. However, there are several obstacles to overcome. These include: 1) material compatibility that was overcome with custom air cylinders made of glass with graphite pistons; 2) lack of stiffness or instability that was overcome with the development of a pneumatic brake that locks the cylinder's rod during needle insertion; and 3) difficult control that was ameliorated by using high-speed valves and shortening pneumatic hose lengths by designing an MRI-compatible controller. Pneumatic actuation seems to be an ideal solution for this robotic system, and it allows the robot to meet all of the design requirements. Further, MR compatibility of the system including the robot and controller is excellent with no more that a 5% loss in average SNR with the robot operational.

The pneumatic and piezoelectric systems have been evaluated in a variety of tests. The MR compatibility has shown to be sufficient for anatomical imaging using traditional prostate imaging sequences. Communications between all of the elements, including the robot, the low level controller, the planning workstation, and the MR scanner real-time imaging interface, are in place. Initial phantom studies validated the workflow and the ability to accurately localize the robot and target a lesion.

The pneumatic actuation and piezoelectric actuation presents complementary advantages and disadvantages. Pneumatic actuation is considered as the "ultimate solution" for MRI applications in the sense that these system can completely get rid of electrical and magnetic fields by using dielectric materials. However, due to the promising result implemented in the piezoelectric robot and the scalability, simplicity, size and inherent robustness of electromechanical systems present a clear advantage over pneumatically actuated systems. The piezoelectric actuation is considered as the rule of thumb actuation method that has been implemented as the slave robot presented here. A future development would be augmenting the slave robot to a teleoperated master-slave system. The slave robot would perform the needle placement under the surgeon motion command while the needle insertion force would be measured using the fiber optic force sensors that are being developed [40, 42, 38].

Acknowledgements

This work is supported in part by the Congressionally Directed Medical Research Programs Prostate Cancer Research Program (CDMRP PCRP) New Investigator Award W81XWH-09-1-0191 and Worcester Polytechnic Institute internal funds.

References

1. Standard test method for measurement of magnetically induced displacement force on passive implants in the magnetic resonance environment. F2052, vol. 13.01. American Society for Testing and Materials (ASTM) (2002)
2. Determination of Signal-to-Noise Ratio (SNR) in Diagnostic Magnetic Resonance Imaging. NEMA Standard Publication MS 1-2008. The Association of Electrical and Medical Imaging Equipment Manufacturers (2008)
3. Blumenfeld, P., Hata, N., Dimaio, S., Zou, K., Haker, S., Fichtinger, G., Tempany, C.: Transperineal prostate biopsy under magnetic resonance image guidance: A needle placement accuracy study 26(3), 688–694 (2007)
4. Bone, G., Ning, S.: Experimental comparison of position tracking control algorithms for pneumatic cylinder actuators. IEEE/ASME Transactions on Mechatronics 12(5), 557–561 (2007)
5. van den Bosch, M.R., Moman, M.R., van Vulpen, M., Battermann, J.J., Duiveman, E., van Schelven, L.J., de Leeuw, H., Lagendijk, J.J.W., Moerland, M.A.: MRI-guided robotic system for transperineal prostate interventions: proof of principle. Physics in Medicine and Biology 55(5), N133 (2010)

6. Cavusoglu, M.C., Sherman, A., Tendick, F.: Design of bilateral teleoperation controllers for haptic exploration and telemanipulation of soft environments. IEEE Transactions on Robotics and Automation 18(4), 641–647 (2002)

7. Chinzei, K., Miller, K.: Towards MRI guided surgical manipulator. Med. Sci. Monit. 7(1), 153–163 (2001)

8. Cole, G., Harrington, K., Su, H., Camilo, A., Pilitsis, J., Fischer, G.: Closed-loop actuated surgical system utilizing real-time in-situ MRI guidance. In: 12th International Symposium on Experimental Robotics - ISER 2010, New Delhi and Agra, India (2010)

9. Cole, G., Harrington, K., Su, H., Camilo, A., Pilitsis, J., Fischer, G.: Closed-loop actuated surgical system utilizing real-time in-situ MRI guidance. In: Khatib, O., Kumar, V., Sukhatme, G. (eds.) Experimental Robotics, Springer Tracts in Advanced Robotics. Springer, Heidelberg (2011)

10. D'Amico, A.V., Cormack, R., Tempany, C.M., Kumar, S., Topulos, G., Kooy, H.M., Coleman, C.N.: Real-time magnetic resonance image-guided interstitial brachytherapy in the treatment of select patients with clinically localized prostate cancer. Int. J. Radiat. Oncol. Biol. Phys. 42(3), 507–515 (1998)

11. DiMaio, S.P., Pieper, S., Chinzei, K., Hata, N., Haker, S.J., Kacher, D.F., Fichtinger, G., Tempany, C.M., Kikinis, R.: Robot-assisted needle placement in open MRI: system architecture, integration and validation. Comput. Aided Surg. 12(1), 15–24 (2007)

12. Elhawary, H., Zivanovic, A., Davies, B., Lamperth, M.: A review of magnetic resonance imaging compatible manipulators in surgery. Proc. Inst. Mech. Eng. H 220(3), 413–424 (2006)

13. Elhawary, H., Zivanovic, A., Rea, M., Davies, B., Besant, C., McRobbie, D., de Souza, N., Young, I., Lampérth, M.: The Feasibility of MR-Image Guided Prostate Biopsy Using Piezoceramic Motors Inside or Near to the Magnet Isocentre. In: Larsen, R., Nielsen, M., Sporring, J. (eds.) MICCAI 2006. LNCS, vol. 4190, pp. 519–526. Springer, Heidelberg (2006)

14. Firbank, M.J., Coulthard, A., Harrison, R.M., Williams, E.D.: A comparison of two methods for measuring the signal to noise ratio on mr images. Phys. Med. Biol. 44(12), N261–N264 (1999)

15. Fischer, G., Iordachita, I., Csoma, C., Tokuda, J., DiMaio, S., Tempany, C.M., Hata, N., Fichtinger, G.: MRI-compatible pneumatic robot for transperineal prostate needle placement. IEEE/ASME Transactions on Mechatronics 13(3), 295–305 (2008)

16. Fu, L., Liu, H., Ng, W.S., Rubens, D., Strang, J., Messing, E., Yu, Y.: Hybrid dosimetry: feasibility of mixing angulated and parallel needles in planning prostate brachytherapy. Med. Phys. 33(5), 1192–1198 (2006)

17. Futterer, J.J., Misra, S., Macura, K.J.: MRI of the prostate: potential role of robots. Imaging in Medicine 2(5), 583–592 (2010)

18. Gassert, R., Dovat, L., Lambercy, O., Ruffieux, Y., Chapuis, D., Ganesh, G., Burdet, E., Bleuler, H.: A 2-dof fMRI compatible haptic interface to investigate the neural control of arm movements. In: Proceedings. 2006 Conference on International Robotics and Automation, pp. 3825–3831. IEEE, Piscataway (2006)

19. Gassert, R., Yamamoto, A., Chapuis, D., Dovat, L., Bleuler, H., Burdet, E.: Actuation Methods for Applications in MR Environments. Concepts in Magnetic Resonance Part B: Magnetic Resonance Engineering 29B(4), 191–209 (2006)

20. Goldenberg, A., Trachtenberg, J., Kucharczyk, W., Yi, Y., Haider, M., Ma, L., Weersink, R., Raoufi, C.: Robotic system for closed-bore MRI-guided prostatic interventions. IEEE/ASME Transactions on Mechatronics 13(3), 374–379 (2008)
21. Haker, S.J., Mulkern, R.V., Roebuck, J.R., Barnes, A.S., Dimaio, S., Hata, N., Tempany, C.M.: Magnetic resonance-guided prostate interventions. Top Magn. Reson. Imaging 16(5), 355–368 (2005)
22. Huang, H., Su, H., Chen, H., Mills, J.K.: Piezoelectric driven non-toxic injector for automated cell manipulation. In: Proceedings of MMVR18 (Medicine Meets Virtual Reality), Newport Beach, California, USA (February 2011)
23. Huang, H., Sun, D., Su, H., Mills, J.: Force sensing and control of robot-assisted cell injection. In: Gulrez, T., Hassanien, A. (eds.) Advances in Robotics and Virtual Reality. Springer, Heidelberg (2011)
24. Jemal, A., Siegel, R., Ward, E., Hao, Y., Xu, J., Thun, M.J.: Cancer statistics. CA Cancer J. Clin. 59(4), 225–249 (2009)
25. Krieger, A., Iordachita, I., Song, S.E., Cho, N., Guion, P., Fichtinger, G., Whitcomb, L.: Development and preliminary evaluation of an actuated MRI-compatible robotic device for MRI-guided prostate intervention. In: 2010 IEEE International Conference on Robotics and Automation (ICRA), May 2010, p. 1066–1073 (2010)
26. Krieger, A., Susil, R.C., Meard, C., Coleman, J.A., Fichtinger, G., Atalar, E., Whitcomb, L.L.: Design of a novel MRI compatible manipulator for image guided prostate interventions. IEEE Trans. Biomed. Eng. 52(2), 306–313 (2005)
27. Lee, L., Narayanan, M.S., Mendel, F., Krovi, V.N.: Kinematics analysis of in-parallel 5 dof haptic device. In: 2010 IEEE/ASME International Conference on Advanced Intelligent Mechatronics, Montreal, Canada (2010)
28. Li, M., Kapoor, A., Mazilu, D., Horvath, K.A.: Pneumatic actuated robotic assistant system for aortic valve replacement under MRI guidance. IEEE Transactions on Biomedical Engineering 58(2), 443–451 (2010)
29. Li, M., Mazilu, D., Horvath, K.A.: Robotic system for transapical aortic valve replacement with MRI guidance. In: Proceedings of the 11th International Conference on Medical Image Computing and Computer-Assisted Intervention, Part II, pp. 476–484. Springer, Heidelberg (2008)
30. Masamune, K., Kobayashi, E., Masutani, Y., Suzuki, M., Dohi, T., Iseki, H., Takakura, K.: Development of an MRI-compatible needle insertion manipulator for stereotactic neurosurgery. J. Image Guid. Surg. 1(4), 242–248 (1995)
31. Melzer, A., Gutmann, B., Remmele, T., Wolf, R., Lukoscheck, A., Bock, M., Bardenheuer, H., Fischer, H.: Innomotion for percutaneous image-guided interventions. IEEE Engineering in Medicine and Biology Magazine 27(3), 66–73 (2008)
32. Mewes, P., Tokuda, J., DiMaio, S.P., Fischer, G.S., Csoma, C., Gobi, D.G., Tempany, C., Fichtinger, G., Hata, N.: An integrated MRI and robot control software system for an MR-compatible robot in prostate intervention. In: Proc. IEEE International Conference on Robotics and Automation ICRA 2008 (May 2008)
33. Pondman, K.M., Futerer, J.J., ten Haken, B., Kool, L.J.S., Witjes, J.A., Hambrock, T., Macura, K.J., Barentsz, J.O.: MR-guided biopsy of the prostate: An overview of techniques and a systematic review. European Urology 54(3), 517–527 (2008)

34. Schouten, M.G., Ansems, J., Renema, W.K.J., Bosboom, D., Scheenen, T.W.J., Futterer, J.J.: The accuracy and safety aspects of a novel robotic needle guide manipulator to perform transrectal prostate biopsies. Medical Physics 37(9), 4744–4750 (2010)

35. Shellock, F.G.: Magnetic resonance safety update 2002: implants and devices. J. Magn. Reson. Imaging 16(5), 485–496 (2002)

36. Song, S.E., Cho, N.B., Fischer, G., Hata, N., Tempany, C., Fichtinger, G., Iordachita, I.: Development of a pneumatic robot for MRI-guided transperineal prostate biopsy and brachytherapy: New approaches. In: Proc. IEEE International Conference on Robotics and Automation ICRA (2010)

37. Stoianovici, D., Patriciu, A., Petrisor, D., Mazilu, D., Kavoussi, L.: A new type of motor: pneumatic step motor. IEEE/ASME Transactions on Mechatronics 12(1), 98–106 (2007)

38. Su, H., Camilo, A., Cole, G., Hata, N., Tempany, C., Fischer, G.S.: High-field MRI compatible needle placement robot for prostate interventions. In: Proceedings of MMVR18 (Medicine Meets Virtual Reality), Newport Beach, California, USA (February 2011)

39. Su, H., Cole, G., Fischer, G.: Active needle steering for percutaneous prostate intervention in high-field MRI. In: 2010 Robotics: Science and Systems Conference, Workshop on Enabling Technologies for Image-Guided Robotic Interventional Procedures, Zaragoza, Spain (August 2010)

40. Su, H., Fischer, G.S.: A 3-axis optical force/torque sensor for prostate needle placement in magnetic resonance imaging environments. In: 2nd Annual IEEE International Conference on Technologies for Practical Robot Applications, IEEE, Boston (2009)

41. Su, H., Harrington, K., Cole, G., Wang, Y., Fischer, G.: Modular needle steering driver for MRI-guided transperineal prostate intervention. In: IEEE International Conference on Robotics and Automation, Workshop on Snakes, Worms and Catheters: Continuum and Serpentine Robots for Minimally Invasive Surgery, Anchorage, AK, USA (May 2010)

42. Su, H., Shang, W., Cole, G., Harrington, K., Gregory, F.S.: Haptic system design for MRI-guided needle based prostate brachytherapy. In: IEEE Haptics Symposium 2010. IEEE, Boston (2010)

43. Su, H., Zervas, M., Cole, G., Furlong, C., Fischer, G.: Real-time MRI-guided needle placement robot with integrated fiber optic force sensing. In: IEEE ICRA 2011 International Conference on Robotics and Automation, Shanghai, China (2011)

44. Su, H., Zervas, M., Furlong, C., Fischer, G.S.: A miniature MRI-compatible fiber-optic force sensor utilizing Fabry-Perot interferometer. In: MEMS and Nanotechnology, Conference Proceedings of the Society for Experimental Mechanics Series, pp. 131–136. Springer, Heidelberg (2011)

45. Sutherland, G.R., Latour, I., Greer, A.D., Fielding, T., Feil, G., Newhook, P.: An image-guided magnetic resonance-compatible surgical robot. Neurosurgery 62(2), 286–292 (2008), discussion 292–3

46. Suzuki, T., Liao, H., Kobayashi, E., Sakuma, I.: Ultrasonic motor driving method for EMI-free image in MR image-guided surgical robotic system. In: Proc. IEEE/RSJ International Conference on Intelligent Robots and Systems IROS 2007, pp. 522–527 (2007)

47. Terris, M.K., Wallen, E.M., Stamey, T.A.: Comparison of mid-lobe versus lateral systematic sextant biopsies in the detection of prostate cancer. Urol. Int. 59(4), 239–242 (1997)

48. Tokuda, J., Fischer, G.S.: OpenIGTLink: an open network protocol for image-guided therapy environment. Int. J. Med. Robot. 5(4), 423–434 (2009)

49. Tokuda, J., Fischer, G.S., DiMaio, S.P., Gobbi, D.G., Csoma, C., Mewes, P.W., Fichtinger, G., Tempany, C.M., Hata, N.: Integrated navigation and control software system for MRI-guided robotic prostate interventions. Computerized Medical Imaging and Graphics 34(1), 3 (2010)

50. Tsekos, N.V., Khanicheh, A., Christoforou, E., Mavroidis, C.: Magnetic resonance-compatible robotic and mechatronics systems for image-guided interventions and rehabilitation: a review study. Annu. Rev. Biomed. Eng. 9, 351–387 (2007)

51. Wallner, K., Blasko, J., Dattoli, M.: Prostate Brachytherapy Made Complicated, 2nd edn. Smart Medicine Press (2001)

52. Wang, Y., Cole, G., Su, H., Pilitsis, J., Fischer, G.: MRI compatibility evaluation of a piezoelectric actuator system for a neural interventional robot. In: Annual Conference of IEEE Engineering in Medicine and Biology Society, Minneapolis, MN, pp. 6072–6075 (2009)

53. Wang, Y., Su, H., Harrington, K., Fischer, G.: Sliding mode control of piezoelectric valve regulated pneumatic actuator for MRI-compatible robotic intervention. In: ASME Dynamic Systems and Control Conference - DSCC 2010, Cambridge, Massachusetts, USA (2010)

54. Webster I, R.J., Kim, J.S., Cowan, N., Chirikjian, G., Okamura, A.: Nonholonomic modeling of needle steering. International Journal of Robotics Research 25(5-6), 509–525 (2006)

55. Yang, B., Tan, U., Gullapalli, R., McMillan, A., Desai, J.: Design and implementation of a pneumatically-actuated robot for breast biopsy under continuous MRI. In: IEEE ICRA 2011 International Conference on Robotics and Automation, Shanghai, China (2011)

56. Yeniaras, E., Lamaury, J., Hedayati, Y., Sternberg, N.V., Tsekos, N.V.: Prototype cyber-physical system for magnetic resonance based, robot assisted minimally invasive intracardiac surgeries. International Journal of Computer Assisted Radiology and Surgery (2011)

A Robot for Surgery: Design, Control and Testing

Basem Fayez Yousef

Department of Mechanical Engineering,
United Arab Emirates University, Al-Ain, United Arab Emirates
basem_yousef@uaeu.ac.ae

Abstract. In this chapter, the design and control of a robotic arm for medical applications will be explained in detail. A general purpose gross positioning robot arm for use in robot-assisted medical applications is introduced. The actuated macro manipulator can carry, appropriately orient, precisely position and firmly "lock" in place different types of micro robots and surgical tools necessary for applications in minimally invasive therapy. With a simple manipulation protocol, the clinician can easily operate the robot in manual mode. Also a remote control mode can be enabled for teleoperation of the robot, and a quick homing routine ensures its re-calibration. Robot workspace analysis is investigated by two methods to study the robot's singularity regions where a 3D visualization technique is used to visualize the location of the singularity trajectories.

Performance analysis techniques using CAD software were used to quantify displacement and angular errors. It will be shown how the sophisticated configuration and joint architecture of the arm enable it to perform and interact efficiently with the constrained and limited workspace of surgical environments. The special features of the proposed robot make it well suited for use with new surgical tools and micro robots for a range of medical interventions.

1 Introduction

In recent years, robot-assisted procedures for surgery and therapy have received considerable attention and are often preferred over conventional, manually performed procedures because of a robot's ability to perform consistent precise movements free of fatigue and tremor, and carry out surgical procedures with high dexterity and accuracy beyond those of a surgeon. Extensive surveys of medical robotics and its applications in various medical fields are given in [1, 2].

Several kinds of special-purpose surgical tools and robotic systems have been developed to perform a variety of surgical procedures. These systems can be classified into two categories. The first category consists of robotic systems that comprise a manipulator and a surgical tool integrated into one complete system. Generally, these floor-mounted systems are bulky and occupy considerable space in the limited workspace of surgical environments. The second category adopts the macro-micro assembly architecture, e.g., see [3] and references therein, which has

T. Gulrez, A.E. Hassanien (Eds.): Advances in Robotics & Virtual Reality, ISRL 26, pp. 33–59.

proved successful in overcoming problems associated with the limited workspace inherent in therapy. In such architecture, a small "micro" robot [4] is attached to the end of a macro robot, which in turn is attached to a rigid frame fixed to the operating table. The macro robot is capable of relatively "large-scale" motions, and is used to place the micro robot in an appropriate position and orientation in the neighborhood of the operative field [5]. The compact sized micro robot can perform more precise manipulation of surgical tools/instruments such as needles, catheters and endoscopes.

The advantage of such system is that the potential for injury is reduced since the micro-robot is only capable of very limited motion. Also, it enables the system to achieve greater accuracy in manipulating surgical tools since the micro-robot operates independently of the macro. Davies et al. [6] developed a robotic system known as the PRobot – a robot for prostatectomy. The robot was required to perform transurethral resection of the prostate (TURP) by moving the cutting tool and actively removing tissue. It was mounted on a large floor-standing counterbalanced framework, which could be locked in position using electromagnetic brakes. It was claimed that the PRobot was the first robotic surgery application of the macro-micro concept. The Zeus^{TM} system [7] from Computer Motion Inc, USA, and the da Vinci system [8] designed for minimally invasive surgery have a master-slave structure where the slave can also be considered to have a macro-micro architecture. The Zeus (no longer available commercially) and the da Vinci have been experimentally used in a number of surgical procedures [9, 10].

Some of the disadvantages associated with the floor-standing systems are: 1) the vibration of the operating table can result in the robot losing its target information, thereby degrading the performance and accuracy of the robot; 2) special floor preparation is needed for the base of the robot and 3) the robot occupies a large amount of space, and as a result, the operating room is reserved mainly for the operations that can be performed using that robotic system.

Because of the limited workspace in surgical environments, special purpose micro robots (e.g., a needle insertion/manipulation mechanism) are usually designed to have compact sizes that are not sufficient to enable the robots to reach the surgical workspace from a reachable distance. Therefore, a sturdy macro robot is required to provide a firm movable base, which can be used to position and orient a micro robot, and then firmly lock it at an appropriate location in the proximity of the operative field. In the prostate brachytherapy application, this would be such that the brachytherapy needle touches the patient's peritoneum, where the needle manipulation micro robot can commence performing its intended specific task, such as needle insertion.

Robotic arms of different configurations have been designed and used to manipulate various kinds of micro robots, surgical tools and instruments that are commonly used for minimally invasive therapy. At the Urology Robotics Laboratory at Johns Hopkins University, the Grey Arm [11], a passive device was developed as a positioning and supporting device for surgical robots and instrumentations. In [12], a biopsy robot is described that uses a manually driven platform mounted on a trolley as a supporting base.

Several disadvantages can be associated with passive arms or those without position sensors (encoders) making them unsuitable for autonomous or semi-autonomous image-guided surgery and therapy or for use in procedures which involve the macro-micro architectures. For instance, the clinician positioning a passive arm would be required to not only hold the end-effector, but also support the arm, which may be tiring when manipulating a relatively complex and/or heavy micro-robot. An example of such a robot is a micro robot that operates under ultrasound image guidance to perform needle insertion in prostate brachytherapy as described in [13].

However, automating image-guided therapy and registering a medical image to the patient requires knowledge of the locations of both the medical image source (e.g. ultrasound) and the robot end-effector, with respect to a global coordinate system that is known relative to the patient. To achieve this, it is essential to unify the coordinate systems of the robot end effector frame and the image source frame in order for the robot to know its target point inside the patient through the ultrasound image. For example, referring to Fig. 1 which shows a schematic diagram for an image-guided robotic system for prostate brachytherapy, the robot can find its target point in the goal frame {G} provided that: 1) the relationship between the tool frame {T} and the station frame {S} is known, (note that this can be easily obtained from the kinematic equations of the robot); 2) the goal frame {G} is known with respect to the ultrasound probe frame {P}, (again this can be provided by the ultrasound imaging software); and 3) frame {P} must be known with respect to frame {S}, which is attainable from the kinematic equations of the ultrasound probe holder [14]. It should be noted that satisfying the 1st requirement entails the use of a robotic arm that is equipped with position sensors which can provide the location of the end effector. Thus the use of an encoded arm enables the reposition of a surgical tool during surgery without losing the robot-to-image registration.

However, in surgical environments, it is desirable to use robots with special configurations and joint architectures that can demonstrate high maneuverability since they can reduce the potential for robot-robot and robot-patient collisions as well as self-collisions.

This chapter presents a stand-alone general-purpose, actuated robotic manipulator that can serve as a gross-positioning system and can carry, appropriately orient, precisely position and firmly "lock" in place different types of micro robots and surgical tools such as endoscopes, ultrasound probes, needle insertion mechanisms and other tools necessary for minimally invasive therapy. It explains how the sophisticated design and the simple manipulation protocols of the robot arm can reduce or overcome the aforementioned obstacles associated with other robotic systems. This chapter consists of six sections organized as follows:

Section (2) provides a brief literature review on the important role of robots in surgery and robot assisted therapy; and outlines the objectives and significance of this chapter. Section (3) describes the mechanical design requirements of the robot arm, the link-joint architecture of the proposed macro robot arm that is customized for use in medical applications, and highlights the main features of the spherical work envelop of the manipulator. Also the motions that can be achieved by the robot are described. Moreover, the robot calibration will be covered in this section. Kinematic equations of the robot are derived in Section (3) and two manipulability analysis approaches are used to study the robot's manipulability and to investigate the singularity configurations of the robot throughout its workspace. A 3D visualization technique is introduced to depict the singularity trajectories in the robot's workspace. In Section (4), the dynamic model of the robot is derived where the terms of the dynamic model are utilized to design two manipulation protocols to operate the robot in two modes. Also, the robot "homing" routine that provides quick re-calibration is described. Experimental setup and performance evaluation procedures are described, analyzed, evaluated and discussed in Section (5). Finally, Section (6) summarizes the key conclusions of this project and the contributions of the proposed mechanism to the robot-assisted medical applications. Potential applications for the designed robot are provided.

2   Robot Description

The robot arm architecture comprises two specially customized links connected through a sophisticated structure of joints, three rotational and one prismatic, as shown in Fig. 2. To reduce the weight of the robot, the links and brackets are made of aluminum. The motors are selected with harmonic drive reduction gearing to provide backlash-free rotational reductions. The maximum payload is determined by the payload capability of the motor at joint 4, which can provide a maximum repeated torque of 20.3 N.m that corresponds to 10 kg payload at the end effectors when considering the perpendicular distance between the line of action of the payload at the end effector to the motor axis (note that the torque capacity of the motor at joint 2 is 90 N.m, and 20.3 N.m for each of joints 2 and 3).

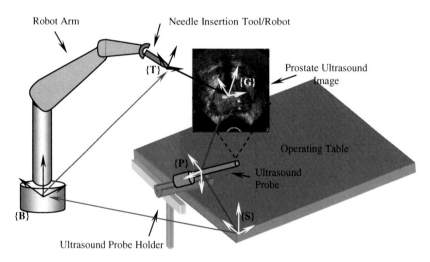

Fig. 1. Schematic diagram of a robotic system for image-guided robot-assisted Brachytherapy.

The design criteria of the robot stem from the drawbacks and limitations associated with other robotic systems, and from the workspace constraints inherent with the surgical environments. Also, the robot will provide the necessary degrees of freedom and a large work envelope sufficient to hold relatively large micro robots. It also can be operated with a simple manipulation protocol and will have an easy-to-reconfigure design for future modifications. Besides, it has to satisfy certain safety requirements.

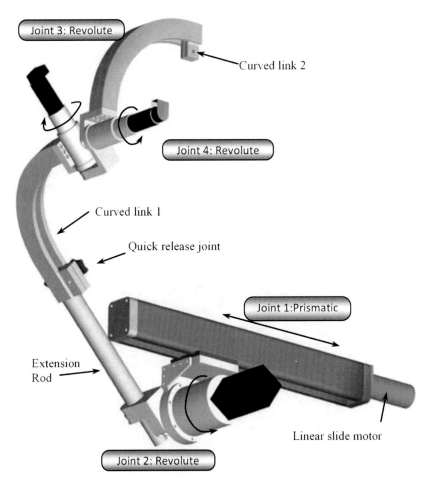

Fig. 2. Macro robot joint structure and components.

2.1 Dexterity and Work Envelope

Robot designers tend to increase the number of joints and links to achieve higher dexterity and improve a robot's maneuverability. This results in cumbersome and bulky robots filling the limited and constrained workspace of the operating room, and increases the chances for collisions during operation. The proposed macro-robot is customized with only two curved links connected by a double-motor joint (i.e. motors 3 and 4 shown in Fig. 2) that enables rotation about 2 perpendicular axes forming a spherical workspace. In addition, the rotational joint formed by motor 2 provides extra height to the end-effector and extends the height of the reachable workspace, which provides the greater space needed for larger and longer end-effectors. Moreover, the prismatic joint formed by the linear slide enables the motion of the end-effector along the operating table providing gross positioning over the patient. This joint-architecture provides the robot with increased dexterity that translates into high maneuverability, thereby enabling it to perform and interact efficiently, in a fairly constrained surgical workspace, with other existing equipment in the operating room and with the patient (Fig. 3). Also, this architecture of the robot links provides the clinician an open field of view to observe the patient from any preferred angle, and enables him/her to easily reach the patient, when needed.

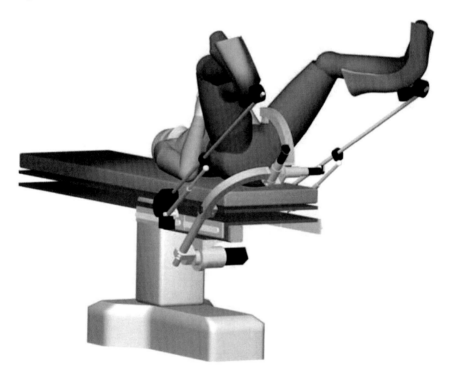

Fig. 3. The macro robot's high dexterity enables efficient functioning in a constrained surgical space such as in prostate brachytherapy.

By attaching the macro robot to the operating table, the following advantages are achieved: 1) any relative motion of the table with respect to the robot arm will be suppressed, i.e., accuracy is maintained; 2) special floor preparation and the need for additional space for the robot are obviated; 3) the robot structure allows for effective space utilization even when it is not in use because its links fold back in-plane to its home position (Fig. 4), thereby occupying minimal space when stored.

Fig. 4. The robot folds back occupying minimal storage space.

The robot's end effector can reach a height of 700 mm above the top of operating table, and covers a width of 560 mm which is close to the standard width of the operating tables (which range from 500-600 mm). However, the distance that the robot can cover along the operating table depends on the length of the linear slide stage at the prismatic joint. The linear stage which is used for the prototype can travel for a distance of 500 mm.

2.2 Safety

The robot is equipped with a "power-fail-safe" braking system, which consists of normally-locked 24V DC brakes at the motor joints to prevent the robot from collapsing in case of a power failure or unexpected power shutdown. The brakes are automatically engaged when the power is switched off. However when the robot is in operation mode, the brakes are easily released/engaged through the controller with a single command upon the clinician's request. A common problem with these kinds of brakes is that they may change the position of the end effector slightly while locking; nonetheless, this does not affect the performance of the robot in locking a micro robot at an approximate initial position.

In addition, to enhance safety, a simple quick-release joint is added between the rod and curved-link 1 (Fig. 2), so that in case of emergency, the clinician can quickly move the robot arm, away from the patient providing him/her full access to the patient. Besides, since it is desirable to be able to remove the mechanism for sterilization or to free the operating table for other procedures and to allow for accommodating different surgical tools/equipment necessary for other procedures, this quick-release joint simplifies the process of attaching the robot arm to and detaching it from the operating table so that, in a short time, the operating room can be made available for other procedures that do not use the robot (Fig. 4).

Detailed specifications of the macro robot are given in Table 1.

Table 1. Macro Robot Design Specifications

Consideration	Specification
Weight	Links and moving parts < 6 kg
Material	Non-corrosive
Payload	10 Kg at the end effector
Architecture	Prismatic-Revolute-Revolute-Revolute (PRRR)
Degrees-of-freedom	4
Braking system	Electrical actuation (24V DC), normally locked
End-effector	Easy attachment
Set-up time	< 1 minute
Workspace envelope (provided by revolute joints 3 and 4)	Spherical with radius = 210 mm
Extended workspace envelope	- 500 mm travel along operating table for gross positioning of the end effector - 600 mm wide - 500 mm above operating table (to accommodate longer surgical tools/micro robots
Operating modes	Manual and autonomous

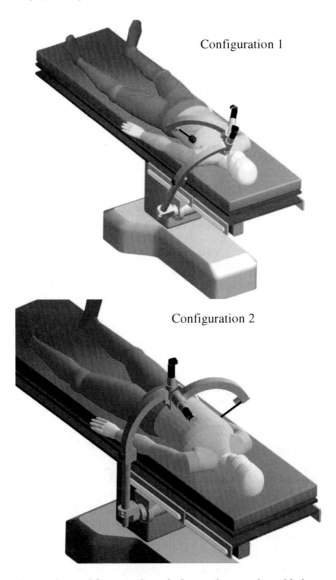

Configuration 1

Configuration 2

Fig. 4. The robot can be used for several surgical procedures such as ablative procedures for lung cancer treatment.

It is important to know that every time the controller is reset, or the control cabinet computer is reset, a "homing" routine must be used to send the robot to a home position and reset the control parameters such that the initial values of the controller parameters reflect the initial configuration of the robot at home position. Also this will ensure that the robot is accurately re-calibrated at set up. Our "homing" routine implements proportional-integral-derivative (PID) velocity control with Coulomb friction compensation. The set up time takes less than one minute and the "homing" routine takes less than one minute to complete.

2.3 Other Features

The robot is designed to serve as a gross positioning system, thus it can be beneficial for several surgical applications that are performed on the lower parts of the body such as prostate brachytherapy (Fig. 3), as well as the upper part of the body such as ablative therapies for lung cancer treatment (Fig. 5). The robot can hold active surgical tools as well as passive tools. Hence, laparoscopes, endoscopes, and needle insertion tools can be attached to the robot's end effector. This broadens the range of medical applications which can benefit from the use of the robot. Of course, a special tool-adapter will be needed to enable attaching the tool to the robot end.

3 Manipulation Protocols

The robot can be operated in two modes: remote and manual control modes. The controllers are designed such that the user can easily switch between the two modes easily. However, the selection of the operating mode depends on the application as explained next.

3.1 Manual Compliant Force Control

Graphical user interfaces (GUIs) are widely used to manipulate robots since they provide simple tools to command the robot. Nevertheless, the clinician needs to undergo intensive training to become familiar with the GUI. The operational protocol presented in this control mode obviates the need for training, thereby making manipulating or commanding the robot very simple. In this intuitive scenario, the clinician drags the end effector with minimal force to the desired position, where the robot controller, in response to the force exerted by the clinician, takes over moving the arm to the intended location so that the surgical tool is placed at the appropriate location. If no force is applied at the end effector, the controller holds the robot arm with the attached tool at its position until the clinician finalizes the robot's pose by locking the brakes with a single push button.

The controller adopts a simple compliant force algorithm that consists of gravity and friction compensation so that the torque τ_j applied at joint j is given by:

$$\tau_j = G_j(\theta) + \tau_{j,friction} \qquad (1)$$

where $G_j(\theta)$ is the gravity term responsible for holding the robot against its own weight (which includes the weight of the attached micro robot), and $\tau_{j,friction}$ is the torque needed to overcome the motor gearhead friction at joint j.

In order to obtain the gravity term, the link-frame assignment shown in Fig. 6 is used in the Newton-Euler formulation to derive the dynamic model of the robot [14]. The gravity term $G_j(\theta)$ in (1) can be extracted from the equation that describes the dynamics of the 4-joint robot given by:

$$\tau = M(\theta)\ddot{\theta} + V(\theta,\dot{\theta}) + G(\theta) + F(\theta,\dot{\theta}) \tag{2}$$

where $M(\theta)$ is the 4×4 mass matrix of the manipulator, $V(\theta,\dot{\theta})$ is the 4×1 vector of centrifugal and Coriolis terms, $G(\theta)$ is the 4×1 vector of gravity terms and $F(\theta,\dot{\theta})$ accounts for all friction forces at the joints.

Since friction is a complicated function of lubrication and other effects, calculating $F(\theta,\dot{\theta})$ can be very difficult. A reasonable model for the friction torque $\tau_{friction}$ is given by [14]:

$$\tau_{friction} = c\,sgn(\dot{\theta}) + v\dot{\theta} \tag{3}$$

where $c\,sgn(\dot{\theta})$ is Coulomb friction which is constant except for a sign dependence on the joint velocity, c is Coulomb friction constant, $v\dot{\theta}$ is the viscous friction and v is the viscous-friction constant. This friction model is depicted in Fig. 7.

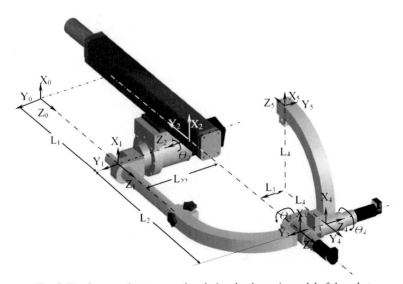

Fig. 5. The frame assignment used to derive the dynamic model of the robot.

However, since for safety, the robot will operate at low velocities, the viscous friction term is ignored and the controller in (2) will only compensate for the Coulomb friction, and $\tau_{j,friction}$ is calculated based on the step function shown in Fig. 8. Thus when the clinician applies a force at the end effector causing the motor shafts to rotate, which in turn will change the encoder readings, the controller uses these changes to calculate the rotational velocities at the motors shafts. When the velocities exceed preset threshold values, this will be interpreted by the controller as a request to move, thus the controller provides joint torques equal to the gravity term $G(\theta)$, plus a friction torque that is equal to the Coulomb friction force gearheads, i.e., $c\,sgn(\dot{\theta})$. As a result, the clinician will be able to move the robot with minimal dragging force exerted at the end effector. On the other hand, if the clinician leaves the robot arm, the calculated torque according to (1) is not enough to move the robot by itself, so the rotary encoder readings remain constant causing the calculated rotational velocity to drop to zero, and thus the controller produces no friction compensation torque $\tau_{friction}$ but only the gravity term that holds the robot without motion. This control scheme enables the user to manipulate and move relatively heavy end effectors without feeling their weight. It's worthwhile mentioning that the software monitors the joint velocities and the torque/voltage on the joint motors. Those values are not allowed to exceed saturation values during operation to ensure low joint velocities even if the algorithm calculations produce higher values.

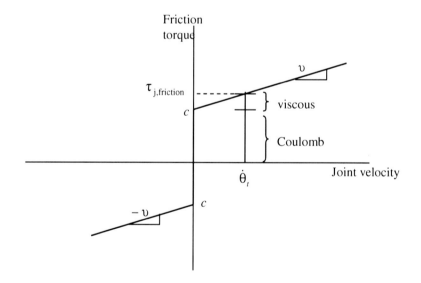

Fig. 6. Friction torque -vs- joint velocity for joint j.

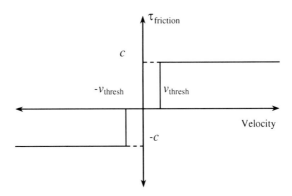

Fig. 7. Joint friction compensation is defined as a function of the joint velocity.

According to (2), it is obvious that if the robot is gravity balanced, i.e., the joint accelerations $\ddot{\Theta}$ and the joint velocities $\dot{\Theta}$ are equal to zero; the torque which will balance the robot and prevent it from collapsing is equal to the gravity term plus the gearhead friction force. Although we obtained the parameters of the arm from CAD, a dynamic identification method of the robot parameters would enable the algorithm to automatically adapt to various changing payloads of the micro robot.

It should be noted that when the controller provides only the Coulomb friction torque in addition to the gravity term, it will not cause any motion at the joints, nonetheless, the joint behaviors will be frictionless and rotation can be caused by minimal force, where the velocity will depend on the viscous force.

For each joint j, $\tau_{j,friction} = c \ sgn(\dot{\theta})$ is obtained experimentally at $\dot{\theta} = 0$ by increasing the torque, i.e., voltage in 0.00122 volt steps while the motor encoder is monitored to determine the torque value that causes the motor to start rotating. This step was repeated 20 times and the average value was used as $\tau_{j,friction}$.

For the prismatic/linear slide joint, the friction force at this joint is mainly caused by the linear slide motor, the ball power screw inside the linear slide, and the sliding carriage. However, in this case the friction compensation term calculated by the controller for this joint exceeds the friction at the joint parts such that it causes the carriage to slide with a desired preset velocity. Accordingly, if the clinician needs to slide the robot over the patient's body, an initial push/pull is needed first to cause a change in encoder value, producing velocity that exceeds a preset threshold value, after which the controller moves the robot by itself at a constant pre-determined velocity in the same direction of the triggering force. If the sliding robot is held by the clinician causing the sliding velocity to drop to zero (or less than the threshold value), the controller stops producing the friction torque and thus, the linear slide stops.

The velocity threshold values are chosen low such that the controller is sensitive enough to respond to the triggering force applied by the clinician, but high enough to filter out any vibrations that may be sensed by the joints' encoders and not caused by the clinician. The choice of these threshold velocity values also determines how much initial force the clinician needs to exert at the robot's end effector before the controller starts producing the appropriate torques at the different joints.

Once the robot is placed at a desired pose (position/orientation) as decided by the clinician, the brakes are applied to lock the robot in place and finalize its pose by a single push button by the clinician after which, the brakes are engaged and the macro robot controller is switched off allowing the attached micro robot or surgical tool to perform its intended task.

One can notice that the control algorithm works in a similar manner to other compliant force controllers that use expensive force sensors and implement complex force feedback loop computations. However, the controller adopted for this manual control mode operates without using any force sensor, hence significantly reducing the cost. Also, since it uses closed form equations, this makes the controller easy to maintain and modify. Besides, the simple, yet intuitive manipulation protocol allows the clinician to operate the robot not only without the need for special training but also without using conventional tools such as joysticks or foot pedals. In addition, another advantage of the controller is that it avoids the robot singularities problem since no Cartesian level control is performed, i.e., the inverse Jacobian matrix [14] does not appear in the control calculations. However, it should be noted that when using a different surgical tool, the controller equations may need to be modified to handle the new weight of the surgical tool and/or the shift of its center of gravity (CG). Hence, in order to handle an occasional change in the surgical tool weight or CG, we developed a Matlab code that generates the necessary controller equations based on the surgical tool's weight and CG.

3.2 Autonomous Remote Control

In some applications, e.g., robot teleoperation, GUIs can be very useful to remotely control the robot whereby a clinician can control the motion of the robot from another location. The robot consists of a wide variety of hardware and software components. These include sensors and their respective control programs, navigational and control strategies and tools for inter-process communication. It is important to develop an intuitive, hierarchical interface which will allow non-technical users to control the robot arm. Hence, the details of the individual programs used to control robot operation are hidden from the user. In these scenarios, the user specifies a desired task, which then becomes the job of the interface to coordinate the actions of the necessary programs in order to perform this task.

For this control mode, the following steps are required:

• Graphical user interface (GUI):

A CAD model of the robot is prepared using SolidWorks, which is used as the Graphical User Interface (GUI). A design spreadsheet table that consists of the joint angles is prepared using the excel tool that is integrated in SolidWorks, and linked with the CAD model such that its values reflect the actual values of the joint angles and offset distances of the robot's CAD model (Fig. 9).

- PID joint controller with friction compensation:

A partitioned control model that implements a proportional-integral-derivative (PID) joint control (Fig. 10) is implemented in C++ to move the robot to track a pre-defined trajectory. In this model, the joint torque τ is given by [14]:

$$\tau = M(\theta)\tau' + V(\theta,\dot{\theta}) + G(\theta) + c\,sgn(\dot{\theta}) \qquad (4)$$

where the terms are as defined in (2) and (3).

$$\tau' = \ddot{\theta}_d + k_p e + k_v \dot{e} + k_I \int e.dt \qquad (5)$$

where the error signal $e = \theta_{desired} - \theta$ and k_p, k_v and k_I are the PID gains.

In this mode, the user drags the end effector of the CAD model and sets the robot in the desired pose/configuration. If SolidWorks detects a potential collision between two parts of the robot or between the robot and the table, the surfaces of the colliding parts are highlighted with a specific color warning the user about the collision. This feature allows the user to confidently make decisions regarding the final robot pose before commanding the controller to move the robot to the corresponding pose.

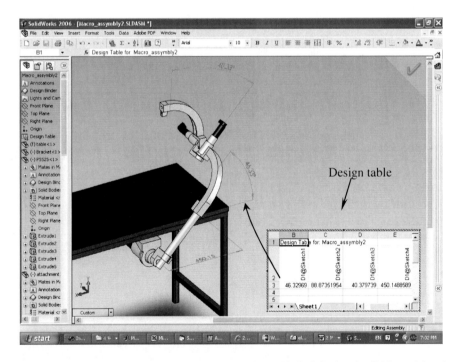

Fig. 8. SolidWorks is used as a GUI whereby a design table is linked to the CAD model and is updated according to the joint angle readings of the CAD model.

After the user chooses the desired robot pose / configuration the "*update*" tool of SolidWorks is used to update the excel spreadsheet values with the final joint angles values. These values are stored in a text file and become the input to the control algorithm as the vector of the final joint angles Θ_f .

After the user finalizes the robots pose in CAD, he or she can instruct the robot to move to the chosen end effector location by running the controller, which will move the robot along a pre-defined trajectory. A quintic polynomial is used as our trajectory generation function since it provides continuous and smooth profiles for Θ_d, $\dot{\Theta}_d$ and $\ddot{\Theta}_d$. Fig. 11a and 11b show the joint-angle tracking response and tracking error of revolute joint 4. The maximum tracking error of $\approx 0.4°$ is attributed to the friction caused by the high gearhead ratio which is 1:100; nevertheless, this does not affect achieving the objective of the controller of generating smooth motion profiles.

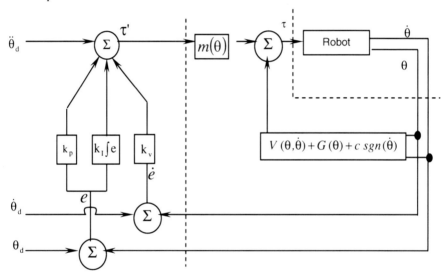

Fig. 9. Partitioned PID joint-level controller model used for autonomous/remote control mode.

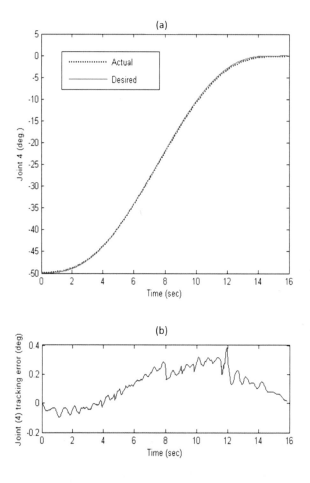

Fig. 10. Joint 4 tracking response under partitioned PID joint control. (a) Joint angle tracking response. (b) Tracking error.

4 Robot Manipulability

The condition number of the robot's Jacobian matrix, $\kappa_i(J)$, which is a performance measure of the robot's manipulability [15] at a particular configuration "i", is defined as the ratio of the maximum singular value, $\sigma_{max}(J)$, to the minimum singular value, $\sigma_{min}(J)$, of the Jacobian matrix J, of the robot in configuration "i".

$$\kappa_i(J) = \sigma_{max}(J) / \sigma_{min}(J) \qquad \kappa_i(J) \in \langle 1, \infty \rangle \qquad (6)$$

However, it is more convenient to use the reciprocal, $1/\kappa_i(J)$, which normalizes the values of this performance measure index between 0 and 1 rather than 1 and ∞. At a point "i", a $1/\kappa_i(J)$ value of zero means that the robot is in a singular configuration, and a value of 1 means that the robot has the best possible kinematic conditioning. Also, if the robot is operating in a region close to singularity, the $1/\kappa_i(J)$ values will be closer to zero.

Before we analyze the robot's manipulability, it should be noted that in order to avoid collisions between the robot and the operating table, the linear slide should not operate unless the robot parts, i.e., the curved links and motors 3 and 4, are above the table's top. Accordingly, the prismatic joint must be operated independently from the 3 revolute joints. Therefore, we excluded the prismatic joint from the kinematic analysis. This allows for the reduction of the Jacobian matrix dimension to become 3×3 making the task of performing matrices operations on the 3×3 square matrix more convenient.

To analyze our robot's performance throughout its entire workspace, the Jacobian matrix was derived considering only the revolute joints. The workspace formed by the three revolute joints consists of a sphere formed by joints 3 and 4, and this sphere rotates in an orbit about the axis of joint 2. The determinant of the 3×3 Jacobian matrix is found to be a function of joint angles θ_3 and θ_4, and not of joint angle θ_2, implying that the singularities for the manipulator are independent of θ_2. Therefore, using the forward kinematic equations, the coordinates of the end effector that correspond to each pair (θ_3, θ_4) are calculated, thus resulting in a 3D spherical subspace which is formed by joints 3 and 4. Also, $1/\kappa_i(J)$ values were calculated for each point on the sphere and these values which range from 0 to 1 were mapped into a $\langle 0,255 \rangle$ grayscale range. Then the result was displayed in 3D as shown in Timeswherein each point on the sphere corresponds to a target point "i" and its shade indicates its $1/\kappa_i(J)$ value where white corresponds to $1/\kappa_i(J)=0$ and the black to $1/\kappa_i(J)=1$. In other words, the mechanism is in singular configuration when the end effector passes through the white regions of the sphere.

It is obvious from Fig. 12 that the robot is in singular configuration at the unusable outer boundaries or unachievable poses of the spherical workspace, e.g., when the curved links overlap or when curved link 2 rotates against the holding brackets. It should be noted that in the previous analysis, the sphere represents the theoretical workspace that includes unachievable and/or unusable poses of the robot, e.g., the lower hemisphere.

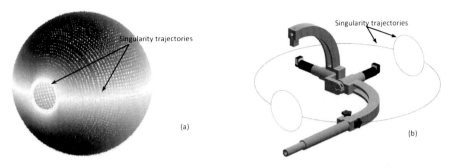

Fig. 11. Robot manipulability analysis using (a) condition number (b) Jacobian matrix determinant.

5 Performance Evaluation

5.1 Unloaded Macro Robot Testing

The first set of experiments involved evaluating the accuracy and controller performance of the unloaded macro robot. The displacement and angular accuracies of the robot were quantified by locking the robot at arbitrary targets and at each target, comparing the actual location and orientation of the end effector against the "theoretical" location. To obtain the actual location of the end effector, the robot was dragged and locked at an arbitrary target "i" (Fig. 12a). Using a caliper of 0.01 mm accuracy, the coordinates (x_{ij}, y_{ij}, z_{ij}) of 5 selected points typically corners, on the end effector were measured relative to an origin on the operating table (Fig. 13a, b) (where $i = 1, 2, \ldots n$, target points, and $j = 1, 2, \ldots, 5$, arbitrary points/corners on the end effector). Level gauges and a precision square were used to assure perpendicularity of the measured dimensions to the reference planes on the table.

On the other hand, the "theoretical" coordinates of the selected points can be obtained either using the forward kinematic equations of the robot or directly from a CAD software such as SolidWorks. For the first method, after reading the joint angles from the encoders, the end effector location was directly calculated using the forward kinematics equations of the robot which can be derived using the link-frame assignments shown in Fig. 6 as explained in [14]. In the second method, a CAD model of the robot was developed and the robot was moved in the CAD environment to the target location "i" again using the joint angles provided by the encoders. Then the theoretical coordinates of the 5 selected points (Fig. 13b) were measured directly from the CAD model. For this set of experiments, the second method was used to obtain the theoretical coordinates.

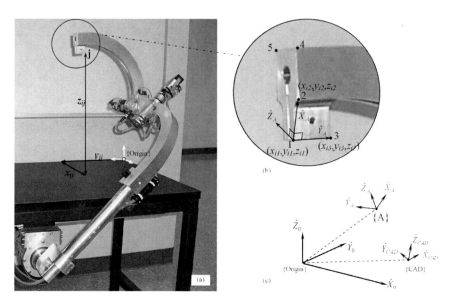

Fig. 12. (a) The macro arm positioned at a target point "i" and the {origin} reference frame is attached to the table; (b) For each target point "i", the coordinates of 5 arbitrary points were determined, where 3 points were used to define a frame {A} that describes the actual orientation of the end effector; (c) The relationship between the actual location frame {A}, the theoretical location frame {CAD} of the end effector, and the reference frame {Origin}.

The displacement error, E_i, for the robot at target "i" is calculated by

$$E_i = \sqrt{\overline{(\Delta x_i)}^2 + \overline{(\Delta y_i)}^2 + \overline{(\Delta z_i)}^2} \qquad (7)$$

where $\overline{(\Delta x_i)}^2 = \dfrac{1}{5} \displaystyle\sum_{j=1}^{5} \left(x_{ij} - x_{ij\,CAD} \right)$, is the mean of the error in the x-direction

for the 5 selected points on the end effector for target "i"; x_{ij} is the x coordinate of point "j" for robot target "i"; and $x_{ij\,CAD}$ is the x coordinate of point "j" for robot target "i" obtained from the CAD model. The mean error, $\overline{E}_t$, for the robot tested at $n = 7$ target points was calculated using:

$$\overline{E}_t = \dfrac{1}{n} \displaystyle\sum_{i=1}^{n} E_i \qquad (8)$$

The data acquired as described before were used to quantify the angular error as follows:

1) Frame {A} is attached to the end effector with the origin at point 1, where the x-axis, $\hat{X}_A$ points in the direction from point 1 to 2, the y-axis, $\hat{Y}_A$ points from point 1 to 3 and the z-axis direction, $\hat{Z}_A$ can be obtained from the cross product of the vectors defining $\hat{X}_A$ and $\hat{Y}_A$ (Fig. 13b). Similarly, frame {CAD} is attached to point 1 on the CAD model, and the axes $\hat{X}_{CAD}$, $\hat{Y}_{CAD}$ and $\hat{Z}_{CAD}$ of the frame are defined in the same way {A} is defined. It can be seen that frames {A} and {CAD} define the actual and the theoretical orientations, respectively, of the end effector for the robot at an arbitrary target "i".

2) The rotation matrices $_A^oR$ and $_{CAD}^oR$ that describe the orientations of frames {A} and {CAD} respectively, with respect to the reference frame {origin} (Fig. 13a), can be obtained by:

$$_A^oR = [\,\hat{X}_A \quad \hat{Y}_A \quad \hat{Z}_A\,] \tag{9}$$

$$_{CAD}^oR = [\,\hat{X}_{CAD} \quad \hat{Y}_{CAD} \quad \hat{Z}_{CAD}\,] \tag{10}$$

Accordingly, the rotation matrix, $_A^{CAD}R$, that describes the orientation of frame {A} with respect to frame {CAD} is given by:

$$_A^{CAD}R = {_{CAD}^oR}^{-1}\,{_A^oR} \tag{11}$$

3) There are many methods [14] that can be used to quantify the angular error of frame {A} with respect to frame {CAD}. We used the "X-Y-Z fixed angles" method, sometimes called the "roll-pitch-yaw angles" in which frame {CAD} is fixed and frame {A} is rotated about $\hat{X}_{CAD}$ by an angle γ, then about $\hat{Y}_{CAD}$ an angle β, and finally about $\hat{Z}_{CAD}$ an angle α. In order to obtain the angles α, β and γ, let the elements of the rotation matrix $_A^{CAD}R$ obtained in equ(11) be:

$$_A^{CAD}R = \begin{bmatrix} r_{11} & r_{12} & r_{13} \\ r_{21} & r_{22} & r_{23} \\ r_{31} & r_{32} & r_{33} \end{bmatrix} \tag{12}$$

then the angles can be calculated by [14]:

$$\beta = \text{Atan2}\left(-r_{31}, \sqrt{r_{11}^2 + r_{21}^2}\right) \tag{13}$$

$$\alpha = \text{Atan2}\left(r_{21}/\cos(\beta), r_{11}/\cos(\beta)\right) \tag{14}$$

$$\gamma = \text{Atan2}\left(r_{32}/\cos(\beta), r_{33}/\cos(\beta)\right) \tag{15}$$

The robot was tested under its own payload of 5.4 kg, which consists of the weight of the links, brackets and the two motors at joints 3 and 4. Table 2 shows that the robot operates within a mean displacement error, $\overline{E_t}$ =0.58 mm and mean angular errors $\overline{\alpha}$ =0.26°, $\overline{\beta}$ =0.26°, and $\overline{\gamma}$ =0.38°.

Also, the controllers performed well in remotely controlling the robot in teleoperation mode, and in providing adequate torques under compliant force control calculated as explained earlier. It is worthwhile mentioning that it is safer if the controller provides the friction torque $\tau_{j,friction}$ that is equal to the Coulomb friction such that it will not cause motion of the robot on its own when a triggering force is removed from the end effector. Therefore, our manual control mode only helps the user to move the robot without feeling neither the friction force, nor the actual weight of the robot arm and the attached micro robot regardless of their weights, but will not move the robot after the triggering force is removed. To illustrate this, consider the case of attaching a micro robot at the end effector and assume that the initial elevation of the micro robot is very close to the operating table surface. If the controller is programmed to move the robot once the end effector is tapped downwards even after removing the force, unexpected motion of the robot may crush the micro robot against the table. This may damage the micro robot or impair its calibration. Hence the controller must help the user to move the robot with minimal force at the end effector, but not be able to move the robot by itself if no force is exerted at the end effector. Male and female users who operated the robot for testing reported that the robot can be moved easily with minimal force without feeling the total payload (i.e. robot and dead weight). This indicated consistency in the controllers' performance under different payloads.

Table 2. The Mean Displacement Error and Angular Errors for the Unloaded Macro Robot

i	$\overline{\Delta x_i}$	$\overline{\Delta y_i}$	$\overline{\Delta z_i}$	E_i (mm)	$\alpha°$	$\beta°$	$\gamma°$
1	0.13	0.24	0.28	0.42	0.50	0.40	0.40
2	0.12	0.56	0.20	0.61	0.30	0.09	0.11
3	0.37	0.23	0.21	0.39	0.15	0.25	0.83
4	0.67	0.12	0.15	0.70	0.10	0.49	0.63
5	0.39	0.50	0.27	0.70	0.20	0.34	0.15
6	0.14	0.50	0.24	0.59	0.30	0.09	0.11
7	0.34	0.49	0.11	0.63	0.29	0.18	0.45
Mean				**0.58**	**0.26**	**0.26**	**0.38**

5.2 Loaded Macro Robot Testing

The second set of experiments involved evaluating the accuracy and controller performance of the loaded macro robot with a dead weight of 6 kg attached at the end effector (Fig. 14).

Again, the angular and translational accuracies are quantified by the same method used for the unloaded robot. Thus, the robot is moved and locked at arbitrary positions and at each position, the actual location and orientation of the end effector are compared against the corresponding "theoretical" location and orientation.

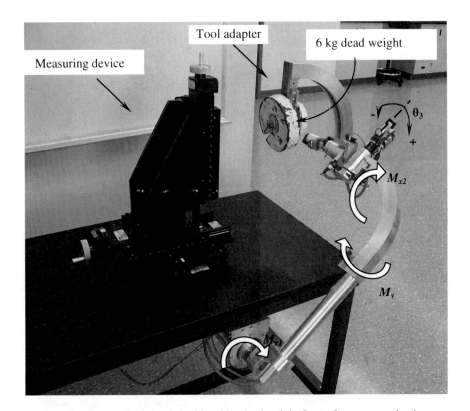

Fig. 13. Macro robot is loaded with a 6 kg dead weight for performance evaluation.

To obtain the actual location of any point on the end effector, a three-axis stage (from Parker Hannifin Co., Irwin, PA) (Fig. 14) which has an accuracy of 2 μm and a workspace envelope of 150 x 150 x 150 mm, was used to measure the coordinates (x_{ij}, y_{ij}, z_{ij}) of 3 selected points on the end effector relative to a chosen origin that was selected in the middle of the work envelope of the measuring device.

Again, the three points are chosen so that the two edges/lines connecting between them form a right angle, where these lines are used to define a coordinate frame {act} that is affixed to the end effector and is used to define the robot end effector's actual orientation as explained earlier and as shown in Fig. 13a, 13b.

On the other hand, the corresponding theoretical values can be obtained as follows: After reading the joint angles from the encoders, using the link frame assignment shown in Fig. 6 and the DH parameters of the robot, it is easy to derive both the forward kinematics equations [14] that provide the theoretical locations of the selected points on the bracket based on the encoders' angles, and the rotation matrix which can be used to describe the orientation of a frame {theo} with respect to the origin frame.

The rest of the mathematical calculations are performed as explained in the unloaded robot tests and the subscript *CAD* is replaced by *theo* to denote the parameters that are obtained using the kinematic equations and the orientation matrices.

Referring to Fig. 14, the payload of the attached weight causes bending moments M_{x1}, M_{x2} and M_y. These moments result in degrading of the accuracy of the robot since the resulting deflections of the robot arm due to these moments are not accounted for by the encoders' readings. Also, they cause flexing in the harmonic drive gearheads which encounter reduced stiffness when loaded. Therefore, the performance evaluation experiments for the loaded macro robot are split into two groups. In the first group, revolute joint #3 is locked at the zero position (as shown in Fig. 14) to reduce flexing in its gearhead due to the effect of the bending moment M_y, while the other joint angles are changed. Table 3 shows that under the constraint of fixing joint #3, the robot operates within a mean displacement error, $\overline{E_t}$ =0.45 mm and mean angular errors $\overline{\alpha}$ =0.15°, $\overline{\beta}$ =0.84°, and $\overline{\gamma}$ =0.19°. One can notice that the angular error $\overline{\beta}$ increases when loading the robot because both the flexing of the gearhead and the deflection of the arm increase due to the effect of M_y. Moreover, in the second group of experiments when lateral motion is involved by moving joint # 3, the flexing in the gearhead of joint #3 becomes larger due to the increased effect of M_y, thus the angular error as can be seen in Table 4 increases ($\overline{\beta}$ =0.92°). Therefore, heavier payloads cause larger deflections/errors. This implies that in the case of micro robots of weights < 6 kg, the error of our robot will be smaller than 1.23 mm (Table 4). Recall that in general, surgical micro robots are designed with compact sizes and lighter weights because of the limited and constrained work space in surgeries. It should be noted that both the maximum translational and angular errors are experienced at the extremes of the workspace.

Furthermore, the controller performed successfully in helping the user to manipulate the loaded robot with ease and simplicity without feeling its weight of 11.4 kg (5.4 kg robot arm + 6 kg dead weight). Again, for the compliant force control mode, the controller is programmed to move the robot only if a force is applied at the end effector, and to stop once the force is removed. One potential disadvantage of a robot with one-sided support is the lower accuracy due to the deflection at the end effector. However, this can be compensated for by an offset with the higher accuracy of the micro. Also, since the main source of inaccuracy is due to the bending moments that cause deflections, higher accuracy can be achieved by increasing the cross section of the robot..

Table 3. Mean Displacement Error and Angular Errors for the Loaded Macro Robot When Joint # 3 is locked at its zero Position

i	$\overline{\Delta x_i}$	$\overline{\Delta y_i}$	$\overline{\Delta z_i}$	$E_i(mm)$	α°_{macro}	β°_{macro}	γ°_{macro}
1	0.73	0.06	0.40	0.83	0.19	0.81	0.23
2	0.55	0.22	0.08	0.60	0.15	0.80	0.02
3	0.24	0.20	0.08	0.33	0.29	0.88	0.28
4	0.66	0.09	0.30	0.73	0.01	0.65	0.12
5	0.07	0.11	0.31	0.34	0.00	0.92	0.08
6	0.27	0.12	0.22	0.37	0.17	0.94	0.49
7	0.44	0.09	0.05	0.46	0.02	0.75	0.16
8	0.12	0.08	0.11	0.18	0.55	0.68	0.28
9	0.15	0.33	0.12	0.38	0.03	1.09	0.12
10	0.20	0.18	0.15	0.31	0.20	0.83	0.07
Mean				**0.45**	**0.16**	**0.84**	**0.19**

Table 4. Mean Displacement Error and Angular Errors for the Loaded Macro Robot When all Joints are moved

i	$\overline{\Delta x_i}$	$\overline{\Delta y_i}$	$\overline{\Delta z_i}$	$E_i(mm)$	α°_{macro}	β°_{macro}	γ°_{macro}
1	0.48	0.73	0.10	0.88	0.05	0.81	0.28
2	0.99	0.93	0.44	1.43	0.15	0.55	0.13
3	0.24	1.26	0.23	1.31	0.18	0.84	0.09
4	0.22	0.40	0.37	0.59	0.08	0.92	0.30
5	1.54	0.59	0.89	1.87	0.22	1.27	0.01
6	1.38	0.03	0.32	1.42	0.01	1.06	0.24
7	1.03	0.37	0.32	1.14	0.63	0.96	0.52
Mean				**1.23**	**0.19**	**0.92**	**0.22**

6 Conclusions

A dexterous robot arm with a sophisticated configuration and joint-structure has been designed and built to perform as a gross positioner that can carry, accurately position and orient different types of surgical tools and end effectors for image-guided robot-assisted therapy. While the robot's normally-locked braking system and quick-release joint enhance its safety features, its compliant force controller provides a simple and intuitive manipulation procedure with no training needed. Also, the remote control mode can be enabled for teleoperation of the robot. Robot workspace analysis showed that the singularity regions are outside the usable work envelope of the robot. A reliable validation technique using CAD models was used to evaluate the performance of the robot. The compact-sized robot demonstrated high accuracy and maneuverability and proved suitable for use in constrained surgical environments. The special features of the proposed robot make it well suited for use with new surgical tools and micro robots for a range of medical interventions such as those described in [3, 13, 16].

Acknowledgments

This research was supported by the Natural Sciences and Engineering Research Council (NSERC) of Canada under the Collaborative Health Research Project Grant #262583-2003 (PI: R.V. Patel).

References

1. Cleary, K., Nguyen, C.: State of the art surgical robotics: Clinical applications and technology challenges. Computer Aided Surgery 6(6), 312–328 (2001)
2. Stoianovici, D., Webster, R., Kavoussi, L.: Robotic Tools for Minimally Invasive Urologic Surgery. In: Complications of Urologic Laparoscopic Surgery: Recognition, Management and Prevention (2002),
   http://pegasus.me.jhu.edu/~rwebster/index_files/pub_files/chapter.pdf
3. Yousef, B.: Design of a Robot Manipulator and an Ultrasound Probe Holder for Medical Applications. Dissertation, The University of Western Ontario, London, Ontario, Canada (2007)
4. Bassan, H.: Design, Construction and Control of a Micro Manipulator for Prostate Brachytherapy. Dissertation, The University of Western Ontario, London, Ontario, Canada (2007)
5. Yousef, B., Patel, R.V., Moallem, M.: Macro-Robot Manipulator for Medical Applications. In: IEEE International Conference on Systems, Man and Cybernetics, pp. 530–535 (2006)
6. Davies, B., Hilbert, R., Timoney, A., Wickham, J.: A surgeon Robot for Prostatectomies. In: Proc. of 2nd Workshop on Medical and Healthcare Robots, Newcastle, pp. 91–101 (1989)

7. Marescaux, J., Leroy, J., Rubino, F., Smith, M., Vix, M., Simone, M., Mutter, D.: Transatlantic Robot-Assisted Telesurgery. Nature 413, 379–380 (2001)
8. Guthart, G., Salisbury, J.: The intuitive telesurgery system: Overview and application. In: Proc. of the IEEE International Conference on Robotics and Automation (ICRA 2000), San Francisco, CA, pp. 618–621 (2000)
9. Eto, M., Naito, S.: Endouronocology. In: Robotic Surgery Assisted by the ZEUS System, pp. 39–48. Springer, Tokyo (2006)
10. Advincula, A.P., Song, A.: The role of robotic surgery in gynecology. Current Opinion in Obstetrics and Gynecology 19(4), 331–336 (2007)
11. Lerner, A.G., Stoianovici, D., Whitcomb, L.L., Kavoussi, L.R.: A passive positioning and supporting device for surgical robots and instrumentation. In: Taylor, C., Colchester, A. (eds.) MICCAI 1999. LNCS, vol. 1679, pp. 1052–1061. Springer, Heidelberg (1999)
12. Phee, L., Di, X., Yuen, J., Chan, C.F., Ho, H., Thng, C.H., Cheng, C., Ng, W.: Ultrasound guided robotic system for transperineal biopsy of the prostate. IEEE Trans. on Robotics and Automation 18, 1315–1320 (2005)
13. Bassan, H., Patel, R.V., Moallem, M.: A Novel Manipulator for 3D Ultrasound Guided Percutaneous Needle Insertion. In: IEEE International Conference on Robotics and Automation (ICRA 2007), pp. 617–622 (2007)
14. Craig, J.J.: Introduction to Robotics: Mechanics and Control, 3rd edn. Pearson Education, Upper Saddle River (2005)
15. Gosselin, C., Angeles, J.: A Global Performance Index for the Kinematic Optimization of Robot Manipulators. ASME Trans. Journal of Mechanical Design 113, 220–226 (1991)
16. Yousef, B., Patel, R.V., Moallem, M.: An Ultrasound Probe Holder for Image-Guided Surgery. ASME Journal of Medical Devices 2(2), 021002-8 (2008)

Force Sensing and Control in Robot-Assisted Suspended Cell Injection System

Haibo Huang[1,*], Dong Sun[2], Hao Su[3], and James K. Mills[4]

[1] Robotics and Micro-systems Center, Soochow University, SuZhou, China
   hbhuang@suda.edu.cn
[2] Department of Manufacturing Engineering and Engineering Management,
   City University of Hong Kong, Hong Kong
   medsun@cityu.edu.hk
[3] Automation and Interventional Medicine (AIM) Laboratory,
   Department of Mechanical Engineering, Worcester Polytechnic Institute,
   Worcester, MA, 01609, USA
   haosu@wpi.edu
[4] Department of Mechanical and Industrial Engineering, University of Toronto. 5
   King's College Road, Toronto, ON, Canada M5S 3G8
   mills@mie.utoronto.ca

Abstract. Stimulated by state-of-the-art robotic and computer technology, cell injection automation aims to scale and seamlessly transfer the human hand movements into more precise and fast movements of the micro manipulator. This chapter presents a robotic cell-injection system for automatic injection of batch-suspended cells. To facilitate the process, these suspended cells are held and fixed to a cell array by a specially designed cell holding device, and injected one by one through an "out-of-plane" cell injection process. Starting from image identifying the embryos and injector pipette, a proper batch cell injection process, including the injection trajectory of the pipette, is designed for this automatic suspended cell injection system. A micropipette equipped with a PVDF micro force sensor to measure real time injection force, is integrated in the proposed system. Through calibration, an empirical relationship between the cell injection force and the desired injector pipette trajectory is obtained in advance. Then, after decoupling the out-of-plane cell injection into a position control in $X - Y$ horizontal plane and an impedance control in the Z- axis, a position and force control algorithm is developed for controlling the injection pipette. The depth motion of the injector pipette, which cannot be observed by microscope, is indirectly controlled via the impedance control, and the desired force is determined from the online $X - Y$ position control and the cell calibration results. Finally, experimental results demonstrate the effectiveness of the proposed approach.

1   Introduction

Since its invention during the first half of the last century, biological cell injection has been widely applied in gene injection, in-vitro fertilization (IVF) [26],

---

* Corresponding author.

T. Gulrez, A.E. Hassanien (Eds.): Advances in Robotics & Virtual Reality, ISRL 26, pp. 61–88.
springerlink.com   © Springer-Verlag Berlin Heidelberg 2012

intracytoplasmic sperm injection (ISCI) [28] and drug development . The cells to be injected in biological engineering are classified as either adherent or suspended cells [11] corresponding to two distinct biomanipulation tasks. Currently, commercial devices such as those provided by Eppendorf, Narishige and Cellbiology Trading [1] are available for automation of adherent cell injection tasks. In contrast, development of methodologies for the automatic injection of suspended cells, have been the focus of a number of research groups. Due to the freedom of movement of cells in the solution, the automatic suspended cell injection process will be more complex and time consuming. Collision between cells and manipulation tools, or the motion of cells themselves, can easily lead to the failure of this process. Only until very recently, most suspended cell injection operations have been performed manually.

In this chapter, a robotic cell injection system for automatic injection of batch of suspended cells is developed. The operation follows certain sequence and the same processing procedure, and is called batch biomanipulation. The proposed injection system utilizes cells arranged and held in circular arrays for insertion. A cell holding device is fixed on an actuated rotary plate, permitting cells to be transported, one by one, into the injection site field of view. This holding device is designed utilizing the ideas of both the hemispherical hole shaped cell holder [7] and the groove shaped cell holder [20]. The hole array can readily immobilize the cells, while the grooves make the embryo pick-and-place process more readily possible. It is essential that the micro-pipette is held at an angle of attack with respect to the horizontal plane. If this was not the case, the injection pipette would simply brush the cells out of the holding devices, failing to pierce the cell wall.

Furthermore, we decouple the out-of-plane cell injection task into two relatively independent control processes: the position control in the horizontal $(X-Y)$ plane and the force control in the depth direction (Z-axis) [8]. This position and force control method differs from those used for macro applications mainly in two aspects. First, a calibration of the cell membrane mechanical properties must be performed in advance to derive an empirical relationship between position and injection force trajectories. This allows us to infer a desired force input online, based on the actual cell injection trajectory obtained from visible positions in the $X-Y$ plane and the fixed attach angle of the pipette.

Second, when the motion of the pipette in the $X-Y$ plane follows the desired position trajectory, the motion in the Z-axis, which cannot be observed under the microscope, is controlled indirectly by following the desired injection force impedance [25]. The desired force, as said above, is determined based on the actual motion in the $X-Y$ plane as seen by the overhead camera and the calibration results. The actual injection force is measured by the micro-force sensor mounted on the injector pipette. With such a cooperation of position and force control in the three axial directions, the desired injection process can be achieved successfully.

In our previous work [9] [10], a planar visual feedback was used to estimate the depth of penetration of the injector pipette, and hence estimate the injection force. With this method, however, the estimate of the injector penetration distance, and hence the injection force, is not robust to the inevitable uncertainty in the cell dynamic model parameters. In addition, the sampling frequency of the vision system is much lower than that of the force sensor, and hence the injection process cannot be controlled more precisely. Note also in [9], that we do not discuss batch processing of embryo cells, nor present any cell holder to facilitate automation of the process.

The remainder of this chapter is organized as follows. Section 2 presents a background of existing cell injection technologies. In Section 3, the automatic batch suspended cell injection system including the suspended cell holding device and PVDF micro force sensor is introduced. This is followed by the system dynamic modeling in Section 4. In Section 5, the cell injection process is introduced, and a position and force control strategy for out-of-plane cell injection tasks is developed. The results of experiments are reported in Section 6. Finally, conclusions of this work are given in Section 7.

2 Background

Fig. 1(a) shows a popular commercial automatic adherent cell injection system setup (computer assisted microinjection of tissue culture cells (AIS 2)) made by Cellbiology Trading and Fig. 1(b) is the operation interface. This automated cell injection system is mainly composed of an inverted microscope equipped with long-distance phase-contrast objectives for observation, a 3-axes-stepping motorized stage for accurate positioning of the cells, another 3-axes-stepping motorized stage controlling the capillary movement, a piezo drive axial injection movement, an Eppendorf pneumatic microinjector with the model FemtoJet, a video system with a CCD camera and a monitor for observation of cells together with the microprocessor control unit and the required software for controlling the cell injection procedure. With the techniques of an autofocus microscope, the image identification of cells and injector pipette, as well as the image-based visual serving for positioning control and the calibrated injection height control, this computer-assisted and microprocessor-controlled injection system can provide high injection rates with optimum reproducibility of injections, and makes quantitative adherent cell injection possible. The software reliably controls the precise injection process and allows fast axial injection at a rate of up to 1500 cells per hour. Retrieving injected cells for analysis and multiple injections into the same cells are also realized with this system.

In contrast, development of methodologies for automatic injection of suspended cells has been the focus of a number of research groups [4] [14] [17] [21] [18] [19] [36] in recent years. Existing semi-automated or tele-operated commercial systems such as those provided by Eppendorf, Narishige and Cellbiology Trading [2] require highly skilled operators to carry out suspended

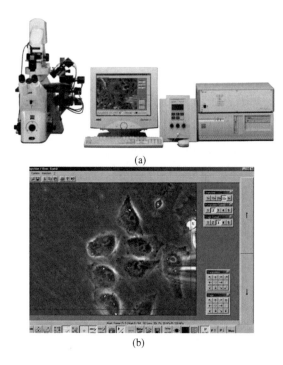

(a)

(b)

Fig. 1. Automated adherent cell injection system (AIS 2). (a) System setup with microscope, motorized stage and micromanipulator, pressure microinjector and video camera system; (b) The friendly user interface makes the microinjection as easy as possible, adapted from [1].

cell injection. A representative automatic suspended cell injection system established in [17] [34] is shown in Fig.2. This kind of automatic suspended cell injection procedure mainly includes the following steps: first, it performs a visual search and recognition of cells, injecting pipette and cell holding tube through image processing; then it drives the holding tube to track and hold the cells and move them to the desired manipulation area; next it guides the injector pipette to inject material into cells and finally, it releases the injected cell in the finishing area and repeats the procedure again until all of the suspended cells have been injected. This is obviously not a reliable control process. Unpredictable situations such as the collision between cells and manipulation tools or the motion of cells themselves can easily lead to the failure of this process.

A number of solutions have been proposed to solve the suspended cell injection problem in a similar manner to the batch adherent cell manipulation. Some of these approaches used chemical or mechanical principles to hold the suspended cells, such as Zhang [37] who utilized fluidic self-assembly technology to position and immobilize Drosophila (fruit fly) embryos, with a thin film

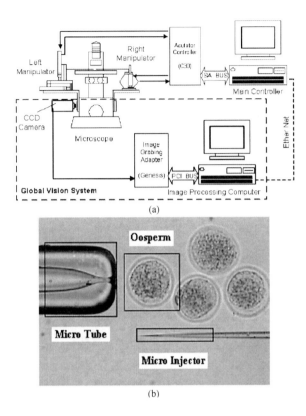

Fig. 2. Traditional automatic suspended cell injection system. (a) sketch of the whole system; (b) operation area under the microscope, adapted from [17] [34].

of polychlorotrifluoro-ethylene oil, and then made parallel injection of genetic material into two-dimensional embryo arrays (as shown in Fig. 3). Wang [30] presented a vacuum-based embryo holding device (as shown in Fig. 4), where evenly spaced through-holes were connected to a vacuum source via a backside channel and a suction pressure, enabling each through-hole to trap a single embryo. Rather than increasing the complexity of the manipulation system, the method utilized natural adhesion forces to fix the cells during the cell injection process. Fujisato [7] developed a cell holder made of microporous glass (MPG) with a glass or stainless coating layer, on which sandblasted micro-pocket holes were formed. Lu [20] reported the development of a prototype micromanipulation system for automatic batch microinjection of Zebrafish embryos using a gel cell holder with several parallel V-grooves (as shown in Fig. 5). All above methods attempt to fix suspended cells into arrays permitting batch cell injection to be performed easily. In particular, the results presented by Fujisato [7] and Lu [20] form the basis of our work: a novel cell holding device that integrates both the merits of hole and grooved

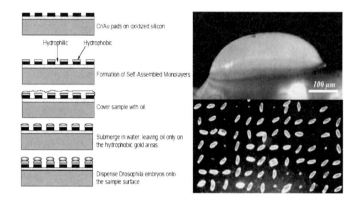

Fig. 3. Fluidic self-assembly technology to position and immobilize Drosophila (fruit fly) embryos, adapted from [37].

shaped cell holders, which makes the suspended cells easily picked and placed, and injected from all directions.

Image-based visual servoing is the dominant control method in micromanipulation, but is subject to the limitation of small depth of field of the microscope used. In recent years, micro force sensors were used for real-time force sensing in cell injection process. Research on bonding a microinjection pipette on the tip of a PVDF sensor or piezoresistive force sensor to detect the injection forces in fish egg biomanipulation was performed in [8] [10] [13] [15] [20] [23] [35] (as shown in Fig. 6). Sun et al. [27] developed a MEMS-based two-axis cellular force sensor to investigate the mechanical

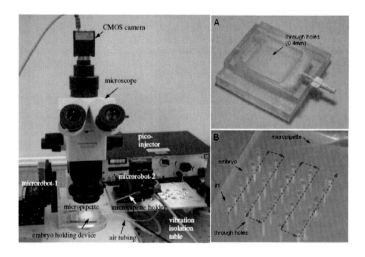

Fig. 4. Vacuum-based embryo holding device, adapted from [30].

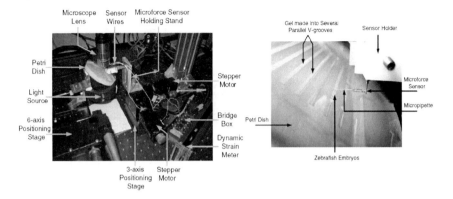

Fig. 5. Gel-made cell holder with several parallel V-grooves, adapted from [20].

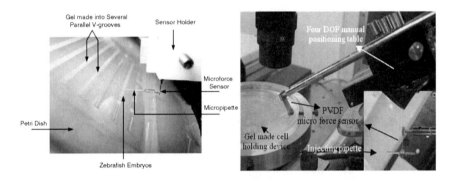

Fig. 6. Bonding a microinjection pipette on the tip of a piezoresistive force sensor (left) or PVDF sensor (right) to detect the injection forces, adapted from [10] [20].

properties of mouse oocyte zona pellucida (ZP) (as shown in Fig. 7). Zhang et al. [37] developed a micrograting-based injection force sensor with a surface micromachined silicon-nitride injector pipette (as shown in Fig. 8). Arai et al. [5] developed a micro tri-axial force sensor with piezoresistive strain gauge which is installed near the tip of the end-effector (as shown in Fig. 9). With development of micro force sensors, several force and vision control approaches were proposed in micromanipulation [38], utilizing sensed force information to guide the injection process at a high level. The work of addressing injection force control using real-time force feedback [31] [32] [33] is both valuable and necessary to our work presented here. With the detected injection force, the cell properties can also be characterized through a mechanical modeling approach [29].

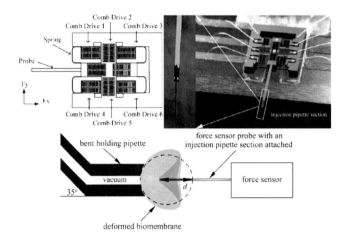

Fig. 7. Multi-axis cellular force sensor with an attached injection pipette section, adapted from [27].

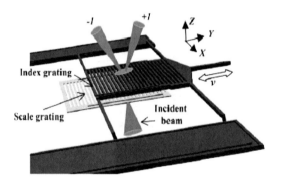

Fig. 8. Optical MEMS force sensor with a surface micro-machined silicon-nitride injector, adapted from [37].

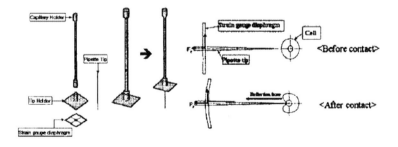

Fig. 9. Strain gauge type cell injection force sensor, adapted from [5].

3 System Design

3.1 System Setup

An automatic suspended cell injection system as shown in Fig. 10(left) was developed in the laboratory for this research. Fig. 10(right) shows a close-up view of the injection area. According to our previous work [9], this system is designed to simulate automatic cell injection of large batches of suspended cells (such as fish embryos) in biological engineering processes. To achieve the purpose of batch injections, the system is designed for the use of cells arrays instead of individual cell holder. A cell holding device, designed specifically for this purpose, is fixed on the actuated rotary plate, permitting batches of cells to be transported one by one, into the field of view of the microscope, for injection. The injection process is achieved with the injector pipette tilted out of the focal plane of the microscope, as shown in Fig. 10(right). It is essential to carry out the insertion with the micro-pipette held at an angle of attack with respect to the horizontal plane in which the cells are held. If this was not the case, the injection pipette would simply brush the cells out of the holding devices, failing to pierce the cells at all. Further, a PVDF (Polyvinylidene fluoride) polymer film force sensor is used to directly sense injection contact force, instead of relying only on the visual estimation in [9].

Fig. 11 illustrates a schematic of the proposed cell injection system, which contains three modules. The executive module consists of a $X-Y-\theta$ positioning table with the injector mounted on the $Z-$axis. The cells to be injected are manually placed on a specially designed cell holding device, formed in a Petri dish, from agarose-gel, shown in Fig. 12(right). Mounted on the positioning table, the cell holder centre is coincident with the θ rotation axis. Coordinated motion of the $X-Y-\theta$ table and the $Z-$axis is required to perform the cell injection task. The $X-Y$ stage has a workspace of 60 $mm \times 60\ mm$, with a positioning resolution of $0.3175\mu m$ in each axial direction. The Z-axis stage has a workspace of 50 mm , and a positioning resolution of $0.3175\mu m$, the rotation axis θ is driven by a precision stepping motor with an rotation resolution of $0.045°$. The pose of the injection pipette is precisely adjusted with a four degree of freedom ($X-Y-\varphi-\beta$) manual positioning table mounted on the $Z-$axis with a work space of ($13mm \times 13mm \times \pm20° \times \pm45°$), where φ is the tilt angle of the injector, β is the angle between the injector and the X-axis, and both φ and β are fixed. A hydraulic CellTram vario (Eppendorf) microinjector pipette is driven by a stepper motor permitting a minimum volume resolution of 2 nl to be delivered into the cell. The sensory module contains a vision system that includes an optical microscope, lighting system, CCD camera, PCI image capture and processing card, and image processing computer. The lighting system is suitable for observing transparent biotechnology objects. The working plate is made of glass. A very thin chip LED back lights (model LDL-TP-27 27-SW) provides illumination of the working area from behind to create a silhouette image. The control module consists

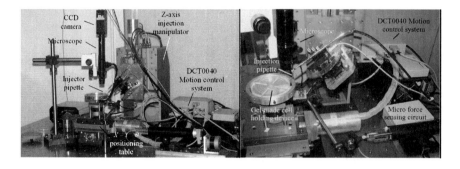

Fig. 10. Photographs of the new laboratory test-bed suspended cell injection system. (left) whole system setup; and (right) close-up.

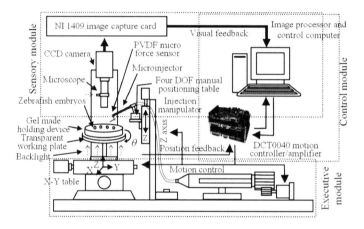

Fig. 11. Schematic of the cell injection system. ©2009 IEEE

of a 2.8 GHz host computer and a DCT0040 motion control/drive system, with further details in [3].

3.2 Suspended Cell Holding Device

Experiments show that both the hemispherical hole shaped cell holder [7] and the groove shaped cell holder [20] have advantages and disadvantages. The array of holes exhibits good capability to hold cells and significantly, the cells can be injected from all directions when they are positioned in this holder. However, placing cells in the individual holes is laborious work for batch injection. A far simpler and expedient approach is to place cells in the half-circle shape grooves. The injection direction is then restricted to along the radial direction of these grooves. Injection attempted in other directions may cause the cells to move during the injection process. Combining these

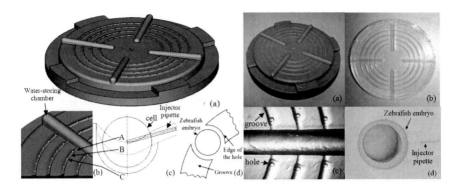

Fig. 12. (Left) CAD prototype of mould for cell holding device. (Right) Improved cell holding device. (a) mould; (b) agarose-gel made cell holder; (c) cell holding device under the microscope; and (d) zebrafish embryo placing in cell holder.

two designs, a new prototype was fabricated in this research, as shown in Fig. 12(left). An array of holes is positioned on a circular profile groove which is centered about the geometric centre of the holder. The diameter of holes varies according to different applications. It is 1.4mm in our prototype for injection of fish embryos. As seen in Fig. 12, the dash-dot line A, dotted line B and solid line C represent the cross section profiles of edge of each groove (namely the groove adjacent to the water-storing chamber, as seen in Fig. 12(left)(b), grooves and the holes, respectively.

The main advantages of our new design are threefold. First, the grooves make the pick-and-place process of embryos more easily. A number of embryos can be placed in the grooves and rolled in the holes one by one along the grooves. Second, the holes array can readily immobilize the cells, which allows the injection to be done from all directions, as seen in Fig. 12(left)(d). Third, with the help of edge of each groove, fluid remains in the grooves during the injection period, keeping the cells wet, and the redundant fluid flows across the edge of grooves and into the water-storing chamber, as shown in Fig. 12(left)(b). Further, the quality of images becomes good, as shown in Fig. 12(right)(c) and (d).

The cell holding device is fabricated from low-melting point agarose gel, a material commonly used in biological research. With soft, hydrous and transparent characteristics, agarose gel is well suited for fabricating the proposed cell holder. To make this cell holding device, the gel placed in Petri dish is made from 5% agarose by dissolving and heating it in a microwave oven. Next, a mould is pressed onto the gel surface and removed after the agarose-gel has cooled. Fig. 12(right)(a) illustrates a cell holder mould which is fabricated using three-dimensional printing technology. Fig. 12(right)(b) shows the agarose-gel cell holding device made by the mould. Procedures to make this cell holding device are shown in Fig. 13.

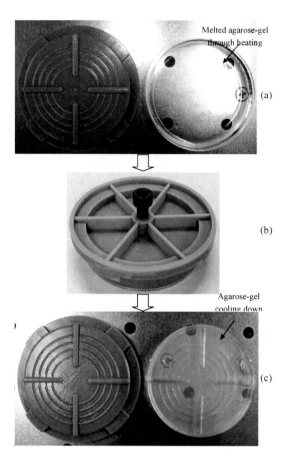

Fig. 13. Procedures to make cell holding device. (a) dissolving and heating low-melting point agarose gel in a microwave oven; (b) pressing the mould on the gel surface; and (c) removing the mould after the agarose-gel has cooled.

3.3 PVDF Force Sensor

Cell injection forces vary significantly from one species to another. For example, mouse embryos require a puncture cell injection force of about 10 μN, while zebrafish embryos require a puncture force of about 1000 μN. Based on factors of measurement range, control precision and the type of cells, a PVDF (Polyvinylidene fluoride) polymer film (MSI Co., Ltd., model LDTC-Horizontal) is used as the micro force sensor to measure real-time cell injection forces during the cell injection process in our research. The PVDF polymer sensor has desirable characteristics of high linearity, wide bandwidth, and high signal-to-noise (S/N) ratio. Compared to other types of micro force sensor [5] [27] [37], PVDF micro force sensor is more suitable for force sensing in our cell injection of fish embryos, in terms of the measurement range

($\mu N \sim mN$), the resolution (submicro-Newton), and the relatively simple structure. Many successful applications have demonstrated the effectiveness of this kind of sensor in cell injection force detection [13] [15] [23] [35].

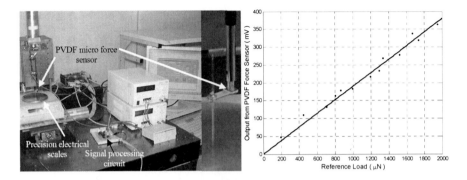

Fig. 14. (Left) Experimental set-up for PVDF micro force sensor calibration. (Right) Force calibration result of cellular PVDF force sensor.

The PVDF sensor used in our research has the dimension (width and height) of 6 $mm \times 12.8\ mm$, and the thickness of 0.1 mm. The sensor is calibrated by moving the sensor, mounted on the injector pipette fixed on the Z-axis, to contact with a precision electrical scale, which is used as a reference sensor [6], to reach different pressures with low speed. Fig. 14(left) illustrates the setup for calibration of the PVDF force sensor. Fig. 14(right) illustrates the calibration results of the sensor outputs via the corresponding pressures measured by the scale, which demonstrates good linearity of the PVDF sensor, with the estimated sensitivity of 0.1901 $mV/\mu N$.

4 System Modeling

Fig. 15 illustrates the configuration of the basic microinjection system. Similar work can be found in [5] [12] [16]. Define $O - XYZ$ as the coordinate frame whose origin is located at the center of the working plate. A microscope mounted on a visual servoing system is positioned on a work table. The optical axis of the microscope coincides with the Z-axis, as shown in Fig. 15. Define $O_c - X_c Y_c Z_c$ as the camera coordinate frame, where O_c is located at the center of the camera, and Z_c coincides with the optical axis of the microscope. Define $O_i - uv$ as the coordinate frame in the image plane, with three coordinates u, v and the rotary angle θ_i. The origin is located in the optical axis, and the axes u and v are within the camera image plane perpendicular to the optical axis.

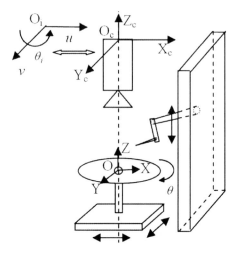

Fig. 15. Configuration of the cell injection system.

The relationship between the image coordinates $\begin{bmatrix} u, v, Z, \theta_i \end{bmatrix}^T$ and the positioning table coordinates $\begin{bmatrix} X, Y, Z, \theta \end{bmatrix}^T$ is given as [9]

$$
\begin{bmatrix} u \\ v \\ Z \\ \theta_i \end{bmatrix} = T \begin{bmatrix} X \\ Y \\ Z \\ \theta \end{bmatrix} + \begin{bmatrix} f_X d_X \\ f_Y d_Y \\ 0 \\ 0 \end{bmatrix} \tag{1}
$$

where $T \in \Re^{4 \times 4}$ is the transformation matrix between the image and positioning table frames. Since both the frames are fixed, T is time-invariant. f_X and f_Y are the display resolutions of the vision system in X and Y coordinate directions. d_X and d_Y are distances in two directions between the two frames.

Using Lagrange's equation of the motion, the dynamics of the four DOF motion stage is given as [9]

$$
M \begin{bmatrix} \ddot{X} \\ \ddot{Y} \\ \ddot{Z} \\ \ddot{\theta} \end{bmatrix} + N \begin{bmatrix} \dot{X} \\ \dot{Y} \\ \dot{Z} \\ \dot{\theta} \end{bmatrix} + \begin{bmatrix} 0 \\ 0 \\ G_z \\ 0 \end{bmatrix} = \tau - \begin{bmatrix} J^T f_e \\ 0 \end{bmatrix} \tag{2}
$$

where: $M = \begin{bmatrix} m_X + m_Y + m_p & 0 & 0 & 0 \\ 0 & m_Y + m_p & 0 & 0 \\ 0 & 0 & m_Z & 0 \\ 0 & 0 & 0 & I \end{bmatrix}$ denotes the inertia matrix

of the system, and m_X, m_Y and m_Z are masses of the X, Y and Z positioning tables, m_p is the mass of the working plate, I_p is the inertia of the rotational axis and the working plate; N denotes the system damping and viscous friction effects; $G_Z = -m_Z g$ is the gravitational force vector; $\tau = \begin{bmatrix} \tau_X, \tau_Y, \tau_Z, \tau_r \end{bmatrix}^T$ denotes the torque inputs to the driving motors; $J = diag\{J_X, J_Y, J_Z\}$ is the Jacobian matrix relating task and joint coordinates of the axes; and $f_e = \begin{bmatrix} f_{eX}, f_{eY}, f_{eZ} \end{bmatrix}^T$ is the external force applied to the motors, and $f_e = 0$ if the injector does not contact the cells. Define $q = \begin{bmatrix} u, v, Z, \theta_i \end{bmatrix}^T$ as the generalized coordinate. We then obtain the following dynamics in terms of q

$$MT^{-1}\ddot{q} + NT^{-1}\dot{q} + G = \tau - \begin{bmatrix} J^T f_e \\ 0 \end{bmatrix} \tag{3}$$

5 Cell Injection Control Design

5.1 Image Processing

The automatic cell injection process begins with identification of positions of the manipulated object and injector pipette by using image processing technology. Much work on image processing and recognition algorithms for detecting the cell and injecting pipette has been done (i.e., [22] [30]), with satisfactory results. With our specially designed cell holding device, only the cell embryo and the injector pipette need to be identified, and the cell injection visual environment, as shown in Fig. 16(a), is simplified.

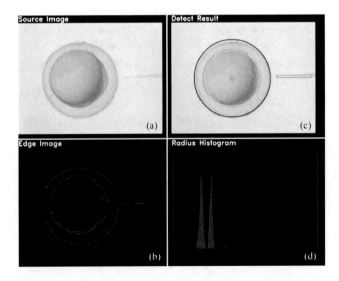

Fig. 16. A typical image grabbed and processed by vision system, (a) Source image; (b) Edge detection; (c) Detection results; and (d) Zebrafish embryos radius histogram. ©2009 IEEE

The image processing procedure is as follows. First, pre-processing is conducted to obtain relatively noise-free binary images. An image is first convolved with a low-pass Gaussian filter for noise suppression. Then the gray-level image is filtered to find the edge of the cell chorion, cytoplasm and injector pipette, using the Canny edge detection algorithm, as shown in Fig. 16(b). To track the movement of cell embryos, a traditional Hough transform has the limitation of long calculation time and relative low precision. A faster and more precise method called the chord midpoint Hough transform (CMHT) [24] is utilized in our research to improve the efficiency and accuracy of ellipse detection in an image. As for the injector pipette, the correlation-based pattern matching method is used to determine its position. Fig. 16(c) shows the image processing results. Fig. 16(d) shows the radius histograms of the zebrafish embryo with and without chorion.

Note that the image processing technology is used for detecting the cell diameter, e.g. the diameter of chorion, the diameter of yolk and outside of the cell, and then guiding the injector pipette towards the desired starting position with visual servoing. When the pipette contacts the cell and starts to inject, the position feedback obtained by the motor encoder, instead of the image feedback, will be used in the position control loop for higher processing speed and accuracy.

5.2   Injection Process

A batch of cells to be injected are placed in the hole array as shown in Fig. 12(c). The cells are observed by the microscope and the position of the cell is adjusted to a proper injection pose by pinching a silastic tube. Fig. 17 illustrates the cell injection process. Since the injector pipette may be located out of the $X - Y$ plane such that it cannot be observed by the microscope,

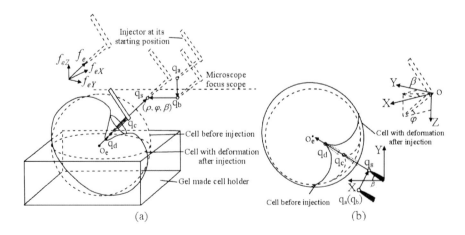

Fig. 17. Cell injection process: (a) side view; (b) top view.

the Z-axis of the injector pipette must move first so that the injector pipette can be viewed by the microscope. Denote q_a as the position of the head of the injector pipette at the beginning of manipulation. The injector pipette moves to the starting injection height that is within the focal range of the microscope, and stops at the position q_b, using a vision guided method. An image-based visual servoing methodology is then used to guide the injector pipette towards the starting injection position q_s. When the injector pipette arrives in the position q_s, the control scheme switches to the more precise cell injection controller, similar to the control scheme proposed by Sun et al. [26].

The precise cell injection control process is designed as follows. First, the so-called pre-piercing step is initiated by accelerating the injector pipette from the starting injection position q_s towards the cell wall. To ensure that the injector pipette is directed toward the centre of the cell, denoted by o_e, the X, Y, and Z axes must move simultaneously, with the desired trajectory of the injector pipette expressed using the spherical polar coordinates. As shown in Fig. 17, when the injector pipette reaches the position q_c located on the surface of the cell membrane, the velocity is designed to be at its maximum, since a sufficiently high injection velocity is required to pierce the cell and cause minimal harm. Next, the injector pipette is decelerated to approach the desired injection position q_d and pierce the cell. The injection distance must be long enough so that the cell can be pierced, which can be determined by calibration experiment. After injecting the genetic material into the cell at q_d, the injector pipette is accelerated again in the opposite direction for removal from the cell membrane. The working plate holding the cells is then rotated to position another cell within the injection area, and the cell injection is repeated.

5.3 Cell Injection Control Strategy

During the out-of-plane cell injection, the depth motion of the injector pipette cannot be observed by the microscope. We therefore propose to utilize a force control in the $Z-$ axis direction to implicitly control the depth motion. With an integration of a micro-force sensor feedback and position feedback from the motor encoder, a position and force control methodology is developed as follows.

First, through cell injection task calibration carried out in the $X-Y$ plane ($\varphi = 0$), i.e., injector pipette is positioned in the same plane as the microscope focal plane, the relationship between the cell biomembrane deformation and the cell injection force can be derived. This geometry is selected so that the injector pierces the cell at its vertical midpoint, such that the $X-Y$ plane deflection observed by the camera is the correct value. Here we assume that batches of the same kind of cells have approximately the same membrane dynamic characteristics. Under this assumption, a similar injection force trajectory will be obtained for the same kind of cells when the injector pipette injects into the cells under the desired velocity and acceleration profile as reported in [9]. Through calibration experiments, an empirical relationship between, what will become the desired cell injection force F^d, and the desired injector pipette trajectory ρ^d, can be obtained, i.e.,

$$F^d = U(\rho^d) \tag{4}$$

In the out-of-plane injection, the desired injection force, corresponding to the desired position trajectory, can be estimated from the cell biomembrane deformation in the $X - Y$ plane and the known angle φ the injector pipette makes relative to the $X-Y$ plane, leading to equation 4. This desired injection force will become the input to the force controller to be designed subsequently.

Second, the out-of-plane cell injection task is decoupled into two relatively independent control process: the position control in horizontal $(X - Y)$ plane and the force control in depth (Z-axis). With this idea, we control the position of the injector in the $X - Y$ plane, which can be observed with the microscope, and simultaneously control the injection force using the force sensor in the Z-axis direction, in which depth information cannot be known. As indicated above, there exists an empirical relationship between the desired cell injection force and trajectory. As a result, the depth movement of the injector can be indirectly controlled via a force control.

5.4   Position and Force Control Design

During the cell injection process, the motion planning of the injector pipette can be simplified if the moving coordinate is represented in a spherical polar coordinate frame. Define $o_e - \rho\varphi\beta$ as the spherical polar coordinate frame whose origin o_e is located at the center of the cell, as shown in Fig. 17. The desired trajectory of the injector pipette is then expressed as

$$\begin{bmatrix} X^d \\ Y^d \\ Z^d \end{bmatrix} = \begin{bmatrix} \rho^d \sin\varphi \cos\beta \\ \rho^d \sin\varphi \sin\beta \\ \rho^d \cos\varphi \end{bmatrix} \tag{5}$$

where ρ^d denotes the desired injector pipette trajectory, and both angles β and φ do not change during the cell injection process.

The dynamics equation (2) during the injection process can be partitioned into two distinct equations of motion, namely

$$\begin{cases} \tau_{XY} = M_{XY} \begin{bmatrix} \ddot{X} \\ \ddot{Y} \end{bmatrix} + N_{XY} \begin{bmatrix} \dot{X} \\ \dot{Y} \end{bmatrix} + \begin{bmatrix} J_X^T f_{eX} \\ J_Y^T f_{eY} \end{bmatrix} \\ \tau_Z = M_Z \ddot{Z} + N_Z \dot{Z} + G_Z + J_Z^T f_{eZ} \end{cases} \tag{6}$$

where τ_{XY} and τ_Z are the torque distributions in $X - Y$ plane and Z- axis, respectively, M_{XY} and N_{XY} are sub-matrices of M and N corresponding to the X and Y axes, M_Z, N_Z and G_Z are sub-matrices of M, N and G

corresponding to the Z axis, and f_{eX}, f_{eY} and f_{eZ} are the actual external forces along the three axes, measured with the PVDF force sensor.

Define the position error in the $X - Y$ plane as $e_{XY} = \rho^d \sin \varphi \begin{bmatrix} \cos \beta \\ \sin \beta \end{bmatrix} - \begin{bmatrix} X \\ Y \end{bmatrix}$. A computed torque controller with measured force compensation is designed below for position control in $X - Y$ plane:

$$\tau_{XY} = M_{XY} \left(\begin{bmatrix} \ddot{X}^d \\ \ddot{Y}^d \end{bmatrix} + k_v \dot{e}_{XY} + k_p e_{XY} \right) + N_{XY} \begin{bmatrix} \dot{X} \\ \dot{Y} \end{bmatrix} + \begin{bmatrix} J_X^T f_{eX} \\ J_Y^T f_{eY} \end{bmatrix} \quad (7)$$

where k_v and k_p are positive control gain matrices. The controller (7) leads to the closed loop dynamics $\ddot{e}_{XY} + k_{v1} \dot{e}_{XY} + k_{p1} e_{XY} = 0$, which implies $e_{XY} = 0$.

As for the depth movement of the injector, an impedance control algorithm is utilized to control the position of the injector in $Z-$axis indirectly. The contact space impedance control is given as:

$$m \ddot{e}_Z + b \dot{e}_Z + k e_Z = \Delta f_{eZ} \quad (8)$$

where m, b and k are the desired impedance parameters, $e_z = Z^d - Z = \rho^d \cos \varphi - Z$, representing the position error of the injector in $Z-$axis. The actual position Z can be obtained from the encoder of the drive motor. $\Delta f_{ez} = f_{eZ}^d - f_{eZ}$ is the force error in the $Z-$axis. The desired force f_{eZ}^d is obtained based on the actual injector position in $X - Y$ plane and the calibrated relationship between the injection position and the force. In other words, f_{eZ}^d corresponds to Z^d closely. $f_{eZ} = f_e \cos \varphi$ is the projection of the cell injection force in $Z-$axis, measured by the PVDF force sensor.

We now solve for $\ddot{Z}$ from equation (8) as follows:

$$\ddot{Z} = m^{-1} \left(m \ddot{Z}^d + b \dot{e}_Z + k e_Z - \Delta f_{eZ} \right) \quad (9)$$

Substituting equation (9) into (6) yields the $Z-$axis torque controller as follows:

$$\tau_Z = M_Z \left(\ddot{Z}^d + m^{-1} (b \dot{e}_Z + k e_Z - \Delta f_{eZ}) \right) + N_Z \dot{Z} + G_Z + J_Z^T f_{eZ} \quad (10)$$

Under (10), the cell injection force can be regulated to meet the desired impedance requirement. Since the desired force has an empirical relationship to the desired position trajectory, the injector position in $Z-$axis is controlled indirectly.

The block diagram of this position and force control is shown in Fig. 18. In the experiments conducted, the high update rate of the motor encoder and force sensor (i.e., 1kHz in our system) conflicts with the relatively slow image capture rate of the CCD camera (i.e., 60fps for a 768x574 pixel image with pixel resolution of 3 μm), when implementing the controllers (7) and (10). Further, cell injection is a rapid motion process (i.e., 1 second in our experiments), which requires high injection speed to produce less damage to the cells. Therefore, we propose to feed back the actual positions X and Y with motor encoders in the position control loop, and feed back the actual forces f_{eX}, f_{eY} and f_{eZ} obtained with the micro force sensor. Note that during the process of cell injection, the vision system still works as a monitor, to inspect the difference between the data obtained from image processing and the motor encoder. When the difference is larger than a pre-defined threshold, a warning signal will be produced.

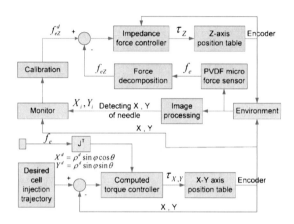

Fig. 18. Control block diagram. ©2009 IEEE

6   Experiments

To verify the effectiveness of the proposed approach, experiments were performed using the setup as shown in Fig. 10 and 11. The modeling parameters of the system were determined in advance. For simplicity, the rotation angle between the image frame and the stage frame is set to zero ($\theta = 0°$) during the injection. The displacement between origins of the two frames is $d = [0, 0, 30mm]^T$. The magnification factor of the microscope objective is $\lambda = 30$. The dynamic model inertia and damping matrices, and the gravitational force vector have been estimated and reported in [9], hence are not given here. The angles between the injector pipette and the X- and Z-axis are $\beta = 45°$ and $\phi = 35°$, respectively.

   The cells selected for injection were Zebrafish embryos. Zebrafish embryos are commonly chosen as an animal model in biomanipulation because of their similarity in genetic makeup to higher vertebrates, are easily obtained, have short generation time, and external fertilization and translucent embryos [23]. The diameter of the Zebrafish egg is approximately 1.15-1.25 mm (including chorion). The radius of the injector pipette is $c = 7.5\mu$m.

6.1 Force Calibration

Experiments on the horizontal cell injection will be conducted first in the $X-Y$ plane ($\varphi = 0$), for calibration purpose, as shown in Fig.19. The injector pipette is positioned in the same plane as the microscope focal plane. Calibration experiments were performed to obtain an empirical relationship between the desired cell injection force F^d and the desired injector pipette trajectory ρ^d. To achieve this goal, the cell injection force was measured using the PVDF micro force sensor while recording the desired injector pipette trajectory ρ^d. To obtain consistent calibration results, calibration tests used pipettes of identical radius, and identical velocity and acceleration profiles, as in the out-of-plane cell injection. During the pre-piercing period ($0\,\text{sec} < t < 0.5\,\text{sec}$), the injector was accelerated to move a distance of 317.5 μm to reach the maximum velocity of 1270 $\mu m/s$ when it contacted the cell membrane. During the next piercing period ($0.5\,\text{sec} < t < 1\,\text{sec}$), the pipette was decelerated and pierced the cell membrane, moving 317.5 μm within the cell to arrive at the yolk's position. Then, it took about two seconds to inject the genetic materials into the yolk. Finally, the injector pipette was extracted from the cell ($3.00\,\text{sec} < t < 5.23\,\text{sec}$), using a lower speed than during the piercing period, where the maximum velocity was 635 $\mu m/s$.

Previously presented results have reported the variations of the mechanical properties on different embryos, where the membrane has been deformed with the pipette. Data has been presented which highlights the variations in pipette penetration force due to variations in mechanical membrane properties of different embryos. A number of zebrafish embryos were used in these calibration experiments. Fig. 20(left) illustrates three typical results obtained in planar cell injection experiments. These forces were measured in the same injection task with different embryos. Note that the force threshold for penetration of the cell membrane differs from embryo to embryo due to variation of cell properties. This difference in penetration forces is not problematic as the micro-force sensor is sensitive to the sudden large force changes as the micro-pipette penetrates the cell. The averages of these forces are plotted in Fig.20(right). Through curve fitting, the relationship between the cell

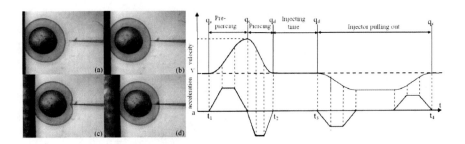

Fig. 19. Planar cell injection experiment for calibration.

injection force and the desired injector pipette trajectory ρ^d, the injection force F was estimated using the following equation:

$$F = 0.0001694 \cdot (r - \rho^d)^2 + 0.3987 \cdot (r - \rho^d) - 8.389$$
$$r > \rho^d > r - 317.5\mu m$$

(11)

where r is the radius of injected cells.

This force-position curve would be subsequently used to guide the insertion of the injector pipette. The process is to move the X and Y axes simultaneously, according to the desired injector pipette trajectory ρ^d, and then based on the actual X and Y to determine the desired injection force at the time instant for Z-axis force control. During the pre-piercing period, the micro-force sensor is used to detect whether the pipette and the embryo make contact. When the contact is confirmed, the controller switches to the position (vision) and force controller as shown in Fig.18.

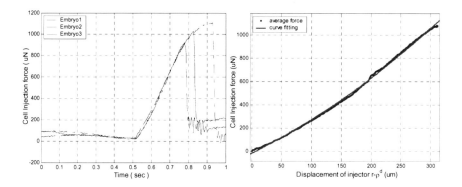

Fig. 20. (Left) Cell injection force calibration. (Right) Curve fitting result of desired force. ©2009 IEEE

6.2   Experimental Results

Although each test used a different embryo cell, the mechanical properties of all cell biomembranes were uniform since they were all collected in accordance with the standard embryo preparation procedures, and all were injected after being placed at room temperature (22-24 °C) for 2 hours after fertilization (blastula stage).

Fig. 21 illustrates the injection procedures including pre-piercing (a), piercing (b), arrival of the desired injection position (c), injecting material (d), and pulling out the pipette from the cell (e) and (f).

The computed torque control gains were $k_v = diag\{ 21.645, 28.87, 31.75 \} \times 10^3 s^{-1}$, $k_p = diag\{ 27.42, 43.4, 66.4 \} \times 10^6 s^{-2}$. The impedance control gains were $m = diag\{0.330, 0.165, 0.1635\}\ N \cdot s^2/m$, $b = diag\{7.143, 4.763, 5.191\}$ $\times 10^3 N \cdot s/m$, and $k = diag\{9.049, 7.144, 10.855\} \times 10^6 N/m$. Experimental results for one of the injected embryo cells, during the piercing period of the injection process ($0\sec < t < 1\sec$), are provided below.

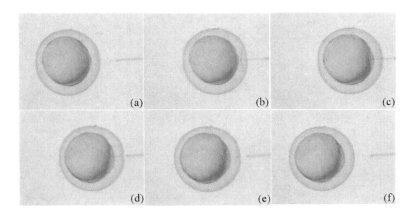

Fig. 21. Out-of-plane cell injection process with position and force control. ©2009 IEEE

Fig. 22(left) and (right) illustrates position tracking results in $X-$axis (a) and $Y-$ axis (b), respectively. It is seen that the actual position could follow the desired position very well, and the maximum position tracking error in each axis was only around 1.5 μm. This demonstrates that the position control algorithm exhibits satisfactory tracking performance for such micro manipulation. Good accuracy of the position control in $X-Y$ plane ensures the force control in the Z- axis to succeed, since the desired force in the Z-axis is based on the actual position feedback in the $X-Y$ plane.

Fig. 23(left) illustrates the impedance control results in the $Z-$axis, and the depth motion error. During the pre-piercing period ($0\sec < t < 0.5\sec$), the injection force was zero since the injector did not contact the cell. When the injector pipette contacted the cell after the injector moved for 0.5 second, the injection force, measured by the force sensor, started to increase. The desired force was determined based on the actually positions in the $X-Y$ plane and the calibration relationship (11). The maximum force error along the $Z-$axis was around 60 μN. The maximum position error in the $Z-$axis was around 6 μm, which was larger than that in the $X-Y$ plane, since it was controlled indirectly via the impedance control.

Fig. 23(right) illustrates both the position and force tracking errors after combining the results in the $X-, Y-$ and $Z-$axes. The position error denotes the difference between each actual position and the desired position in 3D space, calculated by the desired pipette trajectory ρ^d and the fixed angles β

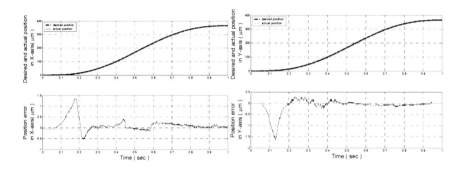

Fig. 22. Position tracking results in $X - Y$ plane.(left) Position tracking results in X-axis; (right) Position tracking results in Y-axis. ©2009 IEEE

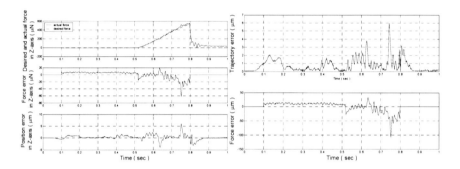

Fig. 23. (Left) Injection force control results in Z- axis and depth motion error. (Right) Combined position and force tracking errors. ©2009 IEEE

and ϕ. The force error was detected along the pipette axis by the installed PVDF force sensor. It is seen that the maximum position error corresponds to the maximum force error. This is because that the depth control follows the force control in $Z-$axis, and the force control performance has great influence on the depth control and hence the whole position control performance.

The same experiment was repeated to simulate out-of-plane cell injection of a batch of 20 zebrafish embryos. Fig. 24 shows the maximum force error and the largest position trajectory errors of these 20 embryos. The maxim force error standard deviation and maxim position error standard deviation are 7.3119 μN and 1.5119 μm, respectively. The statistical data such as the mean error, the standard deviation and the maxim value of these experimental results show that the proposed control method exhibits stable and uniform repeatable performance.

The operation speed of the automated system is about 10 embryos (with chorion) per minute, and the overall success rate in all trials is approximately 98%. This demonstrates that the automated system is capable of repeatedly

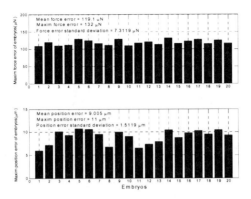

Fig. 24. Evaluation of batch injection performance. ©2009 IEEE

piecing the zebrafish embryos with a high success rate. Rare failure to penetrate an embryo occurs when an embryo having significantly different mechanical properties such as dead or broken embryos, is ignored by the injector pipette rather than piercing the embryo. Compared to a manual operation, the system does not suffer from large failure rate due to technician fatigue and proficiency differences amongst technicians [20]. Experiments demonstrate that the automated microrobotic system is a reliable tool for cell injections.

Note that while experiments were carried out only with Zebrafish embryos in this chapter, the proposed system design concept and the control approach may be applied to other types of cells. For mammalian embryos, the cell holding device and the micro force sensors may need to be modified to suit to the smaller scale of the cells.

7   Conclusion

The development of an automatic cell injection system is a multidisciplinary cross research, in which many technologies in robotics system design, automation, image processing, controls algorithm, MEMS, together with bioengineering are integrated. The special living and soft manipulation objects cultivated in solution have introduced much difficulty in transferring the cell injection process into automation. Compared to the rapid development of abundant research on micromanipulation technologies in microelectronics or micro-electro-mechanical systems, the research on micro bio-manipulation, in contrast, is rare and cannot fully meet the demand by the biological industry. Only until very recently, most of the cell injection operations are still being performed manually, which makes the relevant biological industry labor-intensive.

In this chapter, a new robotic biomanipulation system for automatic injection of large batches of suspended cells is developed. A cell holding device, fixed on the rotary plate, permits cells to be transported, one by one, into the injection field of view. With the controller designed to decouple the out-of-plane cell injection task into a position control in the horizontal plane and an impedance control in the -axis, a position and force control algorithm is developed. Note that the force control in Z-axis indirectly controls the depth motion of the injector pipette. Finally, the effectiveness of the proposed approach is demonstrated experimentally. The technology has great potential for applications. The near term impact is to largely improve work efficiency and success rate of many biological injection applications such as microRNA and DNA injections. Commercialization and use of the proposed technology may lead to growth of high-quality and affordable automatic cell-manipulation industry to support fast growing studies on systems biology.

Acknowledgements

The authors thank Professor Shuk Han Cheng, Ms. Tse Ho Yan of the Department of Biology and Chemistry, City University of Hong Kong, for assistance in preparation of Zebrafish embryos.

References

1. Accessed (2011), `http://www.test.org/doe`
2. Accessed (2011), `http://www.eppendorf.com`
3. Accessed (2011), `http://www.dynacitytech.com`
4. Ammi, M., Ferreira, A.: Realistic visual and haptic rendering for biological-cell injection. In: Proceedings of the 2005 IEEE International Conference on Robotics and Automation, ICRA 2005, pp. 918–923 (2005)
5. Arai, F., Sugiyama, T., Fukuda, T., Iwata, H., Itoigawa, K.: Micro tri-axial force sensor for 3d bio-manipulation. In: Proceedings. 1999 IEEE International Conference on Robotics and Automation (1999)
6. Cho, S.Y., Shim, J.H.: A new micro biological cell injection system. In: Proceedings. 2004 IEEE/RSJ International Conference on Intelligent Robots and Systems, IROS 2004 (2004)
7. Fujisato, T., Abe, S., Tsuji, T., Sada, M., Miyawaki, F., Ohba, K.: The development of an ova holding device made of microporous glass plate for genetic engineering. In: Proceedings of the 20th Annual International Conference of the IEEE Engineering in Medicine and Biology Society (1998)
8. Huang, H., Sun, D., Mills, J., Cheng, S.H.: Integrated vision and force control in suspended cell injection system: Towards automatic batch biomanipulation. In: IEEE International Conference on Robotics and Automation, ICRA 2008, pp. 3413–3418 (2008)
9. Huang, H., Sun, D., Mills, J., Li, W., Cheng, S.H.: Visual-based impedance control of out-of-plane cell injection systems. IEEE Transactions on Automation Science and Engineering 6(3), 565–571 (2009)
10. Huang, H.B., Sun, D., Mills, J.K., Cheng, S.H.: Robotic cell injection system with position and force control: toward automatic batch biomanipulation. Trans. Rob. 25, 727–737 (2009),
    `http://dx.doi.org/10.1109/TRO.2009.2017109`

11. Kallio, P.: Capillary pressure microinjection of living adherent cells: challenges in automation. Journal of Micromechatronics 3(32), 189–220 (2006)
12. Kawaji, A.: 3d calibration for micro-manipulation with precise position measurement. Journal of Micromechatronics 1(14), 117–130 (2001)
13. Kim, D.H., Yun, S., Kim, B.: Mechanical force response of single living cells using a microrobotic system. In: Proceedings. 2004 IEEE International Conference on Robotics and Automation, ICRA (2004)
14. Kumar, R., Kapoor, A., Taylor, R.: Preliminary experiments in robot/human cooperative microinjection. In: Proceedings. 2003 IEEE/RSJ International Conference on Intelligent Robots and Systems, IROS 2003, vol. 3&4, pp. 3186–3191 (2003)
15. Lai, K., Kwong, C., Li, W.: Kl probes for robotic-based cellular nano surgery. In: 2003 Third IEEE Conference on Nanotechnology, IEEE-NANO 2003, vol. 1&2, pp. 152–155 (2003)
16. Li, G., Xi, N.: Calibration of a micromanipulation system. In: IEEE/RSJ International Conference on Intelligent Robots and Systems (2002)
17. Li, X., Zong, G., Bi, S.: Development of global vision system for biological automatic micro-manipulation system. In: Proceedings. 2001 IEEE International Conference on Robotics and Automation, ICRA (2001)
18. Liu, X., Lu, Z., Sun, Y.: Orientation control of biological cells under inverted microscopy. IEEE/ASME Transactions on Mechatronics (99), 1–7 (2010), doi:10.1109/TMECH.2010.2056380
19. Liu, X., Sun, Y.: Microfabricated glass devices for rapid single cell immobilization in mouse zygote microinjection. Biomedical Microdevices 11(6), 1169–1174 (2009)
20. Lu, Z., Chen, P.C.Y., Nam, J., Ge, R., Lin, W.: A micromanipulation system with dynamic force-feedback for automatic batch microinjection. Journal of Micromechanics and Microengineering 17(2), 314 (2007)
21. Matsuoka, H., Komazaki, T., Mukai, Y., Shibusawa, M., Akane, H., Chaki, A., Uetake, N., Saito, M.: High throughput easy microinjection with a single-cell manipulation supporting robot. Journal of Biotechnology 116(2), 185–194 (2005)
22. Mattos, L., Grant, E., Thresher, R.: Semi-automated blastocyst microinjection. In: Proceedings 2006 IEEE International Conference on Robotics and Automation, ICRA 2006, pp. 1780–1785 (2006)
23. Pillarisetti, A., Pekarev, M., Brooks, A., Desai, J.: Evaluating the effect of force feedback in cell injection. IEEE Transactions on Automation Science and Engineering 4(3), 322–331 (2007)
24. Qu, W.T.: Chord midpoint hough transform based ellipse detection method. Journal of Zhejiang University (Engineering Science) 39(8), 1132–1135 (2005)
25. Sun, D., Liu, Y.: Modeling and impedance control of a two-manipulator system handling a flexible beam. In: Proceedings. 1997 IEEE International Conference on Robotics and Automation, vol. 2, pp. 1787–1792 (1997)
26. Sun, Y., Nelson, B.J.: Biological Cell Injection Using an Autonomous Micro-Robotic System. The International Journal of Robotics Research 21(10-11), 861–868 (2002)
27. Sun, Y., Wan, K.T., Roberts, K., Bischof, J., Nelson, B.: Mechanical property characterization of mouse zona pellucida. IEEE Transactions on NanoBioscience 2(4), 279–286 (2003)

28. Tan, K.K., Ng, D.C., Xie, Y.: Optical intra-cytoplasmic sperm injection with a piezo micromanipulator. In: The 4th World Congress on Intelligent Control and Automation, 2002. Proceedings (2002)
29. Tan, Y., Sun, D., Huang, W., Cheng, S.H.: Mechanical modeling of biological cells in microinjection. IEEE Transactions on NanoBioscience 7(4), 257–266 (2008)
30. Wang, W., Liu, X., Gelinas, D., Ciruna, B., Sun, Y.: A fully automated robotic system for microinjection of zebrafish embryos. PLoS ONE 2(9), e862 (2007)
31. Xie, Y., Sun, D., Liu, C., Cheng, S.H.: An adaptive impedance force control approach for robotic cell microinjection. In: IEEE/RSJ International Conference on Intelligent Robots and Systems, IROS 2008, pp. 907–912 (2008)
32. Xie, Y., Sun, D., Liu, C., Cheng, S.H.: A flexible force-based cell injection approach in a bio-robotic system. In: Proceedings 2009 IEEE International Conference on Robotics and Automation (2009)
33. Xie, Y., Sun, D., Liu, C., Tse, H., Cheng, S.: A force control approach to a robot-assisted cell microinjection system. Int. J. Rob. Res. 29, 1222–1232 (2010), http://dx.doi.org/10.1177/0278364909354325
34. Xudong, L.: Automatic micromanipulating system for biological applications with visual servo control. Journal of Micromechatronics 1(4), 345–363 (2002)
35. Sun, Y., Wejinya, U.C., Xi, N., Pomeroy, C.A.: Force measurement and mechanical characterization of living drosophila embryos for human medical study. Proceedings of the Institution of Mechanical Engineers, Part H: Journal of Engineering in Medicine 221, 99–112 (2006)
36. Yu, S., Nelson, B.: Microrobotic cell injection. In: Proceedings. 2001 IEEE International Conference on Robotics and Automation, ICRA (2001)
37. Zhang, X.J., Zappe, S., Bernstein, R.W., Sahin, O., Chen, C.C., Fish, M., Scott, M.P., Solgaard, O.: Micromachined silicon force sensor based on diffractive optical encoders for characterization of microinjection. Sensors and Actuators A: Physical 114(2-3), 197–203 (2004); Selected papers from Transducers 03
38. Zhou, Y., Nelson, B.J., Vikramaditya, B.: Integrating optical force sensing with visual servoing for microassembly. J. Intell. Robotics Syst. 28, 259–276 (2000)

Advanced Hybrid Technology for Neurorehabilitation: The HYPER Project

Alessandro De Mauro[1], Eduardo Carrasco[1], David Oyarzun[1],
Aitor Ardanza[1], Anselmo Frizera-Neto[2], Diego Torricelli[2],
José Luis Pons[2], Angel Gil Agudo[3], and Julian Florez[1]

[1] eHealth and Biomedical Department, VICOMTech. San Sebastian, Spain
   `ademauro@vicomtech.org`
[2] Bioengineering Group, CSIC. Madrid, Spain
   `{frizera,torricelli,jlpons}@iai.csic.es`
[3] Biomechanics Unit, National Hospital of Paraplegics. Toledo, Spain
   `amgila@sescam.jccm.es`

Abstract. Disabilities that follow cerebrovascular accidents and spinal cord injuries severely impair motor functions and thereby prevent the affected individuals from full and autonomous participation in activities of daily living. Rehabilitation therapy is needed in order to recover from those severe physical traumas. Where rehabilitation is not enough to restore completely human functions then functional compensation is required. In the last years the field of rehabilitation has been inspired by new available technologies. An example is given by rehabilitation robotics where machines are used to assist the patient in the execution of specific and physical task of the therapy. In both rehabilitation and functional compensation scenarios, the usability and cognitive aspects of human-machine interaction have yet to be solved efficiently by robotic-assisted solutions. Hybrid systems combining exoskeletal robots (ERs) with motor neuroprosthesis (MNPs) emerge as promising techniques that blends together technologies that could overcome the limitations of each individual one. Another promising technology which is rapidly becoming a popular application for physical rehabilitation and motor control research is Virtual Reality (VR). In this chapter, we present our research focuses on the development of a new rehabilitation therapy based on an integrated ER-MNP hybrid systems combined with virtual reality and brain neuro-machine interface (BNMI). This solution, based on improved cognitive and physical human-machine interaction, aims to overcome the major limitations regarding the current available robotic-based therapies.

Introduction

Cerebrovascular accidents (CVA) and spinal cord injuries (SCI) are the most common causes of paralysis and paresis with reported prevalence of 12,000 cases per million and 800 cases per million, respectively. Disabilities that follow CVA or SCI severely impair motor functions (e.g., standing, walking,

T. Gulrez, A.E. Hassanien (Eds.): Advances in Robotics & Virtual Reality, ISRL 26, pp. 89–108.
springerlink.com

reaching and grasping) and thereby prevent the affected individuals from healthy-like, full and autonomous participation in daily activities.

The main goal of neurorehabilitation is to favor the re-learning process of the Central Nervous System (CNS) in the execution of coordinated movements. This process of neural reorganization involves complex sensorimotor mechanisms, which are nowadays in the focus of intense investigation in (the fields of) neurophysiology. From a therapeutic point of view, the outcome of the neurorehabilitation process depends on two main issues [9, 20]:

- the quality and amount of physical activity performed by the patient.
- the cognitive involvement of the patient in the rehabilitation process.

The traditional rehabilitation therapy is mainly focused on physical exercise, aiming to strengthen the active muscles in several parts of the body. Different modes of exercises are used, ranging from aerobic therapies (e.g. treadmill walking, cycling movement with arm-leg ergometry or seated stepper), and non-aerobic training, to enhance strength or flexibility (e.g. weight machines, isometric exercises, stretching).

In this context, occupational therapy is a specialized profession aimed to train individuals to relearn the tasks of daily living that have personal meaning and value, such as eating, dressing, and grooming.

The latest advancements in robotics and neuroscience have generated a great impact in the field of neurorehabilitation. In the last decade, a number of robot-assisted rehabilitation systems have been developed in order to support and improve the therapist's action by delivering intensive physical therapy and providing objective measures of the patient's performance [16, 5, 51].

In this context, the use of bio-inspired hybrid systems combining exoskeletal robots (ERs) and motor neuroprostheses (MNPs) is emerging. ERs are person-oriented robots, operating alongside human limbs to physically support the function of a limb or to replace it completely [38]. MNPs constitute an approach to restoring neuromotor function by means of controlling human muscles or muscle nerves with Electrical Stimulation. The combination of robotic actuation and bio-electrical stimulation emerges as a promising technique that blends together technologies that could overcome the limitations of each individual one, in both rehabilitation and functional compensation scenarios [15, 33]. The orchestration of the hybrid system with the latent motor capabilities of the patient involves several issues, principally related to the *cognitive* aspects of human-machine interaction: the rehabilitation machine must be capable of deciphering user's volitional commands in a robust manner in order to act reliably and in concert with the subject [21].

In this context, multimodal human-machine interfaces gained relevance as a means to convey a high number of signals of different nature into meaningful feed-forward and feedback information. These signals can be related to muscular activity (EMG), cerebral activity (EEG) and visual, auditory and tactile perception. Virtual Reality (VR) environments have emerged as a powerful tool for integrating such multimodal interfaces within robotic-based rehabilitation scenarios.

In the first two sections of this chapter the authors present the state of the art of the application of Robotics and Virtual Reality in the field of neurorehabilitation. Thereafter, focus is put on the research in the frame of the HYPER project "Hybrid Neuroprosthetic and Neurorobotic devices for Functional Compensation and Rehabilitation of Motor Disorders" [2], which aims at the development of a new hybrid system based on the combined action of NR, MNP and VR, in order to overcome the major limitations of current rehabilitation solutions.

1   Robotics in Neurorehabilitation

In this chapter, first a brief introduction of the state of art in the field of neurorehabilitation is presented. As concrete application of the presented concepts, the HYPER project and the current development is detailed in the following sections. In such project, robotic solutions to assess CVA and SCI (figure 1) are currently being developed by a team of clinical and engineering staff.

In the last decades an increasing number of robotic systems have been made available in the field of rehabilitation. The presence of a robot basically permits to support the therapists in the passive mobilization of the patient's limb. The main advantages of this robotic intervention are twofold: i) therapeutic, in terms of the amount of physical exercise provided during the therapy, and ii) economical, since a reduced number of employees is required to support the patient during the execution of movement.

The recent advances in the fields of human-robot interaction (HRI) [38] have allowed thinking to the machine not only under the point of view of physical interaction, but also as a means to stimulate the mental activity of the subject during the therapy. Here, the cognitive processes involved in the interaction with the machine are crucial. The term cognitive alludes to the close relationship between cognition - as the process comprising high level functions carried out by the human brain, including perception, comprehension, construction, planning, self-monitoring - and motor control.

Cognitive and physical interactions are not independent. On the one hand, a perceptual cognitive process in the human can be triggered by physical interaction with the robot. On the other hand, the cognitive interaction can be used to modify the physical interaction between human and robot, for instance to alter the compliance of an exoskeleton.

1.1   Key Factors in Robot-Assisted Therapy

The main goal of neurorehabilitation is to favor the re-learning process of the Central Nervous System (CNS) in the execution of coordinated movements. This process involves complex sensorimotor mechanisms, which are nowadays

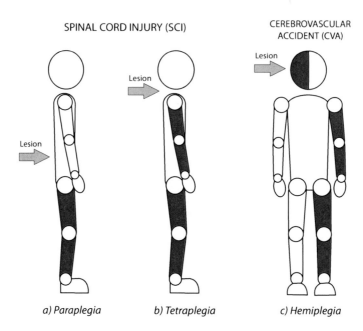

SPINAL CORD INJURY (SCI)   CEREBROVASCULAR
                                            ACCIDENT (CVA)

a) Paraplegia   *b) Tetraplegia*   *c) Hemiplegia*

Fig. 1. Disabilities that follow CVA and SCI. a) Paraplegia refers to the loss of motor and/or sensory function in thoracic, lumbar or sacral segments (SCI). Consequently, the arm function is spared, but the trunk, legs and pelvic organs can be affected. b) Tetraplegia refers to the loss of motor and/or sensory function in the cervical segments of the spinal cord (SCI). It results in an impaired function of the arms, trunk, legs and pelvic organs. c) Hemiplegia is paralysis of the contralateral side of the body occurring after a CVA. It may comprise weakness of the leg on the affected side, where the drop-foot syndrome often prevents walking.

in the focus of intense investigation in neurophysiology. There is no consensus on what are the most adecuate intervention for neurorehabilitation [25, 56]. Nevertheless, some key factors (KFs) for successful robotic-assisted therapy can be identified:

KF1 - Active Role of the Patient. Brain activity plays a fundamental role on the modulation of the neural mechanisms that generate movement [8, 29]. Passive, repetitive training is very likely to be suboptimal, as it leads to the phenomena of "learned helplessness", that is, the lower spinal cord becomes habituated to the training action, decreasing the performance of the therapy [57]. Thereby, active cooperation of the patient is needed to achieve a more functional outcome of the therapy. A solution to this problem is to make the system respond to the physical actions of the patient, in order to re-establish the connection between causes (forces) and effects (movements) [39].

KF2 - Motivation. Motivation is one of the most important factors in rehabilitation and it is commonly used as a determinant of rehabilitation outcome [30], since it is strongly correlated with the degree of patient's activity. User's motivation can be achieved by means of various different types of feedback and modes of interaction, so influencing the motor re-learning process at different levels [6].

KF3 - Assist-as-needed. In order to imitate the action of the physical therapist in supporting the movement of the limb, the new-generation of robotic systems have been provided with the so called Assist-as-needed (AAN) paradigm [11]. The AAN paradigm is intended to simultaneously activate the efferent (motor) and afferent (sensory) pathways, by providing only the minimum assistance necessary during the execution of the movement. This mechanism has been proven to favor cortical reorganization [55].

KF4 - Challenge. Contrary to the assistive techniques, which help the user to reach the task, the challenge-based robotic strategies aim at opposing to the user's intention of movement, using resistance or error-amplification strategies. This approach is based on evidences that resistive exercises requiring high cognitive effort can help improve neuromotor functions [53].

KF5 - Biofeedback. Biofeedback is a crucial factor for the success of the therapy, as it informs about the patient's degree of activity and is a key to maintain and encourage the motivation and increasing the active participation of the patient. Currently, biofeedback in rehabilitation relies mainly on a single source of information, i.e. force-based feedback. By combining other forms of feedback, such as brain activity (EEG), muscular activity (EMG) and visual information on limbs motion, a more accurate and effective outcome might be achieved [28].

KF6 - Bioinspiration. Due to the close cooperation between human and robot, it is necessary to know the properties of the human motor system in order to define the design requirements of a rehabilitation device. With the help of a biological model it is possible to predict the system's behavior and optimize the robotic intervention, in terms of adaptability, functionality and energy consumption [37] (see also paragraph 1.3).

As depicted in the figure 2, the above stated key factors are not independent, like so the technological solutions, which are not univocally defined. This multidimensional problem strongly directs to a multimodal approach solution.

1.2   Hybrid Wearable Technology

The use of wearable devices has shown to be a good solution to achieve the mentioned therapy in which the patient plays such an important role. Wearable robots (WRs) [38] are person-oriented robots and therefore more adapted to the interaction with users. They are worn by human operators to supplement the function of a limb, e.g. exoskeletons. WRs exhibit a close interaction with the human user but structurally are similar to robots, i.e. rigid links, actuators, sensors and control electronics. A wide range of prototypes have been proposed.

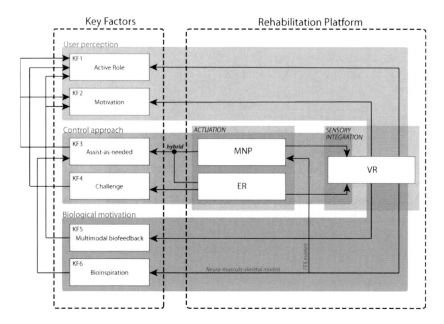

Fig. 2. Cognitive interaction in robotic-assisted rehabilitation. In this scheme, the key factors of neurorehabilitation (KFs) have been classified by their relation with the three main conceptual actors involved in cognitive human-machine interaction: i) user perception, ii) machine control intelligence and, iii) biological motivation. The path followed by the cognitive information responsible of the KFs reinforcement, as well as the interdependencies among the KFs themselves, are shown by arrows.

The Fraunhofer Institute of Berlin developed "Haptic Walker" [43, 42] (Fig. 3 left), a system in which the patient feet are placed on two separated platforms. The patient weight and the device components are supported by a mechanical structure, where the movements of two platforms related to the feet reproduce the kinematics of the leg joints. Haptic Walker is used mainly for gait rehabilitation and includes a module providing VR rehabilitation.

Veneman et al. [51] (Fig. 3 right) have introduced a new gait rehabilitation device that combines a 2-D-actuated and translatable pelvis segment with a leg exoskeleton composed of three actuated rotational joints (two at the hip and one at the knee). Practically it is an exoskeleton that moves in parallel with the legs of a person walking on a treadmill. An interesting outcome of their research is that the system allows both a "patient-in-charge" and "robot-in-charge" mode, in which the robot is controlled either to follow or to guide a patient, respectively.

The Cyberthosis project [31] (funded by the Fondation Suisse pour les Cyberthèses, FSC) combines closed-loop electrical muscle stimulation with motorized orthoses. It is composed of two subsystems (devices): MotionMaker™

[44, 45], which is a couch equipped with two orthoses enabling a controlled movement of the hip, knee and ankle joints (Fig. 4 left) and WalkTrainer™ [47], composed of a body weight support, a gait and pelvic orthosis and several electrodes for electrostimulation (Fig. 4 right).

 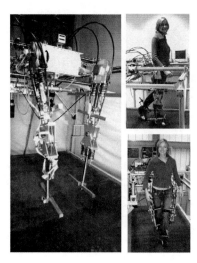

Fig. 3. On the left side: Haptic Walker (Fraunhofer Institute Berlin, Germany). On the right side: the Lopes prototype developed by Jan F. Veneman et al. (Institute for Biomedical Technology, University of Twente, Netherlands).

Together with the mentioned wearable robotic solutions, a different approach in leading to the concept of Soft Robots (SRs). SRs also rely on wearable devices, but use functional human structures instead of artificial counterparts, e.g. artificial actuators are substituted by Functional Electrical Stimulation (FES) of human muscles. The borderline between human and robot becomes fuzzy and this immediately leads to hybrid Human-Robot systems.

It is in this context that Neurorobots (NRs) and Motor Neuroprostheses (MNPs) emerge. They may be considered as a bypass of damaged sensory-motor systems: via Brain-Machine interfaces (BMIs), users can volitionally trigger the device by using the functional part of the body above lesion, e.g. residual muscle activity and/or cerebral activity; MNPs constitute an approach to restoring function by means of artificially controlling human muscles or muscle nerves with Functional Electrical Stimulation (FES). Electrical pulses stimulate motor and/or sensory nerves, thereby generating movement by activating paralyzed muscles.

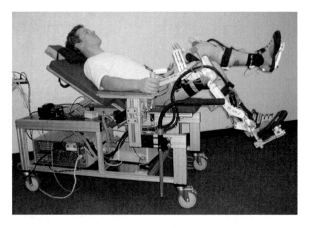

Fig. 4. Prototypes of MotionMaker™ (on the left) and Walktrainer™ (on the right) which are parts of the Cyberthosis project.

A schematic representation of the combined action of MNP and NR on the patient is shown in Fig. 5.

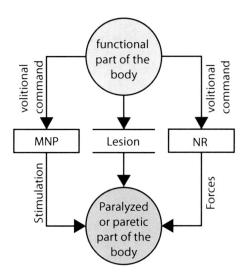

Fig. 5. Hybrid system concept. NRs use volitional commands to drive paralyzed or paretic part of the body, while MNPs bypass the damaged sensory-motor systems using functional electrical stimulation (FES).

NRs use volitional commands for controlling a (mechatronic) WR, which applies controlled forces to drive the affected limbs. Wearable Neurorobotics is more versatile as to the implementation of motor actions: (1) precise kinematics and impedance control is readily available, (2) biomimetic control architectures, e.g. Central Pattern Generators, internal models and reflexes, can be readily applied, and (3) built-in artificial proprioception and vestibular sensors can be used to drive non-volitional motor actions, e.g. control of balance. As a serious drawback, the transmission of motor actions (forces) to the human musculoskeletal system takes place through soft tissues, thus the interaction is to be carefully thought in terms of compatible kinematics and application of controlled forces.

1.3 Bioinspiration

Actual efforts in rehabilitation research are integrating neuroscience knowledge into engineering to develop new effective means for neurorehabilitation, based on a deeper understanding of the human control system. Scientific knowledge in the field of neurophysiology can provide engineers with sources of inspirations for the development of biologically motivated systems and the refinement of human body models. At the same time, mathematical and engineering models of the human body can provide clinicians and neuroscientists with powerful tools to test and simulate a wide range of motor control theories prior to in-vivo experimentation.

The contact between bioinspiration and engineering arose and developed in technical areas which were (and are) not specifically directed to rehabilitation. In particular, the field of human body simulation devoted specific effort to replicate the neuro-musculo-skeletal interactions of walking [50, 34, 35, 19]. Nevertheless, due to the complexity of the problem, further work is needed to translate these simulation results to a real-life environment. In this direction, some interesting approaches have been proposed in the field of bipedal robotics, where the imitation of the structure of the CNS is providing good features for stability enhancement in walking and standing control [27, 14, 46]. An increased interest on the application of biologically motivated methods to rehabilitation robotics and prosthetics is proven by the presence of promising recent research works based on neurophysiological principles [10].

2 Virtual Reality in Neurorehabilitation

VR environments can provide realistic training for the patient in different scenarios and phases of the neurorehabilitation. By using VR in conjunction with Human Computer Interfaces (HCI) the training of daily life activities can be much improved in terms of time and quality. This approach permits a realistic and ergonomic training in a safe, interactive and immersive environment.

Examples of interfaces able to interact with VR are mice, joystick, haptic interfaces with force feedback and motion tracking systems.

Repetition is crucial for the re-learning of motor functions and for the training of the cortical activity. This task has to be connected with the

sensorial feedback on every single exercise. Then again the patient motivation is fundamental and can be improved by assigning a video game format
to the therapy. In this way the training activity becomes more attractive and
interesting [54, 17]. Moreover, VR shows another advantage: the possibility
to be precisely adapted to the patient's therapy and to be specific for each
rehabilitation phase. In addition, it represents a precise tool for the assessment of the therapy during each session. The (tracked/saved) data can be
used by the rehabilitators for monitoring and managing the telerehabilitation
[7]. Several researches have shown that, during VR rehabilitation, the movements are very similar to those used in the traditional therapy. Although
they appear a bit slower and less accurate, [40, 52] they are appropriate for
rehabilitation. Finally, [48, 22, 58, 41, 49, 12] have proven good results in
executing the movements trained in VR in reality.

Some of the significant studies on the application of robotics in conjunction
with VR for rehabilitation purposes shall be introduced briefly. The Rutgers
Arm [24] is one of the first prototypes composed of a PC, a motion tracking
system and a low-friction table for the upper extremity rehabilitation. The
system has been tested on a chronic stroke subject and showed improvements
in arm motor control and shoulder range of motion (Fugl-Meyer [13] test
scores). The same group has developed the Rutgers Ankle [4] for the lower
extremity rehabilitation. It is a haptic/robotic platform, which works with six
degrees of freedom, driving the patient's feet movements (Fig. 6, left). In [32],
the Rutgers Ankle system has been tested. As a result, the group of patients
trained with the robotic device coupled with the VR demonstrated greater
changes in velocity and distance than the group trained with the robot alone.

The InMotion system is a complete robotic system for the rehabilitation
of the upper extremity [7]: arm, shoulder, wrist, and hand. It allows multiple
degrees of movement: pronation, supination, flexion and extension, radial and
ulnar deviation (Fig. 7 left).

In [40], the MOTEK Medical's V-Gait System is described as a tool for clinical analysis, rehabilitation and research. This system combines a self-paced
instrumented treadmill capable of comprehensive measurements of ground
reaction force with a real-time motion capture system and a 3D virtual environment (Fig. 7 right).

The state-of-art in rehabilitation using virtual reality (VR) and robotics is
provided by Lokomat and Armeo (from Hocoma) for the lower and the upper
extremity, respectively (Fig. 6 center and left). These two systems are validated
by the medical community and used in several rehabilitation centers [23].

Lokomat [18] offers a driven orthosis (gait robot) with electrical drives in
knee and hip joints. The orthosis is adaptable to subjects with femur lengths
and an additional body weight support is included. The Armeo [1] provides
support to reacquire and improve motor control for the affected arm and
hand. This support counteracts the effects of gravity. There are three types
of subproducts in the Armeo line: i) the ArmeoPower is a robotic arm exoskeleton, with an electric lifting column for comfortable height and weight

Fig. 6. Successful application of robotics in rehabilitation. From left to right: Rutgers Ankle (Courtesy of Rutgers University), Armeo and Lokomat (Courtesy of Hocoma)

Fig. 7. Examples of products for rehabilitation including VR concepts. On the left: InMotion Arm Robot. On the right: rehabilitation with MOTEK Medical.

adjustment; ii) the ArmeoSpring is an instrumented arm orthosis with a spring mechanism for adjustable arm weight support; and iii) the Armeo-Boom is a simple arm weight support system with low inertia to facilitate functional movements.

All the previous products commercialized from Hocoma are completed by an augmented feedback module which extend the conventional hardware with a computer and a large monitor with acoustic stereo feedback together with a software with interactive training tasks. This option provides various engaging virtual environments to motivate the patients, adjustable level of difficulty and intensity according to the cognitive abilities and the specific needs of each patient.

3   The HYPER Project

3.1   Project Concepts

None of the systems described in the previous section proposes VR in conjunction with an hybrid and wearable MNP-NR system. With the HYPER project (Hybrid Neuroprosthetic and Neurorobotic devices for Functional Compensation and Rehabilitation of Motor Disorders) the authors aim at a breakthrough in the research of neurorobotic and neuroprosthetic devices for rehabilitation and functional compensation. The project focuses its activities on new wearable hybrid systems that will combine biological and artificial structures in order to overcome the major limitations of the current rehabilitation solutions to Cerebrovascular Accident (CVA) and Spinal Cord Injury (SCI). The main objectives of the project are: i) to restore motor functions in SCI patients through functional compensation, and ii) to promote the re-learning of motor control in patients suffering from CVA.

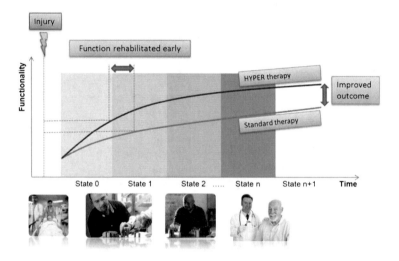

Fig. 8. Rehabilitation using the HYPER system: the therapy is specific for different states. Functions are regained faster and the outcome is improved.

The specific rehabilitation targets of the HYPER project are, on the one hand, to speed up the rehabilitation procedures and, on the other hand, to improve the outcome of the therapy using new paradigms and technologies, as depicted in Fig. 8. This shall be achieved by an integrated use of different interventions comprising multimodal sensing and actuation. Both upper and lower parts of the patient body are assisted. The main emphasis is put on restoring daily life activities, i.e. walking, standing, reaching and grasping. The therapy is subdivided in several states from the state 0 in which the injury happens until the state N in which the patient is fully rehabilitated. The ranges of movements that are significant in the rehabilitation for CVA or SCI patients have been identified by medical doctors. Any of the daily life functions constitutes a combination of these defined movements. In the specific, the upper body joints (and related movements) are:

- shoulder (flexion, extension, abduction, adduction, outward medial rotation, inward medial rotation);
- elbow (flexion, extension, pronation, supination);
- wrist (flexion, extension, abduction, adduction).

Similarly, the lower body joints (and related movements) are:

- hip (flexion, extension, abduction, adduction, medial and lateral rotation);
- knee (flexion, extension);
- ankle (plantar flexion, dorsal flexion, inversion and eversion).

For each of them, both degrees and ranges of movement have been specified in order to assess the patient's skills during the rehabilitation process and to parameterize the rehabilitation training.

Users' groups have been identified in order to adjust therapy and system components to the various therapy needs. Diverse scenarios have been developed and elaborated in detail. Each of them includes a specific configuration of components (MNP, NR, and VR). Some of the components integrated into the HYPER system are presented in Fig. 9.

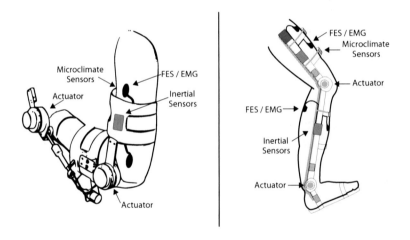

Fig. 9. Components of the HYPER system.

Fig. 10 shows how the inputs and outputs of both actors (human and hybrid device) are interconnected through a multimodal interaction platform. Using a multi-channel acquisition approach, the user's outputs (EEG, EMG, kinetic and kinematic information) serve as inputs to the controller of the hybrid platform. The controller, as in the natural human control system, (re)presents a feed-forward component based on predetermined motion and biomechanical models, and a reactive controller that mimics human neuromotor mechanisms and reflexes. In addition to the limb actuation systems (NR/MNP), a virtual reality system generates visual/auditory feedback to

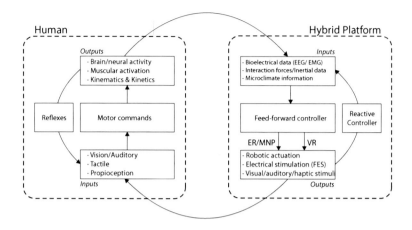

Fig. 10. HYPER human-machine interaction scheme.

increase the user's involvement and immersion, potentiating the cognitive interaction. A Brain Neural Machine Interface (BNMI) is applied to potentiate the cognitive interaction and to drive the hybrid system. BNMIs are able to extract and interpret the volitional commands generated by the humans' neural system, by relying on: (1) direct brain activity monitoring, which might assess user's motor intention; (2) measurement of the peripheral nervous activity, e.g reflexes, that might involuntarily trigger muscle activity; and (3) indirect monitoring of neural activity through EMG, to estimate the fine motor effects of nervous activity. Due to this capacity of extracting relevant information about human motor control mechanism, apt to be fed back to the machine, BNMI might also contribute to new means of improving user-centered strategies (e.g. AAN paradigm).

3.2   The Role of Virtual Reality in HYPER

Virtual reality has the ability to simulate real-life tasks and provides several evident benefits for the new rehabilitation therapy proposed in the HYPER project:

1. specificity and adaptability to each patient and phase of the therapy;
2. repeatability (repetition is crucial for the re-learning of motor functions and for the training of the cortical activity);
3. ability to provide patient engagement (active cooperation of the patient is needed to achieve a more functional outcome of the therapy);
4. capability for precise assessment (data can be used by the rehabilitation specialists for monitoring and managing the therapy);
5. safety (VR provides the user with the possibility to perform tasks with a degree of safety which is normally not possible in the traditional rehabilitation).

We are currently working on providing a VR rehabilitation platform for the HYPER project. The interactive VR environment, an important part of the complex system, was initially based on Radio Frequency (RF) tracking technology. Patient's movements in the execution of the different tasks were initially tracked using RF transmitters positioned on each joint of the patient body (upper/lower part). This solution offers good tracking performances but it suffers from the use of many cables. Considering the patient's needs it has to be considered as a not optimal solution. Therefore we are now exploring a new wireless and inexpensive technology: Kinect® [3]. First results are very promising and even if the accuracy of the tracking has to be measured exactly it seems that for this type of application the needs in terms of accuracy are not highly demanding. Additionally the tracking system appears to be robust enough to track the patient and the related neurorobotic exoskeleton or neuroprosthetic devices on both upper and lower part of the body. A limitation to be considered is that the Kinect® IR camera suffers when the subject is illuminated strongly by the sun light. This is, however, a merely technological limitation which can be overridden.

In order to validate the VR concepts and system accuracy we are currently comparing the two trackers used (RF and Kinect^{TM}) with the Armeo^{TM} measuring the performances in the execution of simple tasks of the upper part of the body. In a similar way, we plan to compare the system with Lokomat^{TM} for the lower part of the patient. Finally, a further part of our research concerns the conjunction between a Brain Computer Interface and virtual reality in order to create a good diagnostic and personalized environment in which it is possible to study the brain signals as answers to external (VR) stimuli or to assess the progress of the patient in the rehabilitation therapy. Several researches [26, 36] have shown how brain-computer interfaces can become a communications medium for VR applications in the range from basic neuroscience studies to developing optimal mental interfaces and more efficient brain-signal processing techniques. Since the engagement and collaboration of the patient is a must in order to obtain good results, an important role can be played by VR during brain signals analysis. A snapshots of simple VR scenes (reaching, moving and grasping a virtual object) is shown in Fig. 11.

Fig. 11. Snapshots of simple VR scenes: reaching, moving and grasping a virtual object.

4 Conclusions

In this chapter we have introduced the main concepts, and the state of the art in the field of neurorehabilitation. In particular we have presented a review of the concepts, advantages and perspectives of robot-assisted therapy, hybrid wearable technology, bio-inspired systems used in CVA and SCI therapies. Additionally, an analysis of the state of art of Virtual Reality based applications in the field of the rehabilitation from motor disorders has been presented focusing the discussion on the potential benefits of the virtual therapy and new available technologies. We have described, as a case study, the first system that combines NR, MNP and VR for neurorehabilitation and functional compensation. The development of the system is still in the preliminary stage, but the application of a new generation of wearable compensational devices together with the integration of the virtual reality environment promises the evolution of (modern/novel) rehabilitation therapies for people suffering from motor disorders.

We believe that this interdisciplinar approach combined with advanced technology will help patients suffering from CVA and SCI in fast re-learning of the skill required to perform autonomously the daily live activities.

Acknowledgment

This research is a part of the HYPER project funded by CONSOLIDER-INGENIO 2010, Spanish Ministry for Science and Innovation. The institutions involved are: Spanish National Research Council (CSIC), Center for Electrochemical Technologies (CIDETEC), Visual Communication and Interaction Technologies Centre (VICOMTech), Fatronik-Tecnalia, University of Zaragoza, University of Rey Juan Carlos, University of Carlos III (UC3M), Bioengineering Institute of Catalunya (IBEC) and Hospital for Spinal Cord Injury of Toledo.

References

1. http://www.hocoma.com/en/products/armeo/
2. http://www.iai.csic.es/hyper/
3. http://www.xbox.com/en-GB/kinect
4. Boian, R., Lee, C., Deutsch, J., Burdea, G., Lewis, J.: Virtual reality-based system for ankle rehabilitation post stroke. In: Proceedings of the First International Workshop on Virtual Reality Rehabilitation, Citeseer, pp. 77–86 (2002)
5. Colombo, G., Jrg, M., Dietz, V.: Driven gait orthosis to do locomotor training of paraplegic patients, pp. 3159–3163 (2000)

6. Colombo, R., Pisano, F., Mazzone, A., Delconte, C., Micera, S., Carrozza, M., Dario, P., Minuco, G.: Design strategies to improve patient motivation during robot-aided rehabilitation. Journal of Neuro Engineering and Rehabilitation 4 (2007)

7. Cano de la Cuerda, R., Muñoz-Hellín, E., Algual-Diego, I.M., Molina-Rueda, F.: Telerrehabilitación y neurología. Rev. Neurol. 51, 49–56 (2010)

8. Dietz, V.: Spinal cord pattern generators for locomotion. Clinical Neurophysiology 114(8), 1379–1389 (2003)

9. Dobkin, B.: Rehabilitation after stroke. New England Journal of Medicine 352(16), 1677–1684 (2005)

10. Eilenberg, M.F., Geyer, H., Herr, H.: Control of a powered ankle-foot prosthesis based on a neuromuscular model. IEEE transactions on neural systems and rehabilitation engineering: a publication of the IEEE Engineering in Medicine and Biology Society 18(2), 164–173 (2010), http://www.ncbi.nlm.nih.gov/pubmed/20071268, doi:10.1109/TNSRE.2009.2039620

11. Emken, J., Bobrow, J.E., Reinkensmeyer, D.: Robotic movement training as an optimization problem: Designing a controller that assists only as needed, pp. 307–312 (2005)

12. Fidopiastis, C., Stapleton, C., Whiteside, J., Hughes, C., Fiore, S., Martin, G., Rolland, J., Smith, E.: Human experience modeler: Context-driven cognitive retraining to facilitate transfer of learning. CyberPsychology & Behavior 9(2), 183–187 (2006)

13. Fugl-Meyer, A., Jääskö, L., Leyman, I., Olsson, S., Steglind, S.: The post-stroke hemiplegic patient. 1. a method for evaluation of physical performance. Scandinavian journal of rehabilitation medicine 7(1), 13–31 (1975)

14. Geng, T., Porr, B., Wörgötter, F.: A reflexive neural network for dynamic biped walking control. Neural Computation 18(5), 1156–1196 (2006)

15. Giszter, S.: Spinal cord injury: Present and future therapeutic devices and prostheses. Neurotherapeutics 5(1), 147–162 (2008)

16. Hesse, S.: Locomotor therapy in neurorehabilitation. NeuroRehabilitation 16(3), 133–139 (2001)

17. Holden, M.: Virtual environments for motor rehabilitation: review. Cyberpsychology & behavior 8(3), 187–211 (2005)

18. Jezernik, S., Colombo, G., Keller, T., Frueh, H., Morari, M.: Robotic orthosis lokomat: A rehabilitation and research tool. Neuromodulation 6(2), 108–115 (2003)

19. Jo, S., Massaquoi, S.: A model of cerebrocerebello-spinomuscular interaction in the sagittal control of human walking. Biological Cybernetics 96(3), 279–307 (2007)

20. Johansson, B.: Brain plasticity and stroke rehabilitation: The willis lecture. Stroke 31(1), 223–230 (2000)

21. Kazerooni, H.: Human-robot interaction via the transfer of power and information signals. IEEE Transactions on Systems, Man and Cybernetics 20(2), 450–463 (1990)

22. Knaut, L., Subramanian, S., McFadyen, B., Bourbonnais, D., Levin, M.: Kinematics of pointing movements made in a virtual versus a physical 3-dimensional environment in healthy and stroke subjects. Arch. Phys. Med. Rehabil. 90, 793–802 (2009)

23. Koenig, A., Wellner, M., Köneke, S., Meyerheim, A., Lunenburger, L., Riener, R.: Virtual gait training for children with cerebral palsy using the lokomat gait orthosis. Medicine meets virtual reality 16, 204 (2008)
24. Kuttuva, M., Boian, R., Merians, A., Burdea, G., Bouzit, M., Lewis, J., Fensterheim, D.: The rutgers arm: an upper-extremity rehabilitation system in virtual reality. In: 4th International workshop on virtual reality rehabilitation, Catalina Islands. Citeseer (2005)
25. Lam, T., Wolfe, D., Eng, J., Domingo, A.: Lower limb rehabilitation following spinal cord injury. In: Eng, J., Teasell, R., Miller, W., Wolfe, D., Townson, A., Hsieh, J., Connolly, S., Mehta, S., Sakakibara, B. (eds.) Spinal Cord Injury Rehabilitation Evidence, Version 3.0, pp. 1–47 (2010)
26. Lécuyer, A., Lotte, F., Reilly, R.B., Leeb, R., Hirose, M., Slater, M.: Brain-computer interfaces, virtual reality, and videogames. Computer 41, 66–72 (2008)
27. Luksch, T.: Human-like control of dynamically walking bipedal robots. Ph.D. thesis, Technischen Universitat Kaiserslautern (2009)
28. Lunenburger, L., Colombo, G., Riener, R.: Biofeedback for robotic gait rehabilitation. Journal of Neuro Engineering and Rehabilitation 4 (2007)
29. MacKay-Lyons, M.: Central pattern generation of locomotion: A review of the evidence. Physical Therapy 82(1), 69–83 (2002)
30. Maclean, N., Pound, P., Wolfe, C., Rudd, A.: Qualitative analysis of stroke patients' motivation for rehabilitation. British Medical Journal 321(7268), 1051–1054 (2000)
31. Metrailler, P., Brodard, R., Stauffer, Y., Clavel, R., Frischknecht, R.: Cyberthosis: Rehabilitation robotics with controlled electrical muscle stimulation. In: Rehabilitation robotics, pp. 648–664. Itech Education and Publishing, Austria (2007)
32. Mirelman, A., Bonato, P., Deutsch, J.: Effects of training with a robot-virtual reality system compared with a robot alone on the gait of individuals after stroke. Stroke 40(1), 169–174 (2009)
33. O'Dell, M., Lin, C.C., Harrison, V.: Stroke rehabilitation: Strategies to enhance motor recovery. Annual Review of Medicine 60, 55–68 (2009)
34. Ogihara, N., Yamazaki, N.: Generation of human bipedal locomotion by a bio-mimetic neuro-musculo-skeletal model. Biol. Cybern. 84(1), 1–11 (2001)
35. Paul, C., Bellotti, M., Jezernik, S., Curt, A.: Development of a human neuro-musculo-skeletal model for investigation of spinal cord injury. Biol. Cybern. 93(3), 153–170 (2005),
    http://dx.doi.org/10.1007/s00422-005-0559-x,
    doi:10.1007/s00422-005-0559-x
36. Pfurtscheller, G., Leeb, R., Faller, J., Neuper, C.: Brain-computer interface systems used for virtual reality control. In: Kim, J.-J. (ed.) Virtual Reality (2011)
37. Pons, J.: Rehabilitation exoskeletal robotics. IEEE Engineering in Medicine and Biology Magazine 29(3), 57–63 (2010)
38. Pons, J., Corporation, E.: Wearable robots: biomechatronic exoskeletons. Wiley Online Library, Chichester (2008)
39. Riener, R., Lïenburger, L., Jezernik, S., Anderschitz, M., Colombo, G., Dietz, V.: Patient-cooperative strategies for robot-aided treadmill training: First experimental results. IEEE Transactions on Neural Systems and Rehabilitation Engineering 13(3), 380–394 (2005)

40. Rizzo, A., Kin, G.: A swot analysis of the field of vr rehabilitation and therapy. Presence: Teleoperators and Virtual Environments 14(2), 119–146 (2005)
41. Rose, F., Attree, E., Brooks, B.: Virtual environments in neuropsychological assessment and rehabilitation. IOS Press, Amsterdam (1997)
42. Schmidt, H., Hesse, S., Bernhardt, R., Krüger, J.: Hapticwalker. a novel haptic foot device. ACM Transactions on Applied Perception (TAP) 2(2), 166–180 (2005)
43. Schmidt, H., Sorowka, D., Piorko, F., Marhoul, N., Bernhardt, R.: Control system for a robotic walking simulator. In: Proceedings. 2004 IEEE International Conference on Robotics and Automation, vol. 2, pp. 2055–2060. IEEE, Los Alamitos (2004)
44. Schmitt, C., Metrailler, P., Al-Khodairy, A., Brodard, R., Fournier, J., Bouri, M., Clavel, R.: The motion maker: a rehabilitation system combining an orthosis with closed-loop electrical muscle stimulation. In: Proceedings of the 8th Vienna International Workshop in Functional Electrical Stimulation, pp. 117–120 (2004)
45. Schmitt, C., Metrailler, P., Al-Khodairy, A., Brodard, R., Fournier, J., Bouri, M., Clavel, R.: A study of a knee extension controlled by a closed loop functional electrical stimulation. In: Proceedings, 9th An. conf. of the IFESS, Bournemouth (2004)
46. Shan, J., Nagashima, F.: Neural locomotion controller design and implementation for humanoid robot hoap-1. In: 20th Annual Conference of the Robotics Society of Japan (2002)
47. Stauffer, Y., Allemand, Y., Bouri, M., Fournier, J., Clavel, R., Metrailler, P., Brodard, R., Reynard, F.: The walktrainer a new generation of walking reeducation device combining orthoses and muscle stimulation. IEEE Transactions on Neural Systems and Rehabilitation Engineering 17(1), 38–45 (2009)
48. Subramanian, S., Knaut, L., Beaudoin, C., McFadyen, B., Feldman, A., Levin, M.: Virtual reality environments for post-stroke arm rehabilitation. Journal of Neuro Engineering and Rehabilitation 4(1), 20 (2007)
49. Sveistrup, H.: Motor rehabilitation using virtual reality. Journal of NeuroEngineering and Rehabilitation 1(1), 10 (2004)
50. Taga, G.: A model of the neuro-musculo-skeletal system for human locomotion. i. emergence of basic gait. Biol. Cybern. 73(2), 97–111 (1995)
51. Veneman, J., Kruidhof, R., Hekman, E., Ekkelenkamp, R., Van Asseldonk, E., Van Der Kooij, H.: Design and evaluation of the lopes exoskeleton robot for interactive gait rehabilitation. IEEE Transactions on Neural Systems and Rehabilitation Engineering 15(3), 379–386 (2007)
52. Viau, A., Feldman, A., McFadyen, B., Levin, M.: Reaching in reality and virtual reality: a comparison of movement kinematics in healthy subjects and in adults with hemiparesis. Journal of neuroengineering and rehabilitation 1(1), 11 (2004)
53. Weiss, A., Suzuki, T., Bean, J., Fielding, R.: High intensity strength training improves strength and functional performance after stroke. American Journal of Physical Medicine and Rehabilitation 79(4), 369–376 (2000)
54. Weiss, P., Kizony, R., Feintuch, U., Katz, N.: Virtual reality in neurorehabilitation. Textbook of Neural Repair and Neurorehabilitation 2, 182–197 (2006)
55. Wolbrecht, E., Chan, V., Reinkensmeyer, D., Bobrow, J.: Optimizing compliant, model-based robotic assistance to promote neurorehabilitation. IEEE Transactions on Neural Systems and Rehabilitation Engineering 16(3), 286–297 (2008)

56. Wolfe, D., Hsieh, J., Mehta, S.: Rehabilitation practices and associated out-
    comes following spinal cord injury. In: Eng, J., Teasell, R., Miller, W., Wolfe,
    D., Townson, A., Hsieh, J., Connolly, S., Mehta, S., Sakakibara, B. M. (eds.)
    Spinal Cord Injury Rehabilitation Evidence, Version 3.0 (2010)
57. Wool, R., Siegel, D., Fine, P.: Task performance in spinal cord injury: effect of
    helplessness training. Archives of Physical Medicine and Rehabilitation 61(7),
    321–325 (1980)
58. You, S., Jang, S., Kim, Y., Hallett, M., Ahn, S., Kwon, Y., Kim, J., Lee, M.:
    Virtual reality-induced cortical reorganization and associated locomotor recov-
    ery in chronic stroke: an experimenter-blind randomized study. Stroke 36(6),
    1166–1171 (2005)

Toward Bezier Curve Bank-Turn Trajectory Generation for Flying Robot

Affiani Machmudah, Setyamartana Parman, and Azman Zainuddin

Department of Mechanical Engineering, Universiti Teknologi Petronas,
Bandar Seri Iskandar, Tronoh, 31750 Perak, Malaysia
de_affela@yahoo.com

Abstract. This paper presents a UAV maneuver planning with an objective is minimize a total maneuvering time and a load factor when the UAV follows an agile maneuvering path. The Genetic algorithm (GA) combined by fuzzy logic approach will lead this challenging work to meet the best maneuvering trajectories in an UAV operational. It considers UAV operational limits such as a minimum speed, a maximum speed, a maximum acceleration, a maximum roll angle, a maximum turn rate, as well as a maximum roll rate. The feasible maneuvering path will be searched first, and then it will be processed to generate the feasible flight trajectories. There are two ways to follow the maneuvering path: constant speed strategy or changing speed strategy. The first strategy is simple; however it will involve the higher load factor. The changing speed strategy is very engaging technique to reduce the load factor during the maneuver; however it needs to generate the speed as function of time which is feasible to be conducted. The trajectory planner needs to keep the speed inside the UAV flight zone. This paper also proposed the strategy in generating the feasible flight trajectory. To investigate the performance of the proposed strategy the simulation environment has been built by creating a program in MATLAB software. Results show that the load factor can be reduced into lower value which is very engaging to be conducted on going to an autonomous goal of the UAV which has an ability to perform the agile maneuvering.

Keywords: Bezier curve, maneuver planning, Genetic Algorithm, fuzzy rule.

1 Introduction

Bats which have outstanding maneuver ability have inspired the researchers in all around the world to understand their behavior where they can conduct a complex maneuver successfully in a hard darkness surrounding. They conduct the difficult maneuver by combining a bank-turn maneuver and a yaw turn maneuver [9, 10]. They have the complex flexible wing to support their maneuver. There is an interesting important question in this issue, how can they find the feasible flight trajectories in the blind condition where they always do the maneuver successfully in the dark night environment.

   This paper will show that there are always the ways to find the feasible maneuvering path as long as the obstacle environment has been known since the

T. Gulrez, A.E. Hassanien (Eds.): Advances in Robotics & Virtual Reality, ISRL 26, pp. 109–131.
springerlink.com   © Springer-Verlag Berlin Heidelberg 2012

trajectories generation is the computational optimization problem which does not depend on the lighting conditions. When the feasible collision-free path has been known, it should be processed next in order to discover the feasible flight trajectories. The GA strategy which based on the Darwin natural selection assisted by the fuzzy logic approach will be utilized to search the feasible maneuvering trajectories in the obstacle environment.

There are two-group papers in this research area based on their results. The first group focused on finding the path which satisfies the specific optimization criteria without discussed how to achieve them for control system. The second group has already discussed the strategy to achieve the path where they utilized the speed as function of time which involves the change in acceleration during the maneuver. Unfortunately, the UAV has the speed and acceleration limits which are not easy to be tackled since they have nonlinear relations with the other UAV constraints. They can fall into very low values or very high values which are outside the UAV operational area. Hence, for second group paper, ordinary they executed the re-planning activity when the trajectories are failure and cannot be repaired.

Shanmugavel et al presented a path planning of a group of UAVs for simultaneous arrival on target at free obstacle environment [1]. The quintic Pythagorean hodograph (PH) path was generated by solving the Dubins path first. An iterative method was used to generate the flyable PH path with the path length close to the Dubins path to satisfy the minimum length. They also presented other approach by replacing the arc segment of the Dubins path with clothoid arcs which have ramp curvature profile to satisfy the curvature continuity [5]. The Dubin path which constructed from two arc circle connected with the line will not agile enough for avoiding collision maneuvering path for the bats. Portas et al also conducted research on UAVs cubic spline trajectory planner considering 11 constraints using Evolutionary Algorithm (EA); however they still considered the path without analyzed how to achieve the path for control system. The velocity constraint and the maximum turn rate which always involve in cubic-spline path that is not easy to be achieved did not be explored yet [6].

Other method has been widely used in the UAV path planning, namely a Rapidly-exploring Random Tree (RRT) [2, 3].

To reach the agile maneuver successfully is an important target for UAV control system to mantain the maneuvering motion in the right and optimum way. The trajectory results need to be analyzed for UAV control system reference.

Koyuncu et al presented the problem of generating agile maneuver profiles for Unmanned Combat Aerial Vehicles using the RRT method [15]. They discovered B-spline avoiding collision path utilizing the random node first by executing RRT algorithm. The dynamically feasible B-spline was discovered by checking the first and second derivatives of the B-spline curve which gave the velocities and acceleration, respectively. If the velocities and acceleration are within the flight envelope, the generated path was valid candidate of dynamically feasible. Their method needed to re-plan the trajectory if the dynamic feasible cannot be repaired during maneuver planning .

In this paper, the trajectory planning strategy, namely two-layer optimization, which involves two steps activities: the path planning and the feasible trajectory generation, will be executed. The feasible maneuvering path will be searched with the optimization objective is to minimize the path length subject to the minimum turning radius as well as the collision detection rules constraints. By this strategy, it will guarantee the existance of the constant speed which satisfies the UAV constraints which have the relation with the speed as well as maximum roll angle constraints. The second way to track the maneuvering path is by changing speed maneuver where the UAV speed is changed during the maneuver. This mechanism is very challenging where it can reduce the load factor during the maneuver; however it needs to keep such speed trajectories that other flight trajectories are feasible to be achieved. The changing speed trajectories will be proposed to be analyzed in normal and tangential coordinate where the speed as well as the acceleration trajectories can be considered independent from the curve speed and curve acceleration. The maneuvering path will contribute the information of the turning radius. The speed and the acceleration need to be managed when the UAV follows this turning radius to always satisfy all UAV constraints. The flight trajectories will change continuously; however they should be controlled to always stay in the UAV operational area for safety.

The GA is very enganging computational method to solve the nature as well as the real engineering problem. The second step will convey a multi-objective optimization. Utilizing the GA to optimize the fuzzy system has been widely accepted in the optimization strategy; however this paper will utilize the new strategy by utilizing the fuzzy set to assist the GA to track the optimization target.

The interesting GA behavior has been detected in this research. The GA accept the training from fuzzy membership function to keep itself to achieve the target. The changing speed strategy goal is to reduce the total maneuvering time and the maximum load factor of the constant speed strategy during the maneuver.

2   Problem Statement

The maneuver begins from starting point "A" to end point "B" at a constant altitude and at a static obstacle environment where the obstacles do not move during the entire process. (refer Fig. 1). This paper utilized cubic Bezier curve as the maneuvering path.

Fig. 1. UAV feasible path at static environment

Mathematically a parametric Bezier curve is defined by

$$\begin{pmatrix} x \\ y \end{pmatrix} = \sum_{u=0}^{n} B_i J_{n,i}(u) \qquad\qquad 0 \le u \le 1 \qquad\qquad (1)$$

$$J_{n,i}(u) = \begin{pmatrix} n \\ i \end{pmatrix} u^i (1-u)^{n-i} \qquad , \qquad \begin{pmatrix} n \\ i \end{pmatrix} = \frac{n!}{i!\,(n-1)!}$$

where $J_{n,i}(t)$ is the ith n^{th} order Bernstein basis function, u is curve parameter. To construct an nth degree Bezier curve, it needs n+1 control points (refer Fig. 2).

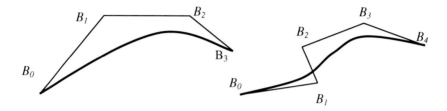

Fig. 2. Bezier curve , Bn is control point

The two-layer optimization is utilized as shown in Fig. 3. There are two steps of the UAV maneuver planning; the first step activity is to find the feasible maneuvering path.

There are two constraints conveyed in the first part, the minimum turning radius or maximum curvature and the collision detection rules. The path result from trajectory planner has satisfied the minimum turning radius for all curve point. Once the feasible path has been discovered, the flight trajectories should be determined so that the UAV can follow the path successfully. There are two options to conduct the agile maneuver: constant speed maneuver or changing speed maneuver.

Some papers has been discussed the agile maneuver for the UAV [1, 2, 3, 5, 6, 15. Some of them just consider the minimum turning radius constraint; however, the flight trajectories also should be guaranteed achievable. It includes speed trajectories, the acceleration trajectories, the roll angle trajectories, the turn rate trajectories, as well as the roll rate trajectories. Since the UAV has many constraints, the flight trajectories generation becomes very complex optimization problem. When the UAV has the maneuver planning that has an intelligent to find the feasible flight trajectories, the autonomous goal of the agile maneuvering can be achieved. This paper works in the issue to generate the feasible flight trajectories when the UAV follows the cubic Bezier curve as the agile maneuvering path.

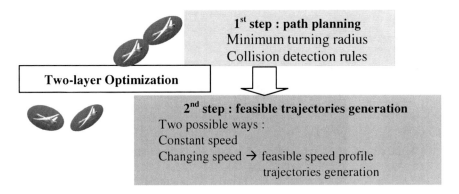

Fig. 3. Two-layer optimization

The equation of a UAV steady level bank-turn is as follows

$$\tan \varphi = v^2/rg \tag{2}$$

where r, g, v, and φ are the turning radius, the gravity acceleration, and the speed, and the roll angle or bank angle, respectively. This equation assumes that the thrust is small as compare to the weight, moreover the angle of attack is also very small and is in radians.

From Eq.(2), the path has satisfied the minimum velocity as well as the maximum roll angle constraint as shown below:

$$r_{min} = \frac{v_{min}^2}{g \ \tan \varphi_{max}} \tag{3}$$

where r_{min}, v_{min}, and φ_{max} are the minimum turning radius, the minimum speed, and the maximum roll angle, respectively.

The load factor which is an important flight characteristic is formulated as follows:

$$n = \frac{1}{\cos \varphi} \tag{4}$$

Thus the maximum load factor constraint is as follows:

$$n_{max} = \frac{1}{\cos \varphi_{max}} \tag{5}$$

where n and n_{max} are the load factor of the UAV and the maximum value of the load factor that can be gained by the UAV, respectively. The value of n of the UAV during the maneuver cannot exeed this n_{max}.

By Eq.(3), it shows that the path has satisfied the minimum speed as well as the maximum roll angle. Consequently, the other UAV constraints which have the relation with both of these constraints have been satisfied by the constant speed maneuver.

For the planar curve, the turning radius will be associated with the curvature as follows

$$K = \frac{\dot{x}\,\ddot{y} - \dot{y}\,\ddot{x}}{\left(\dot{x}^2 + \dot{y}^2\right)^{3/2}} \qquad (6)$$

$$K = \frac{1}{r} \qquad (7)$$

where $\dot{x}, \dot{y}, \ddot{x}, \ddot{y}$, and r are the first derivative of x, the first derivative of y, the second derivative of x, the second derivative of y, and the turning radius, respectively.

The second options to follow the maneuvering path is by changing the speed during the maneuver. The problem is how to generate the speed profile which satisfies all UAV constraints. The difficulty arises when the speed and the acceleration could fall into very low value or very high value which is beyond the UAV operational space. It needs to be managed in order to stay in the allowable zone. This paper consider the velocity constraints which are ordinarily owned by the commercial UAV, $10\text{m/s} \leq v \leq 25\text{m/s}$.

The steady level Bank-turn mechanism will be utilized in these strategies. For angle of attack trajectories, they always can be converted into the lift coefficient. Thus, such constraint is in the lift coefficient form.

3 Constraints

This paper will consider the UAV constraints as indicated in the Table. 1 The trajectory planner needs to lay in these operational limits to achieve the feasibility.

By substituting the values in Table 1 into Eq.(3), the minimum turning radius is as follows

$$r_{min} = \frac{v_{min}^2}{g\,\tan\varphi_{max}} = 5.9 \quad m$$

Other limit values associated with the UAV flight characteristics are determined by substituting the values in Table. 1 into the equation of turn rate and lift force as follows

$$\dot{\psi}_{max} = \frac{g\,\tan\varphi_{max}}{v_{min}} = 1.7 \ \text{rad/s}$$

$$CL_{max} = \frac{2Mg}{\rho\,v_{min}^2\,S\cos\varphi_{max}} = 2.3041 \qquad (8)$$

$$CL_{min} = \frac{2Mg}{\rho\,v_{max}^2\,S} = 0.1843$$

where $\dot{\psi}$, $\dot{\psi}_{max}$, CL_{max} , CL_{min} , ρ , S and M are the turn rate, the maximum turn rate, the maximum lift coefficient, the minimum lift coefficient, an air density, a wing area, and a mass of the UAV and payloads, respectively.

Table 1. The UAV flight characteristics

UAV flight Characteristics	
Max Speed: 25 m/s	Wingspan: 1.50 m
Min Speed: 10 m/s	Length: 1.20 m
Max acceleration=3.5 m/s	Wing Area: 0.55 m^2
Max roll angle : $60^0 = \pi/3$rad	Aircraft Weight: 2.8 kg
Max roll rate : 200^0/s=3.4907rad/s	Payload Capability: 1.14 kg

Since the obstacles are present, the trajectory planner should obey the fuzzy collision detection `rules` constraint as follows

if $\exists$ { (x,y) or (x_{right}, y_{right}) or (x_{left}, y_{left})} $\cap$ obstacle area, then the collision happen, otherwise is free from collisions   (9)

The (x_{right}, y_{right}), (x_{left}, y_{left})} are the offset path which will be explained in the next part. The maneuver planning becomes the fuzzy optimization case by this rules constraint.

4 Offset-Path for Safety

In this paper, to make the path of the UAV is safe, the offset-path is constructed. It should be taken into account since the UAV has a wing span. The safety factor (SF), is constructed to generate the UAV way as illustrated in Fig. 4.

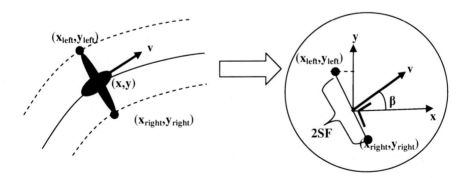

Fig. 4. Offset-path

where, $\tan \beta = \dfrac{dy/du}{dx/du}$, $2SF = $ wing span

The equation of the offset-path is as follows

If $\dfrac{dy}{du} \geq 0$ and $\dfrac{dx}{du} \geq 0$ or $\dfrac{dy}{du} \leq 0$ and $\dfrac{dx}{du} \leq 0$:

$$
\begin{aligned}
x_{right} &= x + SF \; \cos(0.5 \quad \pi - \beta) \\
y_{right} &= y - SF \; \cos(0.5 \quad \pi - \beta) \\
x_{left} &= x - SF \; \cos(0.5 \quad \pi - \beta) \\
y_{left} &= y + SF \; \cos(0.5 \quad \pi - \beta)
\end{aligned}
\tag{10}
$$

Otherwise:

$$
\begin{aligned}
x_{right} &= x + SF \; \cos(0.5 \quad \pi - \beta) \\
y_{right} &= y - SF \; \cos(0.5 \quad \pi - \beta) \\
x_{left} &= x - SF \; \cos(0.5 \quad \pi - \beta) \\
y_{left} &= y + SF \; \cos(0.5 \quad \pi - \beta)
\end{aligned}
$$

5 Fitness Function

The GA is a search technique used to find exact or approximate solutions of optimization and search problems based on genetic theory and Darwin evolution theory.

The GA will start by choosing the population randomly. Then the fitness value of the individual will be evaluated to choose the parent for creating mating pool. The crossover and mutation are performed to born new offspring. The process is repeated till the best individual is found [16]. Unknown variables will be considered as the individuals with specific genes.

The fitness function which is the important tool in discovering the best individual will be presented in this section.

5.1 Minimum Path Length

The minimum path length equation for the parametric curve is as follows

$$
L = \min \; s = \min \int_{0}^{1} \left(\sqrt{\left(\frac{dx(u)}{du}\right)^2 + \left(\frac{dy(u)}{du}\right)^2} \right) du
\tag{11}
$$

where s, x(u) and y(u) are the path length, the equation of Bezier curve in x-component and y-component at Eq. (1), respectively.

The optimum curve is the one with the highest fitness value, thus the first objective function can be evaluated as follows

$$F_1 = \frac{1}{L} \tag{12}$$

The unknown variables should be coded into the genes [16]. Since the first control point, B_0, and the fourth control point, B_3, are known, thus for the maneuvering path planning, there are four genes in the chromosome as follows

B_{1x}	B_{1y}	B_{2x}	B_{2y}

B_{nx} and B_{ny} are the nth Bezier curve control point of x-axis and y-axis, respectively.

5.2 Fuzzy Rules Fitness Function

The UAV will operate in the cluttered environment where it needs to discover the feasible maneuvering path as fast as possible when the UAV experiences a dangerous condition. Considering many strict constraints as well as an injury computational time, the fuzzy logic approach which has an intelligent mechanism i.e. the brain-like human thinking, will be adopted in the GA way. Based on these two requirements, the intelligence and the fast way, the death penalty has been chosen in the GA natural selection method.

The intelligent fuzzy death penalty rules for path planning can be described as follows

1. If any constraint does not satisfy by the path, then the path is failure
2. Otherwise, the path is success (13)

To follow the fuzzy rules, the membership functions have been composed to solve the first maneuvering step as follows

$$
\text{Fitness Function}
\begin{cases}
= 0 & \exists \{K_p < - K_{max} \text{ or } K_p > K_{max} \text{ or } [(x,y) \\
& \text{or}(x_{right}, y_{right}) \text{ or } (x_{left}, y_{leftt})] \cap \text{ obstacle area}\} \\
\\
= \dfrac{1}{\left(\displaystyle\int_0^1 \sqrt{\left(\dfrac{dx(u)}{du}\right)^2 + \left(\dfrac{dy(u)}{du}\right)^2} \; du \right)} & \text{Otherwise}
\end{cases}
\tag{14}
$$

6 Constant Speed Strategy

The previous section has discussed the strategy to find the best maneuvering path. After the feasible path has been known, the associated time needs to be discovered to compose the complete trajectories. This section will discuss the second step of the maneuver planning, utilizing the constant speed strategy.

The dynamics of the moving object in the curvilinear motion at constant speed is as follows

$$s = v\,t \tag{15}$$

where s is the path length, v is the speed, and t is the time, respectively. By Eq.(15), the minimum time is achieved by minimizing the path length, s.

The constant speed conveys the new maximum allowed speed according to the best maneuvering path. The maximum speed probably cannot be achieved anymore.

The new maximum allowed speed can be known as follows

$$v_{max\_} = \sqrt{r_{min}\ g\ \tan\ \varphi_{max}} \tag{16}$$

where the $v_{max\_}$, is the new maximum allowed speed.

The above equation is the maximum speed that can be used to satisfy the UAV speed constraint. From Eq. (16), it shows that the maneuvering path resulted from the first step take an effect in this $v_{max\_}$ in the form of r_{min}.

It still needs to satisfy the roll rate constraint since this paper approximates it by an average roll rate.

The GA needs to check the roll rate constraint fulfillment as follows

$$\text{Fitness Function} \begin{cases} = 0 & \exists\ \{\ \dfrac{\Delta\varphi}{\Delta t} > \left.\dfrac{\Delta\varphi}{\Delta t}\right|_{max}\ \text{or}\ \dfrac{\Delta\varphi}{\Delta t} < -\left.\dfrac{\Delta\varphi}{\Delta t}\right|_{max}\ \} \\ = 1 & \text{Otherwise} \end{cases} \tag{17}$$

where $\dfrac{\Delta\varphi}{\Delta t}$ is the average roll rate for each time discretization. When the $v_{max\_}$ does not satisfy the roll rate constraint, the GA can be executed to find the proper $v_{max\_}$.

There is a new range of the feasible constant speed to conduct the agile maneuvering path. It is the speed zone from the minimum speed to the new maximum allowable speed, $v_{max\_}$, as follows

$$v_{min} \le v \le v_{max\_} \tag{18}$$

The problem becomes the decision to choose the speed in this zone. Thus, there is inheritance limitation in this case. From Eqs.(2, 4, 15), when the minimum speed is chosen, it will have the minimum load factor; however it will take the maximum total maneuvering time.

The speed utilized during the maneuver can be considered as function of the decision as follows

$$v_- = \begin{cases} 0.1(v_{max\_} - v_{min}) + v_{min} & \text{if } m=10\% \\ 0.5(v_{max\_} + v_{min}) + v_{min} & \text{if } m = 50\% \\ m\ (v_{max\_} + v_{min}) + v_{min} & \text{Otherwise} \end{cases} \tag{19}$$

where m and v_- are the percentage of the minimum time from decision maker system and the constant speed utilized during the maneuver, respectively.

7   Speed as Function of Time Strategy

7.1   Normal and Tangential Coordinate

When the motion is analyzed in the normal and tangential coordinate, the speed trajectories profile is independent of the cubic Bezier curve speed profile. Fig. 5 gives the illustration of the speed, the curvature radius, and the acceleration when this coordinate is utilized.

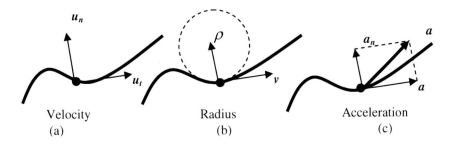

$$
\begin{array}{ccc}
\text{Velocity} & \text{Radius} & \text{Acceleration} \\
\text{(a)} & \text{(b)} & \text{(c)}
\end{array}
$$

Fig. 5. Normal and tangential coordinate illustration

The problem is to find the speed trajectories profile which satisfies all UAV constraints when the UAV tracks the maneuvering path from the first step.

Consider n^{th} degree polynomial speed profile which is a composite function of the speed profile and the linear time-scale as follows

$$v(t) = v(r) \circ r(t)$$
$$= \left(a_n\ r^n + a_{n-1}\ r^{n-1} + \ldots\ldots\ \ldots + a_1 r + a_0\right) \circ r(t) \tag{20}$$

$$r(t) = \frac{t}{T}$$
$$0 \le r \le 1$$

where v(r) and r(t) are the speed profile and the linear time-scale from parameter 0 to 1 with t is the time variable and T is the total maneuvering time, respectively.

The UAV tangential acceleration is as follows

$$a(t) = \dot{v}(r) \ \frac{1}{T} \tag{21}$$

where $\dot{v}(r)$ is the first derivative of $v(r)$ with respect to r.

There is the additional constraint regarding the maximum acceleration limit as follows

$$T_{min} = \begin{cases} \dfrac{\dot{v}_{max}(r)}{a_{max}} & \text{If } \dot{v}_{min}(r) > 0 \text{ or } \{ \dot{v}_{min}(r) < 0 \text{ and } |\dot{v}_{min}(r)| < \dot{v}_{max}(r) \} \\[3mm] \dfrac{\dot{v}_{min}(r)}{-a_{max}} & \text{Otherwise} \end{cases} \tag{22}$$

where $\dot{v}_{min}(r)$ and $\dot{v}_{max}(r)$ are the minimum value and maximum value of first derivatives of $v(r)$ with respect to r, respectively.

The above equation is the minimum total maneuvering time requirement for Eq. (20) to satisfy the acceleration constraint.

The motion of the UAV can be analyzed as follows

$$s = \int_0^T v(t) \ dt = \ = T \int_0^1 v(r) \ dr$$
$$T = \frac{S}{\int_0^1 v(r) \ dr} \tag{23}$$

where s and T are the path length and the total maneuvering time, respectively.

The intersection of Eq.(22) and Eq.(23) becomes the additional constraint to satisfy the maximum acceleration limit as follows

$$T_{min} \leq T = \frac{S}{\int_0^1 v(r) \ dr} \tag{24}$$

where T_{min} is determined from Eq. (22).

The total maneuvering time needs to be managed in condition (24) to satisfy the acceleration limit. The Eq.(24) shows that to tackle this problem, the velocity profile, $v(r)$, needs to be controlled to keep the necessary condition.

7.2 Strategy to Generate the Speed Trajectories

The goal of this strategy is to generate the speed as function of time to reduce the load factor of the constant speed strategy. The third degree polynomial function will be utilized as speed profile in this paper. The speed controls points become the variable which should be managed to keep the speed trajectories lay in the feasible zone. The strategy to generate the speed trajectories is by controlling the initial speed and the final speed.

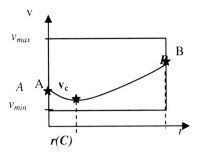

Fig. 6. Control variables

Fig. 6 shows the illustration of this strategy. The initial speed (point A) and final speed (point B) are the control variables that GA should manage them always in the UAV allowable speed zone. Other controlled variables are one floating point which is chosen at the proper point of r, r(C). This floating point is kept to be the optimum point. Hence, there are four variable control points: the initial speed (point A), the final speed (point B), the floating point (point C), and the r(C) such as illustrated in Fig. 6.

For third degree polynomial, the speed equation is in the following

$$v(t) = \left(a_3 r^3 + a_2 r^2 + a_1 r + a_0\right) \circ r(t) \tag{25}$$

where a_n is the nth coefficient of polynomial and r is linear time-scale.

The first control point, i.e. the initial speed, are determined by substituting r = 0 or t = 0 as follows

$$v(0) = \left(a_3 0^3 + a_2 0^2 + a_1 0 + a_0\right) \tag{26}$$

Thus,

$$a_0 = v(0) = v_i \tag{27}$$

where v_i is the initial speed at r = 0.

The second control point, which is the final speed, are determined by substituting r = 1 or t = T as follows

$$v(1) = \left(a_3 1^3 + a_2 1^2 + a_1 1 + a_0\right) \tag{28}$$

Thus,

$$a_3 = v_f - a_2 - a_1 - v_i \tag{29}$$

where v_f is the final speed at r = 1.

Consider the optimum point, r^*, the condition of the optimum point is $\dot{v} = 0$. By taking the derivative of the third degree polynomial, the following equation should be satisfied

$$3a_3 r^{*2} + 2a_2 r^* + a_1 = 0 \tag{30}$$

By substituting a_1 from Eq. (29), the a_3 is as follows

$$a_3 = \frac{v_i - v_f - a_2\left(2r^* - 1\right)}{\left(3r^2 - 1\right)} \tag{31}$$

This optimum point will have the speed as follows

$$v^*(r^*) = \left(a_3 r^{*3} + a_2 r^{*2} + a_1 r^* + v_i \right) \tag{32}$$

By substituting a_1 into the above equation, the equation becomes

$$v^*(r^*) = \left(a_3 r^{*3} + a_2 r^{*2} + (v_f - v_0 - a_2 - a_3)r^* + v_i \right) \tag{33}$$

By substituting a_3 in Eq.(29) into the above equation, the equation becomes

$$a_2 = \frac{\dfrac{\left(v_f - v_i\right)\left(r^{*3} - r^*\right)}{\left(3r^* - 1\right)} + v^* - v_f r^* - v_i\left(1 - r^*\right)}{\left(r^* - r\right) - \dfrac{\left(2r^* - 1\right)\left(r^{*3} - r^*\right)}{3r^* - 1}} \tag{34}$$

Thus, for this scenario the genes are in the following

r^*	v_i	v_f	v^*

The searching areas are $0 \le r^* \le 1$ for the first gene and $10 \le v_f, v_i, v^* \le 20$ for the second gene, the third gene, and the fourth gene, respectively.

8 Reward Fuzzy Membership Function for GA Training

The mission of using the changing speed trajectories is to improve the performance of the constant speed strategy. Such as in the acceleration constraint, giving the constraint in the total maneuvering time, T, and the maximum load factor gained from the trajectories generation is very useful information for the GA.

Such constraints are as follows

$$
F = \begin{cases} 0 & \text{if } T > T_{target} \text{ or } n_{max\_} > n_{max\ target} \\ \\ y \cdot F_2 & \text{otherwise} \end{cases}
\tag{35}
$$

where T_{target}, $n_{max\_}$, and $n_{max\ target}$ are the target of total maneuvering time, the maximum load factor resulted from the trajectories generation step, and the target of the maximum load factor, respectively. In this paper, the target utilized is when the constant speed maneuver has the m=0.5.

y is the reward function which is formulated as follows

$$
\begin{aligned}
y_T &= m_1 T \\
y_n &= m_2 n \\
y &= \text{mean}(y_T \ \ y_n)
\end{aligned}
\tag{36}
$$

y_T and y_n are reward functions of the total maneuvering time and the load factor, respectively. The m_1 and m_2 can be determined as illustrated in Fig. 7.

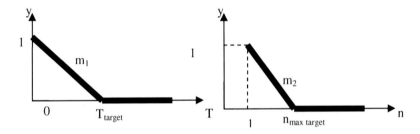

Fig. 7. Reward fuzzy membership function

The aim of the reward function is to keep the GA in the proper way to track the total maneuvering time and load factor targets. y is the reward value which value increases as linear function of the total maneuvering time and the load factor. The mean of the load factor score, y_n, and total maneuvering time score, y_T, is utilized as reward for successful trajectories. The fitness function is the multiplication of the fuzzy reward value with the original fitness function.

F_2 is the original fitness function of the trajectories generation as follows

$$
F_2 = aT + \int (1 - a) \sum_1^n \left(\frac{n}{n_{max}} \right)^2 dt
\tag{37}
$$

where a is the weighting factor which is chosen 0.5 for this paper. The total maneuvering time, T, is determined from Eq.(23). F_2 is the objective function which is balance of the time and the load factor that need to be minimized. It has the dimension of time.

By Eq.(35), the trial and error in deciding the T_{target} and $n_{max\ target}$ can be avoided. This strategy will guarantee that the total maneuvering time as well as load factor maximum at m=0.5 will be reduced by such speed profile. The authors investigated that this is very important information for the GA since the GA probably can be trapped into the unwanted results where the total maneuvering time or the maximum load factor is more than such variables at m=0.5 for constant speed. The detail information of the target is important analysis for guiding the GA to produce the correct pattern to go to the best trajectories generation.

9 Simulation Results and Discussion

The MATLAB environment has been built to simulate the non-steady level Bezier curve bank-turn maneuver utilizing the GA-fuzzy approach. The simulation used 20 individuals in the population, 0.4 mutation rate, 0.5 the selection rate, 15 bits, 200 curve points, and a Simpson's rules as numerical integration to compute the integration in Eq.(14). All results operated when UAV had the maximum payload and maneuvered at very low altitude. For maneuvering path, the 100 generations was used while it utilized 500 generations for maneuvering trajectories.

The simulation was done at small environment since it will convey the small radius which was not modest task to be tackled regarding the UAV constraints. Two cases of point-to-point maneuver planning were simulated with detail information were shown on Table 2.

Table 2. Bezier curve control points results

Case	B_0		B_1 (results)		B_2 (results)		B_3	
I	-30	0	6.42415	14.2186	37.7972	-38.1085	70	20
II	80	-60	8.17591	17.0507	20.9449	-38.9752	0	60

9.1 Maneuvering Path

The fuzzy logic approach was brain-like human thinking where in the human brain the information from all system body such as neuron, cell, as well as other human-biological system was processed by converting in binary code or bit string such as in a computer way, thus the chromosome used in this paper was in binary string.

Fig. 8 and Fig. 9 showed the results of path planning in two cases and the curvature results, respectively. The detail information of the cubic Bezier curve path is on Table 2. After the best path was achieved, the information of the curvature profile can be determined. The results showed that the both constraints were achieved by trajectory planner. The curvatures changed with time continuously without exceeded the maximum allowed value and the path was also safe since it was collision-free.

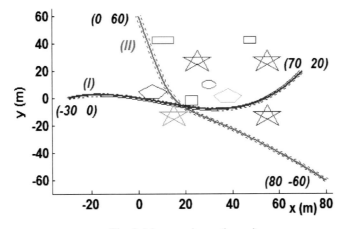

Fig. 8. Maneuvering path results

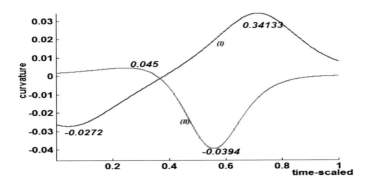

Fig. 9. Curvature results

9.2 Constant Speed Trajectories

The total maneuvering time at constant speed needed to be discovered associated with the curvature information from step 1. The maximum allowed value, $V_{max\_}$, depended on the turning radius of the maneuvering path result. By Eq.(16) the $V_{max\_}$ for case I and case II were 22.311 m/s, and 20.76 m/s, respectively.

For case I, the range of the feasible speed to conduct the maneuver was from 10 m/s to 22.311 /s. When the goal was balance between the minimum time and minimum load factor, i.e. m=50%, using membership function in Eq.(19), the appropriate speed was in the following

$$speed = 0.5(22.311-10)+10 = 16.156 \text{m/s}$$

The constant speed strategy will simplify the maneuvering trajectories problem; however it conveyed the higher load factor. The results of load factor distribution during the motion when the goals were 100% minimum time and 100% minimum load factor, respectively were shown in Fig. 10 bellow

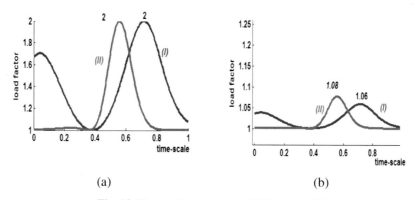

(a)   (b)

Fig. 10. The load factor for m=100% and m=0%

The UAV maximum load factor which was allowed to be utilized in this paper, n_{max}, was 2. Above this value was forbidden since it was very dangerous for the UAV structure. Fig. 10 gave the illustration of the maximum load factor when the speed were $v_{max\_}$ and v_{min}, respectively. It showed that when the goal was the minimum time, or m =100%, the maximum load factor, 2, was reached. The maximum load factor was lower when the goal was the minimum load factor, or m =0%. It was the lower load factor; however it operated in the minimum speed which will take the maximum maneuvering time.

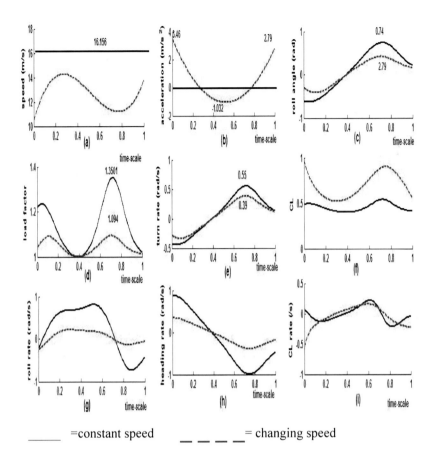

_____ =constant speed _ _ _ _ = changing speed

Fig. 11. The case I trajectories

Fig. 11a, Fig. 11b, Fig. 11c, Fig. 11d, Fig. 11e, Fig. 11f, Fig. 11g, and Fig. 11i showed the trajectories for case I of the speed, the roll angle, the load factor, the turn rate, the lift coefficient, the roll rate, the heading rate, and the CL rate for the constant speed maneuver as well as the changing speed maneuver, respectively. For all trajectories, the constraints were achieved successfully.

Fig. 12a, Fig. 12b, Fig. 12c, Fig. 12d, Fig. 12e, Fig. 12f, Fig. 12g, and Fig. 12i showed the trajectories for case II of the speed, the roll angle, the load factor, the turn rate, the roll rate, the lift coefficient, the roll rate, the heading rate, and the CL rate for the constant speed maneuver as well as changing speed maneuver, respectively. For all trajectories, the constraints were achieved successfully.

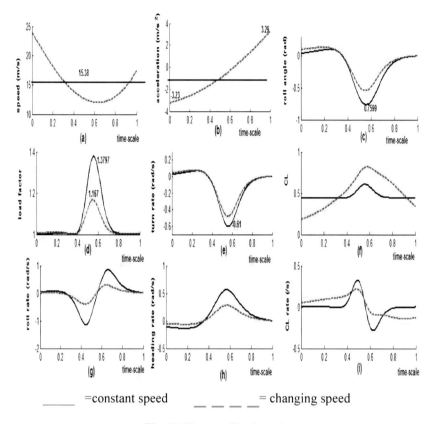

_____ =constant speed   _ _ _ _ = changing speed

Fig. 12. The case II trajectories

9.3   Speed as Continuous Function

The second way to follow the maneuvering path was by changing the speed during
the maneuvering. The goal of this strategy was to find the feasible speed profile
for reducing the load factor and the maneuvering time of constant speed strategy
at m=0.5.

From Fig. 11 and Fig. 12, they showed the comparison of the flight trajectories
between the constant speed and changing speed strategies for case I and case II,
respectively. They showed that the changing speed trajectories had reduced the
maximum load factor during the maneuvering. The coefficient of lift, CL, of
changing speed maneuver was higher than the constant speed maneuver. It was
due to the fact that the when the maneuver was done at constant altitude; it needed
to manage the speed as well as the lift force in such away that the lift force always
balances with the weight force. To perform this task when the speed was changed
conveyed more effort than when it conducted at constant speed.

All constraints were achieved successfully where they were feasible to be done.
For case I, the maximum load factor and the total maneuvering time for constant
speed with m=0.5 were 1.3508 and 11.372 second, respectively, while the

maximum load factor and the total maneuvering time for changing speed were 1.094 and 8.98 second, respectively. For case II, the maximum load factor and the total maneuvering time for constant speed m=0.5 were 1.3797 and 9.809, respectively, while the maximum load factor and the total maneuvering time for changing speed were 1.1671 and 9.8058 second, respectively.

In this paper, the CL rate and heading rate did not included as constraints yet since it want to know these results trajectories for each case without control these trajectories. The CL rate can be converted into an angle of attack rate for elevator deflection reference. Fig. 11 and Fig. 12 showed that the heading rate and the CL rate changed during the motion where their maximum values are in medium values. This fact has shown that the most difficult problem was to find the changing speed trajectories which satisfied the velocity constraint and the acceleration constraint.

There was the challenge to generate the UAV feasible speed profile to improve the constant speed strategy performance as shown in the trajectory results. For both presented cases, the mission target has been achieved where the total maneuvering time and the maximum load factor have been successfully to be reduced. The fuzzy reward function has succeeded in guiding the GA to achieve this target. The authors investigated that without any information about such targets the GA can be trapped into the opposite condition where the reducing conditions did not come.

Conducting the constant speed strategy was simple where the path has been guaranteed to be achieved; however it had inheritance limitation in the load factor distribution during the maneuver. Changing the speed has evolved the constant speed trajectories performance. The load factor has nonlinear relation with the speed, thus by constructing the proper speed trajectories in a proper way; its value can be reduced. For case I, it has been reduced into the lower value with significant reduction; from 1.3501 to 1.094 was very interesting result since the load factor minimum which can be achieved by the UAV was 1 where it was happen in the straight motion. This value was closed to 1.06, the maximum load factor which was achieved when the constant speed was conducted at the minimum speed, 10 m/s^2.

The interesting behavior of the GA has been detected when it solved the agile maneuver planning. The GA accepted the informations which represented the detail work. The GA respons when the information was just to find the polynomial function were failure; however the GA worked well when there was the information that the initial speed and the final speed should be kept in the allowable zones. The GA could accept the guidance from the fuzzy membership function. The punishment values were ordinary utilized in the GA while this paper considered the reward value of the individual which it depended on their achievment.

10 Future Work Issues

The results presented in this paper have proven that it was possible to find the feasible Bezier curve maneuver trajectory for flying robot control system. The

Bezier curve was very agile as path guidance which was suitable as model of bats maneuvering path. Finding the feasible maneuvering trajectories was the problem of computational optimization which did not depend on the condition of environment such as darkness or lightness. As long as the environment model has been known, it was always possible to calculate the maneuvering trajectories by random search techniques. When the path has been already known, the next activity was to process it to meet the feasible maneuvering trajectories.

The flying robot will have high maneuverability performance which will be opposite with the stability achievement. It was not a modest task to be adjusted each other. Thus, to design the control system that suitable for agile maneuvering trajectory was the future important work.

From the results presented in this paper, the bats should have the intelligence maneuver planning system in generating the avoiding collision maneuvering path as well as such maneuvering trajectories that their maneuver can be conducted perfectly in the cluttered dark night environment. They have an ability to collect the environment information to know all obstacles position. Based on the presented results, the feasible maneuvering path can be computed as long as the obstacles environment has been known. After the feasible avoiding collision maneuvering path has been gathered, they have their path guidance that the light existence did not need anymore since the path has saved in their memories. The bats have changed the speed during their maneuver [10]. Thus, probably they have their own intelligent algorithm especially in generating the speed trajectories which should be adjusted with all their flight limitations. They have the complex flexible wing which was very outstanding design to support their difficult maneuvering mission; however they should generate the maneuvering trajectories which conveyed the fuzzy optimization and multi-objective optimization problems. The ability to generate the feasible maneuvering trajectories which made them the truly maneuver organism was also very amazing, the great creation from the nature behavior.

11   Conclusion

The intelligent GA-fuzzy approach has succeeded to discover the best trajectory which satisfies all UAV constraints when they follow the agile maneuvering path. The changing speed strategy is very challenging to reduce the load factor during the UAV maneuver. To achieve the target mission, the fuzzy membership function can be constructed as the guidance to keep the GA in the right way.

References

[1] Shanmugavel, M., Tsourdos, A., White, B.A., Zbikowski, R.: Differential geometric path planning of multiple UAVs. J. Dyn. Syst. Meas. Control 129, 620–632 (2007), doi:10.1115/1.2767657

[2] Yang, K., Gan, S.K., Sukkarieh, S.: An efficient path planning and control algorithm for RUAV's in Unknown and Cluttered Environment. J. Intell. Robot. Syst. 57, 101–122 (2010), doi:10.1007/s10846-009-9359-1

[3] Neto, A.A., Macharet, D.G., Campos, M.F.M.: On the generation of trajectories for multiple UAVs in environment with obstacles. J. Intell. Robot. Syst. 57, 123–141 (2010), doi:10.1007/s10846-009-9365-3

[4] Sujit, P.B., Beard, R.: Multiple UAV path planning using anytime algorithms. In: Proc. American Control Conference, pp. 2978–2983 (2009), doi:10.1109/ACC.2009.5160222

[5] Shanmugavel, M., Tsourdos, A., White, B.A., Zbikowski, R.: Co-operative path planning of multiple UAVs using Dubins paths with clothoid. Control Eng. Pract. 18, 1084–1092 (2010), doi:10.1016/j.conengprac.2009.02.010

[6] Portas, E.B., Torre, L., Cruz, J.M.: Evolutionary trajectory planner for multiple UAVs in realistic scenarios. IEEE Trans. Robotics 26, 619–632 (2010), doi:10.1109/TRO.2010.2048610

[7] Senthilkumar, K., Bharadwaj, K.K.: Hybrid genetic-Fuzzy approach to autonomous mobile robot. In: Proc. IEEE Int. Conf. on Industrial Elect. Control and Instrumentation, pp. 29–34 (1993), doi:10.1109/TEPRA.2009.5339649

[8] Aydin, S., Temeltas, H.: Time optimal trajectory planning for mobile robots by differential evolution algorithm and neural networks. In: Proc. IEEE Int. Symp. On Industrial Electronics, pp. 352–357 (2003), doi:10.1109/ISIE.2003.1267273

[9] Diaz, J.I., Swartz, S.M.: Kinematics of slow turn maneuvering in the fruit bat Cynopterus brachyotis. J. of Exp. Biol. 211, 3478–3489 (2008), doi:10.1242/jeb.017590

[10] Aldridge, H.D.J.N.: Turning flight of bats. J. of Exp. Biol. 128, 419–425 (1987)

[11] Konak, A., Coit, D.W., Smith, A.E.: Multi-objective optimization using genetic algorithms: A tutorial. J. Reliab. Eng. and Syst. Saf. 91, 992–1007 (2006), doi:10.1016/j.ress.2005.11.018

[12] Michalewicz, Z.: A survey in constraint handling techniques in evolutionay computation methods. In: Proc. of the 4th Annual Conference on Evolutionary Programming, pp. 135–155 (1995)

[13] Jimenez, F., Cadenas, J.M., Verdegay, J.L., Sanchez, G.: Solving fuzzy optimization problem by evolutionary algorithms. J. of Inf. Sci. 152, 303–311 (2003), doi:10.1016/S0020-0255(03)00074-4

[14] Jimenez, F., Cadenas, J.M., Verdegay, J.L., Sanchez, G.: Multi-objective evolutionary computation and fuzzy optimization. J. of Approx. Reason. 91, 992–1107 (2006), doi:10.1016/j.ress.2005.11.018

[15] Koyuncu, E., Ure, N.K., Inalhan, G.: Integration of path/maneuver planning in complex environment for agile maneuvering UCAVs. J. Intell. Robot. Syst. 57, 143–170 (2010), doi:10.1007/s10846-009-9367-1

[16] Goldberg, D.E.: Genetic algorithms in search, optimization & machine learning. Kluwer Academic Publishers, Boston (1989)

Part II

Enabling Human Computer Interaction

Audio-Visual Speech Processing for Human Computer Interaction

Siew Wen Chin, Kah Phooi Seng, and Li-Minn Ang

The University of Nottingham Malaysia Campus
keyx8csw@nottingham.edu.my, kezkps@nottingham.edu.my,
kezklma@nottingham.edu.my

Abstract. This chapter presents an audio-visual speech recognition (AVSR) for Human Computer Interaction (HCI) that mainly focuses on 3 modules: (i) the radial basis function neural network (RBF-NN) voice activity detection (VAD) (ii) the watershed lips detection and H∞ lips tracking and (iii) the multi-stream audio-visual back-end processing. The importance of the AVSR as the pipeline for the HCI and the background studies of the respective modules are first discussed follow by the design details of the overall proposed AVSR system. Compared to the conventional lips detection approach which needs a prerequisite skin/non-skin detection and face localization, the proposed watershed lips detection with the aid of H∞ lips tracking approach provides a potentially time saving direct lips detection technique, rendering the preliminary criterion obsolete. Alternatively, with a better noise compensation and a more precise speech localization offered by the proposed RBF-NN VAD compared to the conventional zero-crossing rate and short-term signal energy, it has yield to a higher performance capability for the recognition process through the audio modality. Lastly, the developed AVSR system which integrates the audio and visual information, as well the temporal synchrony audio-visual data stream has proved to obtain a significant improvement compared to the unimodal speech recognition, also the decision and feature integration approaches.

1 Introduction

Recent years, due to the increasing demands for the symbiosis between human and robots, humanoid robots are expected to offer the perceptual capabilities as analogous as human being. In particular, hearing capabilities are essential for social interaction. Unfortunately, most of the audio-only speech recognition (ASR) technologies possess a general weakness; they are vulnerable to cope with the audio corruption, the performance would dramatically degrade under the real-world noisy environments. This undesired outcome stimulates a flourish development of audio-visual speech recognition (AVSR). With the aid of additional speech information from the visual modality, the humanoid

T. Gulrez, A.E. Hassanien (Eds.): Advances in Robotics & Virtual Reality, ISRL 26, pp. 135–165.
springerlink.com   © Springer-Verlag Berlin Heidelberg 2012

robots would integrate both audio and visual information and be able to make a more precise decision dealing with the incoming command.

There are numerous literatures available about the HCI which uses the AVSR as the communication pipeline. Takami Yoshida et al. [1] proposed a two-layered AV integration framework that consists of audio-visual VAD based on a Bayesian network and AVSR using a missing feature theory to improve performance and robustness of ASR. The ASR system is implemented with the proposed two-layered AV integration framework on HARK, which is the open-sourced robot audition software. L. Guan et al. [2] discussed the issues and challenges in the design of HCI through two types of examples, i.e., emotion recognition and face detection. The audio-visual based bimodal emotion recognition, face detection in crowded scene, and facial fiducial points detection are the focus in [2]. The integration of these systems is claimed to provide more natural and friendly man-machine interaction. Furthermore, various aspects of research leading to a text-driven emotive audio-visual avatar are presented in [3]. A general framework of emotive text-to-speech (TTS) using a diphone synthesizer is implemented and integrated into a generic 3-D avatar face model. Alternatively, Beom-Cheol Park et al. presented a sound source localization (SSL) based on audio-visual information with robot auditory system for a network-based intelligent service robot. The audiovisual-based human-robot interaction (HRI) components are combined to naturally interact between human and robot for SSL. The sound is first localized based on the GCC (generalized cross-correlation)-PHAT (phase transform) by frequency characteristics and subsequently the robot moves forward to the caller based on the face detection.

The AVSR system essentially consists of three major parts, i.e.: (i) audio front-end processing which includes voice activity detection and audio feature extraction (ii) visual front-end processing that involves lips detection, lips tracking, and subsequently the visual feature extraction (iii) back-end processing which classifies and integrates the information from both audio and visual modalities to produce the final recognized speech. The most challenging task to develop a robust AVSR system would be the way of attaining the recognition performance equally or better than the audio-only or visual-only speech recognition models under different kind of noisy circumstances. Under the low noise level condition, the acoustic modality would have a better recognition performance compared to visual modality. Therefore, the AVSR in this case should offer at least an equivalent performance as the ASR. On the other hand, as the noise condition increased to certain level where the visual-only speech recognition (VSR) outperform ASR, the AVSR which integrates the information from both the audio and visual modality should have the same or better recognition capability than the VSR. The aforementioned challenge refers to the bound of achieving the minimum recognition performance.

2   Audio-Visual Speech Processing

The robustness of the AVSR is highly dependent to three essential factors: (i) the performance of the ASR, (ii) the efficiency of the VSR, and also (iii) the effectiveness of audio-visual integration. In this chapter, the proposed AVSR would be focused onto the pin pointed factors and the overall proposed architecture is illustrated in Figure 1. The audio and visual data streams are first obtained from the input speech signal and pumped into the respective modality for further processing. For the audio front-end processing, the RBF-NN VAD is proposed to extract the voice speech signal. Once the voice signal is detected, the subsequent audio feature extraction would only be triggered. The most representative audio features, which are the Mel frequency cepstral coefficient (MFCC) would be then sent to the back-end processing to merge with the visual features for final recognition process.

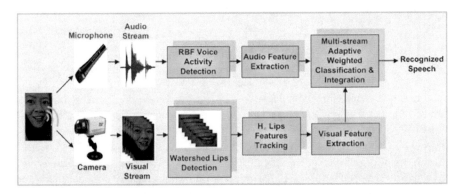

Fig. 1. The architecture of the proposed AVSR system.

On the other hand, the visual stream data is first sent to the watershed segmentation and lips detection by using the pre-defined lips colour cluster boundary. Subsequently, with the obtained lips position, the H∞ filtering is applied as the predictor to predict the lips location in the succeeding frame. For the incoming frame, the watershed lips detection process would focus on the small window size image set around the predicted location. The full size screening would only be going through again if the lips position is predicted wrongly and the lips failed to be detected. With the aid of lips tracking system, the processing time for the overall lips detection process is able to be reduced and it is apparently a credit for the hardware implementation.

Lastly, once the audio and visual features are ready at the back-end platform, the proposed multi-stream classification is developed to classify and integrate the incoming audio and visual features to produce the final recognized speech. The overall proposed back-end processing is compared with decision and feature level integration, the simulation results have disclosed the importance of the temporal synchrony information between the audio and visual stream offered by the proposed AVSR while remaining the precious asynchronous characteristic. It has shown an improvement of the overall AVSR performance not only under the noisy condition but also the clean ambient.

This chapter is organized as: Section 3 discusses about the audio front-end processing. Subsequently, Section 4 explain some of the theory needed for the lips segmentation and detection process, the lips tracking system, the visual feature extraction and also the audio-visual back-end processing in Section 5. The proposed voice activity detection with radial basis function neural network and continuous wavelet transform is demonstrated in Section 6. On the other hand, Section 7 discusses the proposed watershed lips detection and tracking with H∞. The multimodal audio-visual speech classification and integration of the aforementioned audio and visual modalities are explained in Section 8. Lastly, some simulation results are demonstrated in Section 9 followed by the conclusion in Section 10.

3 Audio Front-End Processing

The main concern of the ASR system is how well the acoustic speech could be most represented by the feature vectors and classified at the back-end processing. The general architecture of the ASR system is demonstrated in Figure 2. The input audio stream is first sent to the audio front-end processing which includes the process of voice activity detection (VAD) and feature extraction. Voice activity detection (VAD) is a crucial pre-processing for the audio speech processing, the failure of the VAD to precisely extract the presence of speech signal would directly affecting the performance of the subsequent applications. As a result, a robust VAD for speech/non-speech localization, particularly under the noisy condition, has become an essential research topic throughout the decades. There have various kinds of VAD approaches been reported. The zero-crossing rate and short-term signal energy [4, 5] are commonly used as the acoustic feature for the VAD process due to their simplicity.

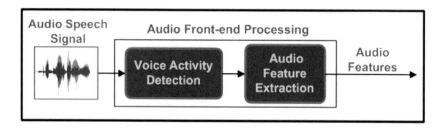

Fig. 2. The architecture of the audio front-end processing.

Once the voice signal is detected by the VAD, the audio feature extraction process would be triggered to extract the most representative audio features from the voice speech signal. There are numerous audio feature extraction approaches being presented in the past [6-9]. Among the audio feature extraction algorithm, Mel frequency cepstral coefficients (MFCC) are the most popular feature extraction approach [10, 11]. For MFCC, the audio input signal is first windowed into the fixed length size. Subsequently, the windowed signal is pre-emphasized for spectrum flattening purpose. In the implemented system, the Finite Impulse Response (FIR) filter [12] as equated below is used for the pre-emphasize process:

$$H(z) = 1 - az^{-1} \qquad (1)$$

where a is normally chosen to be the value between 0.9 to 0.95.

After framing, each of the frames is going for the Hamming filtering to smooth out the signal. In speech analysis, Hamming [13, 14] is one of the most commonly used window functions to reduce the signal discontinuities and spectral distortion in the framed speech. By applying the Hamming window as the aforementioned, the output signal is mathematically represented as:

$$h(n) = g(n)w(n) \qquad (2)$$

where $0 \leq n \leq$ N-1,
    $g(n)$ = segmented voiced signal
    $h(n)$ = output windowed signal

After going through the framing and windowing process for each frame, FFT is used to convert the frame data from time domain into frequency domain. The FFT for the windowed signal, which consists of length N could be mathematically represented as [12]:

$$X(k) = \sum_{j=1}^{N} x(j)\omega_N^{(j-1)(k-1)} \qquad (3)$$

where $\omega_N = e^{-2\pi i/N}$ is an Nth root of the unity.

The decibel amplitude of the spectrum, which is the output from FFT is then mapped onto the Mel scale. The Mel scale is formed with the concept of, the frequency of 1kHz is equal to the Mel-frequency of 1000Mels. The relationship between the Mel-frequency and the frequency is equated as [15]:

$$Mel(f) = 2595 \times log_{10}\left(1 + \frac{f}{700}\right) \qquad (4)$$

After working on the frequency wrapping with the Mel-frequency filter bank, the log Mel-scale is converted back to the time domain by using the

discrete cosine transform (DCT). The MFCC, its delta and double delta coefficients and their re-spective energy components are extracted by using the
equation as shown below in (5)-(8) [16]. The total number of 39 audio feature
coefficients (13 coefficients for each of the respective MFCC, delta and double
delta MFCC) is generated as the most representative audio information for
the subsequent RBF NN speech/non-speech classification.

$$M(j) = \sum_{n=1}^{N} \log \{|m_t(mel)|\} \cos \left\{ \frac{k\pi}{N}(n - 0.5) \right\} \quad k = 0, ..., J \quad (5)$$

where N is known as the number of bandpass filter, and j is the number of
the overlapping frame.

$$DeltaMFCC, \Delta M_t(j) = \frac{\sum_{n=-p}^{p} n \bullet M_{t-n}(j)}{\sum_{n=-p}^{p} n^2} \quad (6)$$

$$DoubleDeltaMFCC, \Delta^2 M_t(j) = \frac{\sum_{n=-p}^{p} n \bullet \Delta M_{t-n}(j)}{\sum_{n=-p}^{p} n^2} \quad (7)$$

$$\widetilde{M_t} = \begin{cases} M_t(j) \\ \Delta M_t(j) \\ \Delta^2 M_t(j) \end{cases} \quad (8)$$

4   Visual Front-End Processing

From Figure 3, it is shown that the visual stream from the incoming speech
signal would be first fed into the visual front-end processing. The role of
visual front-end processing system is to obtain the visual cues, mainly the
lips of a speaker and transforms it into a proper representation of visual data
that is compatible with the recognition module. The most two concerns of the
visual front-end design are: (i) lips detection (ii) lips tracking and (iii) visual
speech representation in a relatively small number of informative features by
applying an appropriate feature extraction methodology.

4.1   Lips Segmentation and Detection

Dealing with the AVSR, the front-end lips detection and tracking is a key
to make the overall AVSR system a success. There are various kinds of image segmentation approaches which could be able to segment the lips region
from the incoming image sequence. Watershed algorithm has become one of
the popular image segmentation tools since it possesses the characteristic of

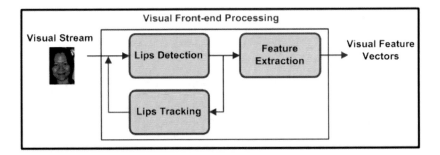

Fig. 3. The architecture of the visual front-end processing.

closed boundary region segmentation. The watershed algorithm is first introduced by [17] with the inspiration from the topographical studies. For a digital image, watershed algorithm is used to evaluate the grayscale image in the form of topographic surface, where each of the pixels is located at certain altitude level depending on its gray level. The pixel in white which has the intensity value of 255 is denoted as the peak or the 0 intensity value is corresponded to the minimum altitude. The other pixels would be assigned to a particular level between these two extremes. For image processing, watershed algorithm applied to split the image into a set of catchment basins, where the catchment basins are formed by the regional minimum. The output from the watershed algorithm, which is the segmented regions are then sent to the lips colour cluster for lips region detection process.

The watershed algorithm in this chapter is based on the rain-flow simulation pre-sented in [18]. The rain-flow simulation employs the falling rain concept where following the steepest descent path, the rain drop falls from the highest altitude to the lowest level, which known as catchment basin. The output of the watershed transformation would be the division of an image into several catchment basins which built by its own regional minimum, and each of the pixels is labelled to a specific catchment basin number.

Figure 4 shows the example of steepest descent path and the catchment basin from part of the pixels value which extracted from the sample image respectively. Referring to Figure 4(left), the steepest descending path which represented by the light grey, is originating from location (1, 1) to the minimum location at (4, 4). Alternatively, for the dark grey path, there are two steepest descending paths which starts from the same point of (9, 6) and goes to two different regional minima. On the other hand, the segmented image is shown in Fgure 4right). The image has segmented into four catchment basins according to their respective steepest descending paths to the minima pixels. The circle from the image represents the regional minima of each catchment basin. The priority of the steepest descending paths which starts at the same beginning point is given from up to the left.

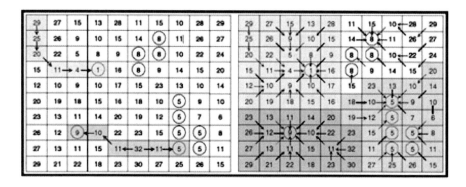

Fig. 4. (left) The example of steepest descending paths, (right) the catchment basin from part of the pixel values [18].

4.2 Lips Tracking

Referring to the process flow of the visual front-end processing in Figure 3, after successfully detecting the lips region described in the previous section, the track-ing system is triggered to predict the lips location of the next incoming source. The purpose of having the tracking system rather than purely applying lips detection for the entire input source is to reduce the computational time. With the estimation of the incoming lips location, the detection process would only focus on the area nearby the predicted point. The detection area is reduced to a relatively small window area instead of the full frame size and the system efficiency is hence increased.

Kalman filtering is widely applied in the tracking system since it possesses the characteristic of minimizing the mean of the system squared error. Kalman filtering nonetheless, would only outperform if the prior knowledge on the system characteristic and the noise statistic are available, which is usually hard to be acquired from the real-world applications. Dealing with the aforementioned matter, an advanced version of the Kalman filtering [19-21], which is the H∞ filtering [22, 23] has become a trend to be an alternative solution. The consideration of the worst case scenario makes the H∞ performs better than Kalman filtering. Nevertheless, the sensitivity tuning of the system parameter is the main issue for the system designer to consider the usage of H∞ as the substitution of the Kalman filtering.

Due to the aforementioned limitation of the Kalman filtering in the real world ap-plication, H∞ has become an alternative solution. The advantages of H∞ filtering over the Kalamn filtering have drawn a considerably attention in recent year. The concept of the H∞ filtering is related to the game theory, where the optimal strategies for the players involving in the multi-players negotiations are studied. The H∞ filtering has been recognized to be useful in the design ofthe robust controller. For a time invariant systems over an infinite time, the game theory has been applied into the H∞ filtering and its cost function could be mathematically defined as below [24]:

$$J = \frac{\sum_{k=0}^{N-1} ||z_k - \hat{z}_k||^2_{S_k}}{||x_0 - \hat{x}_0||^2_{P_0^{-1}} + \sum_{k=0}^{N-1} (||w_k||^2_{Q_k^{-1}} + ||v_k||^2_{R_k^{-1}})} \tag{9}$$

where $\hat{x}_0$ is an a priori estimation of the x_0; $x_0 - \hat{x}_0$, w_k and $w_k \neq 0$; S_k, P_0^{-1}, Q_k^{-1} and R_k^{-1} are the positive definite, symmetric weighting matrices which decided by the designer according to their performance requirement.

For the H∞ filtering game theory concept, the nature goal is to maximize the cost function, J as equation in (9) to worsen the system, therefore the role of the system designer is to, on the other hand minimizes the J by finding the w_k, v_k and x_0 for $\hat{x}_k$ estimation. Referring to (9), it is difficult to obtain a direct cost function minimization solution. As a result, by having the user-defined boundary, $J = 1/\theta$, the H∞ filtering is designed to provide the estimation error as small as possible so that the cost function, J is always bounded by the boundary, $1/\theta$. In the other words, no matter what is the condition of the disturbances energy, even at the worst case scenario, the H filtering possesses the system robustness to the errors by the boundary limitation.

With the user-defined boundary, the cost function equated in (9) could be redefined as following:

$$-\frac{1}{\theta}||x_0 - \hat{x}_k||^2_{P_0^{-1}} + \sum_{k=0}^{N-1} ||z_k - \hat{z}_k||^2_{S_k} - \frac{1}{\theta}(||w_k||^2_{Q_k^{-1}} + ||v_k||^2_{R_k^{-1}}) < 0 \tag{10}$$

As to fulfill the redefined cost function in (10), the state equation of the H∞ filtering is updated as:

$$\overline{S}_k = L_k^T S_k L_k \tag{11}$$

$$K_k = P_k[1 - \theta\overline{S}_k P_k + H_k^T R_k^{-1} H_k P_k]^{-1} H_k^T R_k^{-1} \tag{12}$$

$$\hat{x}_{k+1} = F_k\hat{x}_k + F_k K_k(y_k - H_k\hat{x}_k) \tag{13}$$

$$P_{k+1} = F_k P_k[1 - \theta\overline{S}_k P_k + H_k^T R_k^{-1} H_k P_k]^{-1} F_k^T + Q_k \tag{14}$$

For every time step, k, the following condition has to been fulfilled in order to obtain a true solution for the aforementioned estimator:

$$P_k^{-1} - \theta\overline{S}_k + H_k^T R_k^{-1} H_k > 0 \tag{15}$$

4.3 Visual Feature Extraction

After obtaining the detected lips region, the lips region would be then sent for feature extraction for dimension reduction. By providing an input video of a person talking, the role of the VSR system is to extract the visual speech features that could represent the spoken stream for further uttered words recognition. Visual features could be categorized into three groups: appearance-based, shape-based, and combination of both. The appearance-based feature extraction has been applied by many researchers [25-28]. The main advantage of appearance-based feature is, this approach extracts the entire image information where the ROI (lips or sometimes includes the chin of the speaker) located. From this point of view, the extracted feature vectors would have a large dimensionality which is not appropriate for statistical modelling of the speech classes by using HMM for instance. As the result, the extracted ROI from each input frames is compressed by using Principal Component Analysis (PCA) [29, 30], Discrete Cosine Transform (DCT) [31, 32], Discrete Wavelet Transform (DWT) [32, 33], or Linear Discriminant Analysis (LDA) [34] to obtain a low-dimensional representation.

For the appearance-based feature extraction, discrete cosine transformation (DCT) which known as one of the most commonly used image transformation tool and able to perform well for highly correlated data and provides an excellent performance for the energy compaction [35, 36] is implemented. A brief introduction for the DCT is explained as following:

$$Y(u,v) = \frac{2c(u)c(v)}{\sqrt{MN}} \sum_{i=0}^{M-1} \sum_{j=0}^{N-1} X(i,j) \cos[\frac{(2i+1)u\pi}{2M}] \cos[\frac{(2j+1)v\pi}{2N}] \quad (16)$$

$$c(k) = \begin{cases} \frac{1}{\sqrt{2}} & if\ k = 0 \\ 1 & otherwise \end{cases} \quad (17)$$

5   Audio-Visual Back-End Processing

After obtaining the useful information from both of the audio and visual modality, the feature vectors, which are the representations of these modalities should be then going through the back-end processing to obtain the final recognized speech. In general, the main issue which the researchers would normally look into when dealing with the back-end processing is, the architectures of the audio and visual classification and their integration. There are two main different architectures available in the literature for back-end processing: feature integration [37, 38] and decision integration [39-42].

For AVSR, a number of classification approaches have been proposed in the literature. The most popular classification tools are Dynamic Bayesian Network [40, 43-45], artificial neural network (ANN) [46, 47], and Support Vector Machine (SVM) [48-50]. Hidden Markov Models (HMM), is known as the simply version of the dynamic Bayesian Network. It is one of the most popular classification tools which normally applied not only in ASR, but as well as AVSR system. The system offers a stochastic framework, which is the most commonly used classifier for both audio-only and audio-visual speech recognition [40, 43-45] HMMs assume a class-dependent generative model for the observed features and it is applied to statically model the transition between the speech classes.

As discussed in the previous section, there are two main back-end processing architectures: decision fusion and feature fusion. For feature fusion models which combines the acoustic and visual features into a single feature vectors before going through the classification process. From this case, a single and bimodal classifier, which is a regular left-to-right HMM [51] is required. Alternatively, two parallel unimodal classifiers would be involved in the decision fusion approach. The results from each classifier are then fed into the final decision fusion, such as probabilistic basis to produce the final decision. With the assumption that the acoustic and visual streams are asynchronous, the conventional HMM classifiers fail to perform well in decision fusion. Therefore, instead of using the general HMM classifier, the Coupled HMM (CHMM) [34, 52, 53], Multi-Stream HMM (MSHMM) [54-56], Factorial HMM (FHMM) [57, 58], Product HMM (PHMM)[31, 59] and Independent HMM (IHMM) [60] take over the role from the conventional HMM for decision fusion classification.

6   Voice Activity Detection with Radial Basis Function Neural Network and Continuous Wavalet Transform

Figure 5 illustrates the overview of the proposed RBF-CWT VAD algorithm. The audio input signal is processed within a fixed length of non-overlapped window. Under the supervised learning practice, the MFCC with delta and double delta are extracted as the most representative audio features from a number of audio signals under different environment conditions. Subsequently, the RBF NN is trained by the extracted window-level MFCC to classify the speech and non-speech signal under different level of signal-to-noise ratio. On the other hand, during the VAD operation, the CWT and its energy calculation is triggered once the frame where the interchange of speech to non-speech or vice versa occurred. The starting/ending points of the speech are localized according to the energy threshold predefined by the user and finally the detected speech signal would be sent for further ASR system to obtain final recognized speech.

A general three-layer RBF NN is implemented as illustrated in Figure 6. From the architecture shown, the first layer of the classifier is denoted as the input vector, which is the MFCC features, known as $X(k) = [x_1(k), x_2(k), ..., x_n(k)]^T$. On the other hand, the centre layer includes the

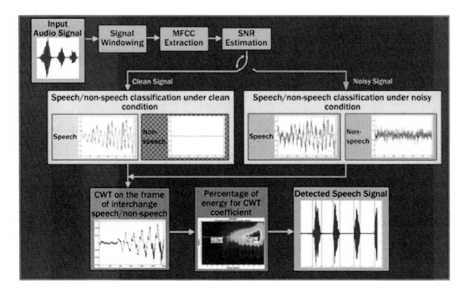

Fig. 5. The overview of the proposed RBF-NN VAD.

parameters of RBF centre, C_n and the Gaussian width, σ_n for each of the RBF unit that represents its corresponding subclass. The output of the RBN NN is the third layer and mathematically represented as following [61]:

$$y(k) = W^T(k)\Theta(k) \tag{18}$$

where,
$$W(k) = [w_1(k), w_2(k), ..., w_N(k)]^T$$
$$\Theta(k) = [\phi_1(k), \phi_2(k), ..., \phi_N(k)]^T$$

The above mentioned output layer could be written in different yet equivalent expression as:

$$y(k) = \sum_{n=1}^{N} W_n(k)\phi_n(k) \tag{19}$$

where ϕ_n could be mathematically represented as:

$$\phi_n(k) = exp(-\frac{||X(k) - C_n||}{\sigma_n^2}) \quad n = 1, 2, ..., N \tag{20}$$

Referring to the expression $|| \bullet ||$ in (20), it is denoted as the Euclidean norm of the input since the RBF NN classification is based on the distance

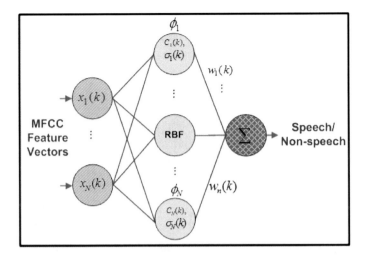

Fig. 6. The architecture of the RBF NN for the speech/non-speech classification.

between the input and the centre values of each subclass. Alternatively, C_n is the RBF centre and the σ_n is the Gaussian width of the nth RBF unit.

One of the results from the RBF NN classifier is depicted in Figure 7. The frame with zero value reflects the non-speech signal while the value of one shows the speech signal. Once the interchange of the speech to non-speech signal or vice versa is triggered, the interchange windows (within the boxes labelled in Figure 7) would be sent to the subsequent CWT speech localization to further localize the starting/ending point.

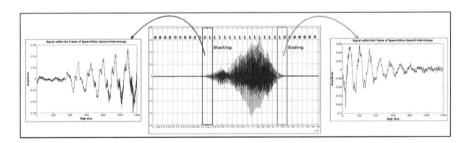

Fig. 7. The example of the output from speech/non-speech classification.

Once the interchange windows of the speech to non-speech and vice versa are triggered by using the aforementioned RBF NN speech classifier, the energy change for the continuous wavelet transform (CWT) coefficients would be calculated. The CWT possesses the capability of illustrating the frequency contents of the sound sources with respect to the time function. The CWT coefficients are generated as following. The CWT is defined as the sum of the signal over all time, $f(t)$ multiplied by the scaled and shifted version of the wavelet function, ψ [62]:

$$Coeff(scale, translation) = \int_{-\infty}^{+\infty} y(t)\psi(scale, translation, t)dt \qquad (21)$$

According the (21), the CWT is the inner multiplication of a wavelet family $\psi_{\vartheta,\varphi}(t)$ with the signal y(t) and it could be expressed as:

$$Coeff(\vartheta, \varphi) = \int_{-\infty}^{+\infty} \frac{1}{\sqrt{\vartheta}}\overline{\psi}(\frac{t-\varphi}{\vartheta})y(t)dt \qquad (22)$$

where ϑ and φ are the parameters for the respective scale and translation; the $\overline{\psi}$ is known as the complex conjugate of the ψ, and the output of the CWT would be the wavelet coefficients denoted as $Coeff$.

After the CWT coefficients are generated, the percentage of the energy change for each of the coefficients is computed as:

$$P = abs(coeff \times coeff) \qquad (23)$$

$$Energy\ Percentage = \frac{P}{Sum(P)} \times 100\% \qquad (24)$$

The percentage of energy change could be graphical represented by the scalogram. From the obtained scalogram, the maximum change of energy is first found. Subsequently, according to the energy which gradually degraded in the contour spreading manner from the coordination where the maximum change of energy took place, the starting/ending point of the speech signal would be localized once the predefined threshold at the same horizontal position as the maximum energy is reached. The coordination observed from the percentage energy change of the CWT coefficients would be the starting and the ending points of the input speech signal after converting back to the time step base.

7   Watershed Lips Detection and Tracking with H∞

The overall lips segmentation and detection is illustrated in Figure 8. Instead of sending the image into the watershed transform directly, some pre-processing steps need to be carried out as to deal with the watershed over-segmentation issue shown in Figure 9. The pre-processing includes edge detection using Sobel filtering, obtaining the background ridge line and foreground object. The output from the watershed algorithm with the aforementioned pre-processing is depicted in Figure 10.

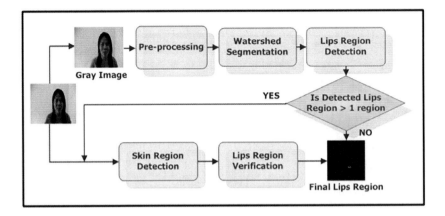

Fig. 8. The overview of the proposed watershed lips segmentation and detection.

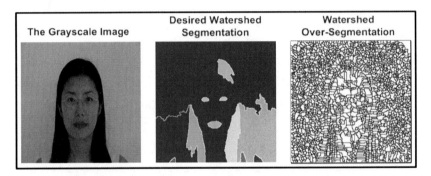

Fig. 9. (a) The grayscale image (b) the desired watershed segmentation and (c) the watershed over-segmentation.

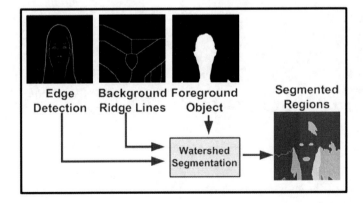

Fig. 10. The outcome of the watershed segmentation with pre-processing.

After the watershed transform, the segmented regions are passed to the lips detection process which contains cubic spline interpolant (CSI) lips colour modelling to obtain the lips region. While modelling the lips colour cluster, the images are first transformed from RGB to YCbCr colour space. Only the chrominance components, Cb and Cr are involved in the system.

For generating the lips colour model, the Asian Face Database (AFD) [63] is used. 6 sets of 6x6 dimensions lips area are cropped from each of the subject in the database. Subsequently, the total of 642 sets lips data is plotted onto a Cb-Cr graph as illustrated in Figure 11(a). Only the heavily plotted pixels are categorized as the lips colour cluster. After morphological closing, the final lip colour cluster is as in Figure 11(b). Moreover, the cluster is encircled using the cubic spline interpolant as equated in (25)-(26) which could be noticed in Figure 11(b). The lips cluster boundary is saved for the lips detection process, the segmented region which falls within the boundary would be detected as the lips region.

$$
S(y) = \begin{cases}
S_1(y) & if \ x_1 \leq x \leq x_2 \\
S_2(y) & if \ x_2 \leq x \leq x_3 \\
\quad \vdots \\
S_{k-1}(y) & if \ x_{k-1} \leq x \leq x_k
\end{cases}
\tag{25}
$$

where S_n is the third degree polynomial which could be described as:

$$
S_n(y) = a_n(y - y_n)^3 + b_n(y - y_n)^2 + c_n(y - y_n) + d_n \quad n = 1, 2, ..., k-1 \tag{26}
$$

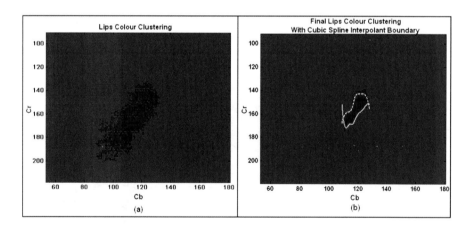

Fig. 11. Lips colour modeling (a) initial cluster (b) the final lips colour cluster encircled with the cubic spline interpolant boundary.

If the number of detected lips region after the aforementioned lips colour modelling is more than one, a further verification process using skin cluster boundary would be triggered. The detected region which as well falls within the skin cluster boundary as shown in Figure 12(b) would be only considered as the final detected lips region. The skin colour boundary is built using the similar methodology as the lips colour boundary. 20 x 20 dimension of the skin area is cropped from the AFD and the skin cluster is plotted as depicted in Figure 12(a).

By using the current detected lips position, the modified H∞ filtering is applied as the predictor to predict the lips location in the succeeding frame. For the incoming frame, the watershed lips detection process would only focus on the small window size image set around the predicted location. The full size screening would only be going through again if the lips position is predicted wrongly and the lips failed to be detected. With the aid of lips tracking system, the processing time for the overall lips detection process is able to be reduced and it is apparently a credit for the hardware implementation. Figure 13 depicts the flow of the overall aforementioned lips detection and tracking system.

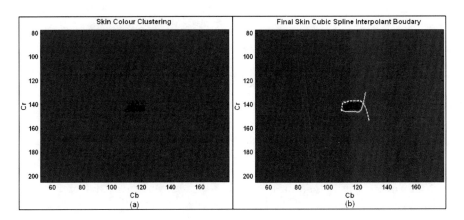

Fig. 12. Lips colour modeling (a) initial cluster (b) the final lips colour cluster encircled with the cubic spline interpolant boundary.

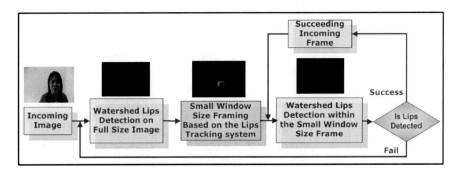

Fig. 13. The process flow of the proposed H∞ lips tracking.

8 Multimodal Audio-Visual Speech Classsifications and Integration

The purpose of proposing multi-stream classification as depicted in Figure 14 which consists of the audio-only model, the visual-only model, and also the audio-visual model is to obtain the advantages provided by the decision-level fusion while enhance its performance. The proposed algorithm remains the asynchronous processing characteristic for the audio and visual stream data, further with the additional benefit of producing the fruitful temporal information between both modalities for the succeeding recognition process which is lack in the conventional decision-level fusion method.

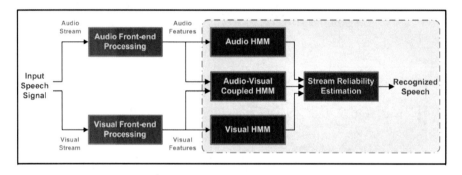

Fig. 14. The proposed multi-stream back-end processing.

As to capture the temporal information between the audio and visual stream, the coupled HMM (CHMM) [64] is developed for the proposed multi-stream integration. The CHMM is used to collect the HMMs from each stream data, where the hidden backbone nodes for each HMM at time duration, t are conditioned by the backbone nodes at time, t-1 for all the associated HMMs. The parameters for the CHMM are denoted as below [65].

Initial state probability distribution:

$$\pi_0^c(i) = P(q_1^c = i) \tag{27}$$

$$b_t^c(i) = P(O_t^c | q_t^c = i) \tag{28}$$

$$a_{i|k,k}^c = P(q_t^c = i | q_{t-1}^a = j, q_{t-1}^v = k) \tag{29}$$

where at time t, the q_t^c is denoted as the state of the couple node for the cth stream.

For a continuous mixture with Gaussian elements, the node observation probability could be written as:

$$b_t^c(i) = \sum_{m=1}^{M_i^c} w_{i,m}^c N(O_t^c, \mu_{i,m}^c, U_{i,m}^c) \tag{30}$$

where $w_{i,m}^c, \mu_{i,m}^c and U_{i,m}^c$ are known as the mixture weight, mean and covariance corresponding to the ith state, the mth mixture and the cth modality respectively; O_t^c is the observation vector at the duration of t to c modality; the M_i^c is denoted as the number of mixtures corresponding to the ith state in the cth modality. After passing through the most representative speech features to the respective audio HMM, visual HMM and audio-visual CHMM, the output, which is the normalized log-likelihood would be sent to the back-end integration platform to produce the final recognized speech.

9   Experimental Results

The proposed AVSR would be analyzed according to the three main modules: (i) the ASR module using the proposed RBF-NN audio front-end processing (ii) the VSR module with the proposed watershed lips detection and tracking with the H filtering and (iii) the multi-stream back-end AVSR processing. These modules are evaluated by applying CUAVE database from Clemson University [66]. From the CUAVE database, it contains 36 speakers uttering continuous, connected and isolated digits. For the experiment setup, the system is evaluated through the method of leave-one-out cross-validation. The experiment would be going through 36 runs, with one speaker reserved as the testing while 35 speakers as the training data for each run. The final recognition rate would be the average of the result of these 36 runs.

9.1   Simulation of the Audio Front-End Processing

For the purpose of analyzing the performance of the proposed RBF-NN VAD approach in terms of speech/non-speech discrimination, the clean uttered data extracted from the CUAVE database is used as the reference decision. As to prefer the reference decision, each utterance is manually hand-marked as either the speech or non-speech frames. The evaluation of the proposed RBF-NN speech detection algorithm is based on the speech hit-rate (SHR) and the non-speech hit-rate (NSHR) as follow [67]:

$$SHR = \frac{N_{speech}}{N_{speech}^{ref}} \tag{31}$$

$$NSHR = \frac{N_{non-speech}}{N_{non-speech}^{ref}} \qquad (32)$$

where N_{speech} and $N_{non-speech}$ are the number of speech and non-speech
frames which have been precisely classified respectively; while N_{speech}^{ref} and
$N_{non-speech}^{ref}$ are the respective number of reference speech and non-speech.

The SHR and the NSHR of the proposed RBF-NN VAD is compared with
the VAD using (i) the MFCC and SVM binary classifier presented in [68]
where the SVM tool used in [68] is similarly downloaded from [69] and (iii)
the zero-crossing rate and energy [4, 5] is plotted in Figure 15 and Figure
16 respectively. From the graphs shown, it could be noticed that with the
proposed RBF-NN VAD, a higher capability of speech signal detection is
obtained by showing higher SHR in Figure 15 while at the same time; the
false detection shown in the NSHR graph (Figure 16) is reduced particularly
under the noisy condition where the SNR is getting lower.

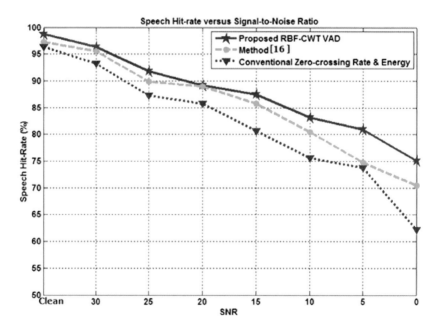

Fig. 15. The speech hit-rate of the (i) proposed RBF-NN VAD, (ii) the SVM
binary classifier [68] and (iii) the zero-crossing rate & energy [4, 5].

The ASR performance evaluation that tabulated in Table 1 is based on the
(i) the proposed RBF-NN VAD (ii) the CMN reported in [70] (iii) the SVM
binary classifier and (iv) the zero-crossing rate and energy. It is observed
that ASR is highly affected by the front-end VAD, a more accurate VAD

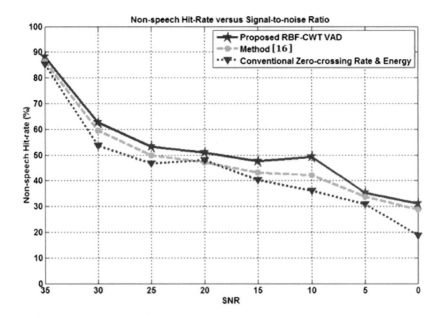

Fig. 16. The non speech hit-rate of the (i) proposed RBF-NN VAD, (ii) the SVM binary classifier [68] and (iii) the zero-crossing rate & energy [4, 5].

would gives higher final speech recognition rate and vice versa. This could be proved by the simulation results shown in Table 1, the ASR which consists of the audio feature post-processing, that is the proposed VAD and the CMN reported in [70] would offer a better recognition rate. Although the VAD could help in good performance for the ASR which shown by the SVM classifier and the zero-crossing rate algorithms, the performance would be still facing a dramatically dropped due to the mismatch matter between the training and testing data, an obvious distinct could be noticed under the low SNR level. With the aid of the proposed RBF-NN VAD, the speech recognition demonstrates a better recognition rate compared to others approaches under both the clean and noisy conditions.

9.2 Simulation of the Visual Front-End Processing

The CUAVE database and AFD are used to evaluate the proposed lips segmentation and detection system. Figure 17(b), Figure 18(b) and Figure 19(b) show some of the watershed segmentation results. Subsequently, the CSI lips colour modelling is used to detect the segmented regions for obtaining the final lips region as in Figure 17(c), Figure 18(d), and Figure 19(c). If the detected region is more than one, the verification process using skin colour modelling is then activated. The segmented region which as well falls within

Table 1. The performance comparison of the ASR with the proposed FCN-RBF VAD to other recently reported algorithm.

SNR(dB)	ASR+Proposed VAD	[70]	ASR+SVM Classifier	ASR+ZCR
Clean	98.47	98.25	96.83	95.02
25	98.03	97.92	94.18	92.64
20	97.81	97.36	93.40	87.91
15	95.03	95.25	80.77	78.54
10	92.14	91.49	63.84	64.06
5	83.61	82.24	41.02	35.88
0	66.83	67.35	15.59	13.95
-5	46.99	42.54	7.18	8.60
-10	24.11	23.18	3.62	2.54

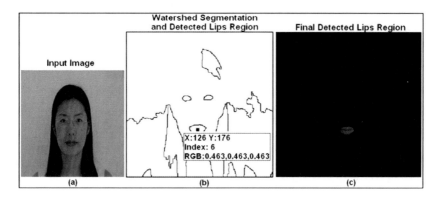

Fig. 17. (a) The input image from AFD (b) the watershed segmentation and lips detection and (c) the final detected lips region.

the skin colour boundary would only be known as the final lips region. The verification using skin colour boundary is depicted in Figure 18(c).

The performance of the proposed parallel watershed lips segmentation and detection system is quantitatively evaluated. The evaluation is based on percentage of overlap (POL) between the segmented lips region X_1 and the ground truth X_2 [71]. The ground truth is built by manually segmenting the lips region. The POL is equated as [71]:

$$POL = \frac{2(X_1 \cap X_2)}{X_1 + X_2} \times 100\% \tag{33}$$

According to the above measurement, the total agreement of the overlap would be 100%. Furthermore, the segmentation error (SE) is also evaluated using the (34) as [71]:

$$SE = \frac{OLE + ILE}{2 \times TL} \times 100\% \tag{34}$$

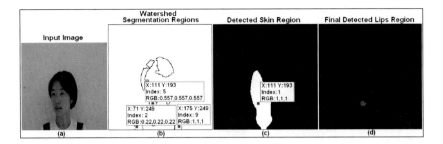

Fig. 18. (a) The input image from AFD (b) the watershed segmentation and lips detection, (c) the lips verification with skin cluster boundary and (d) the final detected lips region.

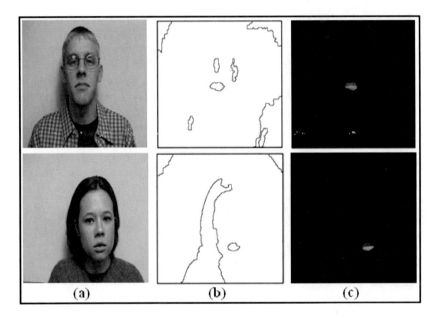

Fig. 19. (a) The input image from CUAVE database (b) the watershed segmentation and detection (c) the final detected lips region.

where the OLE is defined as the outer lips error, meaning the number of non-lips pixels being categorized as lips pixels. The ILE, inner lips error is explained as the number of lips pixels being categorized as non-lips pixels. Besides, TL is denoted as the total lips pixels in the ground truth. The total agreement of the segmentation error would be 0%.

The average of the POL and SE for the respective ASIAN face database and CUAVE database are tabulated in Table 2. The proposed algorithm demonstrates an appreciable detection performance with a high POL and while showing low SE.

Table 2. The average POL and SE for the CUAVE and ASIAN database using the proposed algorithm.

Database	POL (%)	SE (%)
CUAVE	89.72	14.09
ASIAN	89.28	14.82

After obtaining the lips region from the previous watershed lips detection process, the H∞ tracking system is activated to estimate the location of the lips region on the succeeding frame. For the purpose of analyzing the propose tracking algorithm, some in-house video sequences is prepared using Canon IXUS-65 camera. The video sequence is first converted into 256 x 256 dimensional frames. Figure 20, 21 and 22 shows some of the resultant output of the lips detection and lips tracking process respectively. According to Table 1, the proposed modified H∞ tracking system performs better by giving smaller estimation error compared to the conventional H∞ approach. Figure 15 shows the tracking results for every tenth frames in x- and y-position.

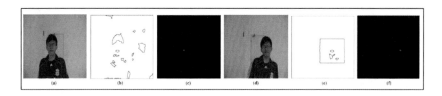

Fig. 20. (a) The first input image (b) full frame watershed segmentation (c) the detected lips region (d) the succeeding input image (e) the small window size segmentation and (f) the detected lips region.

Fig. 21. Resultant output of the lips tracking using in-house video sequence.

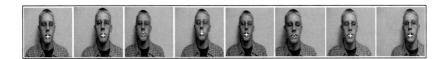

Fig. 22. The resultant output of the tracking using CUAVE database.

Table 3. Average estimation error of tracked pixel in x-position and y-position.

Average Estimation Error (No. of Pixels)	H∞ FIltering	Kalman Filtering
x-position	12.66	16.49
y-position	3.19	5.97

9.3   Simulation of the Audio-Visual Back-End Processing

The simulation results for the ASR, VSR and the proposed AVSR respectively are plotted in Figure 23. From the graph shown, it is noticed that by adding in the complementary information from the visual modality, the performance of the AVSR could be significantly improved compared to the ASR particularly under the noisy circumstances, since the visual stream has provide an extra and more trustable information to the classifier compared to the audio stream which has been severely distorted under the low SNR level.

As to analyze the effect of adding the CHMM to become the multi-stream HMM, the proposed back-end classifiers would be compared with the conventional audio-visual early and late integration. For the late integration, the

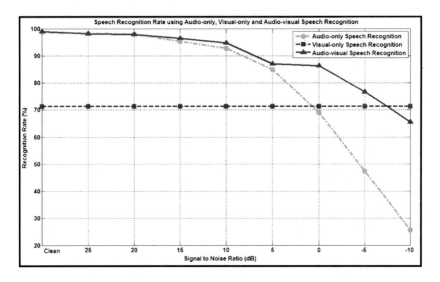

Fig. 23. The performance of the ASR, VSR, and the proposed AVSR.

CHMM would be taken off from the proposed multi-stream classifiers and the rest of the experiment setup remains unchanged. On the other hand, for the feature-level fusion, only single HMM is applied. The audio (44.1kHz) and visual features (29.97fps) are first synchronized into 90Hz and concatenated into a vector before sending into the HMM classifier. The simulation results are demonstrated in Figure 24. The proposed multi-stream classification with the additional fruitful partial temporal synchrony information has shown a better recognition rate compared to others.

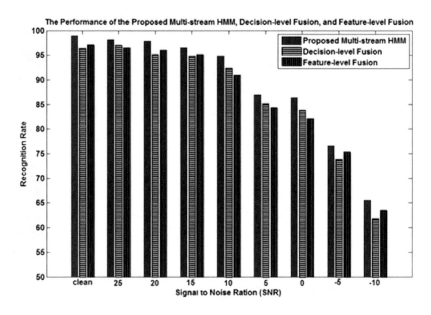

Fig. 24. The performance of the proposed multi-stream HMM, the decision-level HMM, and the feature-level HMM classification.

10   Conclusion

An audio-visual speech processing for the HCI is presented in this paper. Audio-visual speech recognition (AVSR) is a multimodal human-machine interaction technology which uses both acoustic and visual speech information to automatically recognize the incoming speech signal. Three main modules are focused in this chapter: (i) the RBF-NN VAD (ii) the watershed lips detection and tracking with H∞ and (iii) the multi-stream audio-visual back-end processing. The proposed RBF-NN VAD approach demonstrates a more accurate detection compared to the zero-crossing rate and short-term signal energy algorithms not even under the clean condition but as well the noisy circumstances. With a higher capability of speech/non-speech detection offers, the subsequent classification of the audio modality has shown a higher audio recognition rate. Alternatively, a direct lips detection using watershed segmentation and H∞ tracking without the prerequisite face localization is presented. The proposed lips detection successfully obtain a closed boundary lips region while the H∞ tracking helps to reduce the image processing time since only a small window size image need to sent for lips detection in the succeeding image sequence. Lastly, the proposed multi-stream classification is compared with various kinds of classification approaches and it has shown a better recognition rate for the overall AVSR system since the proposed multi-stream AVSR offers not only the asynchronous characteristic but also the fruitful temporal synchrony audio-visual information. The overall proposed AVSR would be fully implemented into the real-world application to finalize the actual efficiency of the system.

References

1. Yoshida, T., et al.: Automatic speech recognition improved by two-layered audio-visual integration for robot audition. In: 9th IEEE-RAS International Conference on Humanoid Robots, Humanoids 2009, pp. 604–609 (2009)
2. Guan, L., et al.: Toward natural and efficient human computer interaction. Presented at the Proceedings of the 2009 IEEE international conference on Multimedia and Expo., New York, NY, USA (2009)
3. Hao, T., et al.: Humanoid Audio \& Visual Avatar With Emotive Text-to-Speech Synthesis. IEEE Transactions on Multimedia 10, 969–981 (2008)
4. Rabiner, L.R., Sambur, M.R.: Algorithm for determining the endpoints of isolated utterances. The Journal of the Acoustical Society of America 56, S31 (1974)
5. Bachu, R.G., et al.: Voiced/Unvoiced Decision for Speech Signals Based on Zero- Crossing Rate and Energy. In: Elleithy, K. (ed.) Advanced Techniques in Computing Sciences and Software Engineering, pp. 279–282. Springer, Netherlands (2010)
6. Chakrabartty, S., et al.: Robust speech feature extraction by growth transformation in reproducing kernel Hilbert space. In: IEEE International Conference on Acoustics, Speech, and Signal Processing, Proceedings (ICASSP 2004), vol. 1, pp. I-133–I-136 (2004)
7. Satya, D., et al.: Robust Feature Extraction for Continuous Speech Recognition Using the MVDR Spectrum Estimation Method. IEEE Transactions on Audio, Speech, and Language Processing 15, 224–234 (2007)
8. Zheng, J., et al.: Modified Local Discriminant Bases and Its Application in Audio Feature Extraction. In: International Forum on Information Technology and Applications, IFITA 2009, pp. 49–52 (2009)
9. Umapathy, K., et al.: Audio Signal Feature Extraction and Classification Using Local Discriminant Bases. IEEE Transactions on Audio, Speech, and Language Processing 15, 1236–1246 (2007)
10. Delphin-Poulat, L.: Robust speech recognition techniques evaluation for telephony server based in-car applications. In: IEEE International Conference on Acoustics, Speech, and Signal Processing. Proceedings (ICASSP 2004), vol. 1, p. I-65–I-68 (2004)

11. Chazan, D., et al.: Speech reconstruction from mel frequency cepstral coefficients and pitch frequency. In: Proceedings. 2000 IEEE International Conference on Acoustics, Speech, and Signal Processing, ICASSP 2000, vol. 3, pp. 1299–1302 (2000)

12. Denbigh, P.: System analysis and signal processing with emphasis on the use of MATLAB. Addison Wesley Longman Ltd, Amsterdam (1998)

13. Zhuo, F., et al.: Use Hamming window for detection the harmonic current based on instantaneous reactive power theory. In: The 4th International Power Electronics and Motion Control Conference, IPEMC 2004, vol. 2, pp. 456–461 (2004)

14. Song, Y., Peng, X.: Spectra Analysis of Sampling and Reconstructing Continuous Signal Using Hamming Window Function. Presented at the Proceedings of the 2008 Fourth International Conference on Natural Computation, vol. 07 (2008)

15. Shah, J.K., Iyer, A.N.: Robust voice/unvoiced classification using novel featuresand Gaussian Mixture Model. Temple University, Philadelphia, USA (2004)

16. Jurafsky, D., Martin, J.: Speech and Language Processing: An Introduction to Natural Language Processing. In: Computational Linguistics and Speech Recognition. Prentice Hall, Englewood Cliffs (2008)

17. Vincent, L., Soille, P.: Watersheds in digital spaces: an efficient algorithm based on immersion simulations. IEEE Transactions on Pattern Analysis and Machine Intelligence 13, 583–598 (1991)

18. Osma-Ruiz, V., et al.: An improved watershed algorithm based on efficient computation of shortest paths. Pattern Recogn. 40, 1078–1090 (2007)

19. Aja-Fern, S., et al.: A fuzzy-controlled Kalman filter applied to stereo-visual tracking schemes. Signal Process 83, 101–120 (2003)

20. Canton-Ferrer, C., et al.: Projective Kalman Filter: Multiocular Tracking of 3D Locations Towards Scene Understanding. In: Machine Learning for Multimodal Interaction, pp. 250–261 (2006)

21. Maghami, M., et al.: Kalman filter tracking for facial expression recognition using noticeable feature selection. In: International Conference on Intelligent and Advanced Systems, ICIAS 2007, pp. 587–590 (2007)

22. Chieh-Cheng, C., et al.: A Robust Speech Enhancement System for Vehicular Applications Using H∞ Adaptive Filtering. In: IEEE International Conference on Systems, Man and Cybernetics, SMC 2006, pp. 2541–2546 (2006)

23. Shen, X.M., Deng, L.: Game theory approach to discrete H∞ filter design. IEEE Transactions on Signal Processing 45, 1092–1095 (1997)

24. Dan, S.: Optimal State Estimation: Kalman, H Infinity, and Nonlinear Approaches. Wiley-Interscience, Hoboken (2006)

25. Eveno, N., et al.: New color transformation for lips segmentation. In: 2001 IEEE Fourth Workshop on Multimedia Signal Processing, pp. 3–8 (2001)

26. Hurlbert, A., Poggio, T.: Synthesizing a color algorithm from examples. Science 239, 482–485 (1988)

27. Yau, W.C., et al.: Visual recognition of speech consonants using facial movement features. Integr. Comput.-Aided Eng. 14, 49–61 (2007)

28. Harvey, R., et al.: Lip reading from scale-space measurements. In: Proceedings. 1997 IEEE Computer Society Conference on Computer Vision and Pattern Recognition, pp. 582–587 (1997)

29. Xiaopeng, H., et al.: A PCA Based Visual DCT Feature Extraction Method for Lip-Reading. In: International Conference on Intelligent Information Hiding and Multimedia Signal Processing, IIH-MSP 2006, pp. 321–326 (2006)

30. Peng, L., Zuoying, W.: Visual information assisted Mandarin large vocabulary continuous speech recognition. In: Proceedings. 2003 International Conference on Natural Language Processing and Knowledge Engineering, pp. 72–77 (2003)
31. Potamianos, G., et al.: Recent advances in the automatic recognition of audio-visual speech. Proceedings of the IEEE 91, 1306–1326 (2003)
32. Seyedin, S., Ahadi, M.: Feature extraction based on DCT and MVDR spectral estimation for robust speech recognition. In: 9th International Conference on Signal Processing, ICSP 2008, pp. 605–608 (2008)
33. Wu, J.-D., Lin, B.-F.: Speaker identification using discrete wavelet packet transform technique with irregular decomposition. Expert Syst. Appl. 36, 3136–3143 (2009)
34. Nefian, A.V., et al.: A coupled HMM for audio-visual speech recognition. In: Proceedings. IEEE International Conference on Acoustics, Speech, and Signal Processing, ICASSP 2002, pp. 2013–2016 (2002)
35. Guocan, F., Jianmin, J.: Image spatial transformation in DCT domain. In: Proceedings. 2001 International Conference on Image Processing, vol. 3, pp. 836–839 (2001)
36. Hao, X., et al.: Lifting-Based Directional DCT-Like Transform for Image Coding. IEEE Transactions on Circuits and Systems for Video Technology 17, 1325–1335 (2007)
37. Kaynak, M.N., et al.: Analysis of lip geometric features for audio-visual speech recognition. IEEE Transactions on Systems, Man and Cybernetics, Part A 34, 564–570 (2004)
38. Meynet, J., Thiran, J.-P.: Audio-Visual Speech Recognition With A Hybrid SVM-HMM System. Presented at the 13th European Signal Processing Conference (2005)
39. Teissier, P., et al.: Comparing models for audiovisual fusion in a noisy-vowel recognition task. IEEE Transactions on Speech and Audio Processing 7, 629–642 (1999)
40. Potamianos, G., et al.: An image transform approach for HMM based automatic lipreading. In: Proceedings. 1998 International Conference on Image Processing, ICIP 1998, vol. 3, pp. 173–177 (1998)
41. Neti, G.P.C., Luettin, J., Matthews, I., Glotin, H., Vergyri, D., Sison, J., Mashari, A., Zhou, J.: Audio-Visual Speech Recognition. The John Hopkins University, Baltimore (2000)
42. Heckmann, F.B.M., Kroschel, K.: A hybrid ANN/HMM audio-visual speech recognition system. In: Presented at the International Conference on Audio-Visual Speech Processing (2001)
43. Yu, K., et al.: Sentence lipreading using hidden Markov model with integrated grammar. In: Hidden Markov models: applications in computer vision, pp. 161–176. World Scientific Publishing Co., Inc., Singapore (2002)
44. Çetingül, H.E.: Multimodal speaker/speech recognition using lip motion, lip texture and audio. Signal Processing 86, 3549–3558 (2006)
45. Yau, W., et al.: Visual Speech Recognition Using Motion Features and Hidden Markov Models. Computer Analysis of Images and Patterns, 832–839 (2007)
46. Yuhas, B.P., et al.: Integration of acoustic and visual speech signals using neural networks. IEEE Communications Magazine 27, 65–71 (1989)
47. Meier, U., et al.: Adaptive bimodal sensor fusion for automatic speechreading. In: IEEE International Conference Proceedings - Presented at the Proceedings of the Acoustics, Speech, and Signal Processing, vol. 02 (1996)

48. Gordan, M., et al.: Application of support vector machines classifiers to visual speech recognition. In: Proceedings. 2002 International Conference on Image Processing, vol. 3, pP. III-129–III-132(2002)
49. Saenko, K., et al.: Articulatory features for robust visual speech recognition. Presented at the Proceedings of the 6th International Conference on Multimodal Interfaces, State College, PA, USA (2004)
50. Zhao, G., et al.: Local spatiotemporal descriptors for visual recognition of spoken phrases. Presented at the Proceedings of the International Workshop on Human-centered Multimedia, Augsburg, Bavaria, Germany (2007)
51. Rabiner, L., Juang, B.H.: Fundamental of speech recognition. Prentice-Hall, Upper Saddle River (1993)
52. Xie, L., Liu, Z.-Q.: A coupled HMM approach to video-realistic speech animation. Pattern Recogn. 40, 2325–2340 (2007)
53. Nefian, A.V., Lu Hong, L.: Bayesian networks in multimodal speech recognition and speaker identification. In: Conference Record of the Thirty-Seventh Asilomar Conference on Signals, Systems and Computers, vol. 2, pp. 2004–2008 (2003)
54. Xie, L., Liu, Z.-Q.: Multi-stream Articulator Model with Adaptive Reliability Measure for Audio Visual Speech Recognition. In: Advances in Machine Learning and Cybernetics, pp. 994–1004 (2006)
55. Luettin, J., et al.: Asynchronous stream modeling for large vocabulary audiovisual speech recognition. In: Proceedings. 2001 IEEE International Conference on Acoustics, Speech, and Signal Processing, ICASSP 2001, vol. 1, pp. 169–172 (2001)
56. Marcheret, E., et al.: Dynamic Stream Weight Modeling for Audio-Visual Speech Recognition. In: IEEE International Conference on Acoustics, Speech and Signal Processing, ICASSP 2007, pp. IV-945–IV-948 (2007)
57. Dean, D.B., et al.: Fused HMM-adaptation of multi-stream HMMs for audiovisual speech recognition (2007)
58. Dean, D.B., et al.: Fused HMM-Adaptation of Synchronous HMMs for Audio-Visual Speech Recognition (2008)
59. Kumatani, K., et al.: An adaptive integration based on product hmm for audiovisual speech recognition. In: IEEE International Conference on Multimedia and Expo, ICME 2001, pp. 813–816 (2001)
60. Lee, A., et al.: Gaussian mixture selection using context-independent HMM. In: Proceedings. 2001 IEEE International Conference on Acoustics, Speech, and Signal Processing, ICASSP 2001, vol. 1, pp. 69–72 (2001)
61. Seng Kah, P., Ang, L.M.: Adaptive RBF Neural Network Training Algorithm For Nonlinear And Nonstationary Signal. In: International Conference on Computational Intelligence and Security, pp. 433–436 (2006)
62. Sinha, S., Routh, P.S., Anno, P.D., Castagna, J.P.: Spectral decomposition of seismic data with continuous-wavelet transforms. Geophysics 70, 19–25 (2005)
63. Lab, I.M.: Asian Face Image Database PF01. Pohang University of Science and Technology
64. Brand, M., et al.: Coupled hidden Markov models for complex action recognition. In: Proceedings. 1997 IEEE Computer Society Conference on Computer Vision and Pattern Recognition, pp. 994–999 (1997)
65. Nefian, A., et al.: A Bayesian Approach to Audio-Visual Speaker Identification. In: Kittler, J., Nixon, M.S. (eds.) AVBPA 2003. LNCS, vol. 2688, pp. 1056–1056. Springer, Heidelberg (2003)

66. Patterson, E., et al.: CUAVE: a new audio-visual database for multimodal human-computer interface research. In: Proceedings. IEEE International Conference on Acoustics, Speech, and Signal Processing, ICASSP 2002, pp. 2017–2020 (2002)
67. Ramírez, J., Górriz, J.M., Segura, J.C.: Voice Activity Detection. Fundamentals and Speech Recognition System Robustness (Robust Speech Recognition and Understanding) (2007)
68. Tomi Kinnunen, E.C., Tuononen, M., Franti, P., Li, H.: Voice Activity detection Using MFCC Features and Support Vector Machine. In: SPECOM (2007)
69. Joachims, T.: SVM light (2008), http://svmlight.joachims.org/
70. Gurban, M.: Multimodal feature extraction and fusion for audio-visual speech recognition. Programme Doctoral En Informatique, Communications Et Information, Signal Processing Laboratory(LTS5), Ecole Polytechnique Fédérale de Lausanne (EPFL), Lausanne, Switzerland (2009)
71. Liew, A.W.C., et al.: Segmentation of color lip images by spatial fuzzy clustering. IEEE Transactions on Fuzzy Systems 11, 542–549 (2003)

Solving Deceptive Tasks in Robot Body-Brain Co-evolution by Searching for Behavioral Novelty

Peter Krčah

Computer Center
Charles University
Prague, Czech Republic
peter.krcah@ruk.cuni.cz

Abstract. Evolutionary algorithms are a frequently used technique for designing morphology and controller of a robot. However, a significant challenge for evolutionary algorithms is premature convergence to local optima. Recently proposed Novelty Search algorithm introduces a radical idea that premature convergence to local optima can be avoided by ignoring the original objective and searching for any novel behaviors instead. In this work, we apply novelty search to the problem of body-brain co-evolution. We demonstrate that novelty search significantly outperforms fitness-based search in a deceiving barrier avoidance task but does not provide an advantage in the swimming task where a large unconstrained behavior space inhibits its efficiency. Thus, we show that the advantages of novelty search previously demonstrated in other domains can also be utilized in the more complex domain of body-brain co-evolution, provided that the task is deceiving and behavior space is constrained.

1 Introduction

Evolutionary algorithms are a frequently used technique for designing morphology and controller of a robot [1, 2, 3, 4, 5, 6, 7, 8, 9, 10, 11, 12]. The advantage of using evolutionary algorithms compared to other optimization methods is that only a high level fitness function and a set of genetic operators (mutation, crossover) need to be specified. While such fitness-driven search can be very successful, a common problem in evolutionary algorithms is premature converge to local optima. This is often caused by too much focus on exploitation of already discovered areas of the search space as opposed to exploration of yet unknown solutions. The trade-off between exploration and exploitation is well understood and several methods for addressing this issue have been proposed [13]. Most methods attempt to increase the rate of exploration by maintaining the diversity of individuals in a population through techniques such as fitness sharing [14, 15], age-layered population structure [16], multi-objectivization [17] or hierarchical fair competition [18].

T. Gulrez, A.E. Hassanien (Eds.): Advances in Robotics & Virtual Reality, ISRL 26, pp. 167–186.
springerlink.com   © Springer-Verlag Berlin Heidelberg 2012

Several of the diversity maintaining methods have also been applied to evolutionary robotics [19, 20, 11].

However, recent works suggest that the underlying problem lies not just in the balance of exploration and exploitation, but in the deceitfulness of the fitness function itself [21, 22, 23, 24, 25]. Novelty search, a recently proposed approach [21], introduces a radical idea that for some problems convergence to local optima can be avoided by simply ignoring the original objective altogether and instead only searching for *any* novel behaviors regardless of their quality with respect to the fitness function. While seemingly counterintuitive, this approach has already been successfully applied to problems in multiple domains: evolving neuro-controllers for robot navigation in a maze [22, 21], evolving genetic programs [25, 23] or evolving learning artificial neural networks for maze navigation [24].

In this work, we use search for behavioral novelty as a method for avoiding premature convergence to local optima in the evolution of body and controller of a simulated robot. Subjects of our experiments are robots placed in a virtual 3D environment with physics simulated using rigid-body dynamics. Both body structure and controller of the robot are subject to optimization by evolution, forming a larger and more complex search space than in domains where novelty search was previously tested. We demonstrate the advantages and disadvantages of novelty search in this domain in two experiments: one with a deceitful fitness function with constrained space of possible behaviors and one with a large behavioral space and a less deceitful fitness function.

We start the chapter by describing the representation of a robot (sections 2.1, 2.2), fitness evaluation (section 2.3) and search algorithms (sections 2.4 and 2.5) used in our experiments. Section 3 then describes the setup of our experiments. Results of experiments along with examples of evolved behavior are provided in section 4. Chapter concludes with the discussion of results (section 5) and conclusions (section 6).

2   Methods

2.1   The Robot

The robot is composed of boxes of varying sizes connected by joints (see figure 1 for examples of evolved robots). Genetic representation of a robot is a directed graph (cycles in the graph are permitted) of nodes representing the body parts. Robot is created from this genome by first adding the body part represented by the root node of the graph and then recursively traversing connections in depth-first order, adding encountered nodes and connections to the robot (see figure 2 for examples of manually constructed robots). To prevent infinite recursion, each node has a *recursive limit* which limits the number of passes through the given node during transcription. Each node can thus be copied multiple (but finite) times to the robot, forming repeated structures. Each genotype connection also has a *terminal flag*. Connection with the terminal flag enabled is applied only when the recursive limit of its source node is reached. Terminal flag can be used to represent structures appearing at the end of chains or repeating units. Transition from a genotype graph with a terminal connection to corresponding phenotype graph is illustrated in figure 3.

Each node also specifies the *size* of the resulting body part (i.e. dimensions of a box) and a type of joint connecting body part to its parent in the resulting tree. Each joint type is defined by a set of rotational constraints imposed on two connected body parts. Different joint types thus have a different number of degrees of freedom (DOF). The following joint types are used: fixed (0 DOF), hinge (1 DOF), twist (1 DOF), hinge-twist (2 DOF), twist-hinge (2 DOF), universal (2 DOF) and spherical (3 DOF). Joint type is subject to mutation and recombination.

Both genotype node and genotype connection have various other properties used for building their phenotype counterparts. Each genotype connection contains information about the position of the child node relative to its parent node. The position is represented by child and parent *anchor points*, relative *rotation*, *scaling factor* and a set of three *reflection flags*, one flag for each major axis. Each enabled reflection flag causes a mirrored copy of the child node to be added to the phenotype graph (along with the original non-mirrored child node). All enabled reflection flags are always applied (if one, two or three reflection flags are enabled, two, four or eight mirrored copies of a child node are created in the phenotype graph, respectively). All geometric transformations (such as scaling, rotation and reflection) are cumulative, i.e. they are applied to the entire subtree of the phenotype graph during its construction. Step-by-step construction of a sample robot is shown in figure 4.

This representation permits very compact encoding of complex body structures allowing for features such as symmetry (using reflection flags) and repetitive segmentation (using recursive transcription).

Although other encodings have been proposed for evolving body and brain of robots [26, 27, 1, 8, 10], encoding used in this work has been successfully used for the evolution of robots in several previous works [7, 28, 29, 4, 30, 6, 2]. Examples of evolved robots are shown in figure 1.

ODE physics engine was used to provide realistic environment for the robot [31]. Water environment was simulated by applying a drag force opposing the movement of each body part and disabling gravity.

2.2   The Controller

Robot's controller is distributed in the body of the robot. Each body part contains a local neuro-controller (an artificial neural network), as well as a local sensor and effector. Artificial neural network was used for controllers. Apart from standard sigmoidal transfer function, oscillatory transfer function was used to enable faster discovery of efficient swimming strategies. A sensor

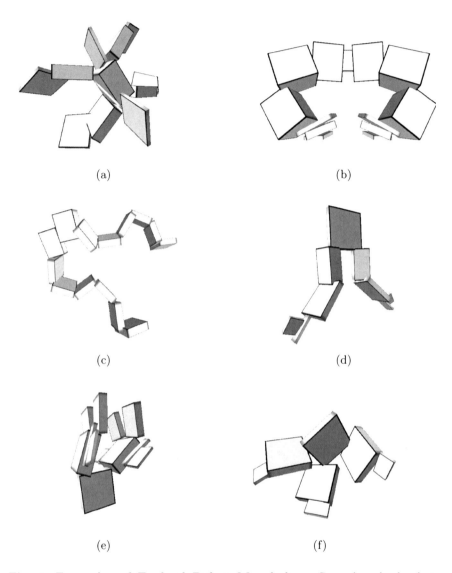

(a)   (b)

(c)   (d)

(e)   (f)

Fig. 1. Examples of Evolved Robot Morphology. Several evolved robots exhibit symmetry (1a, 1b, 1d) and segmentation (1c).

in each body part is measuring current angle of each degree of freedom of a joint.

Effectors allow the robot to move in the simulated world by applying torque to the joints between body parts of the robot. Each degree of freedom of each

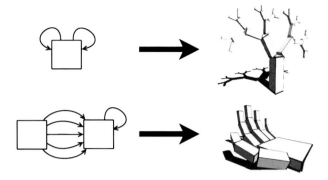

Fig. 2. Manually designed examples of a transcription of genotype to a robot phenotype.

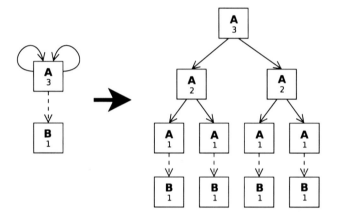

Fig. 3. Robot genotype (left) and phenotype (right). Dashed line represents a terminal connection. Recursive limit of each node is shown under the node mark (A, B).

joint can be controlled separately. The straightforward application of the effector values to physical joints causes undesirable effects and instability in the simulation. Therefore, several transformations are applied to effector values before their application. Effector values are first limited to range $[-1, 1]$ and then scaled by a factor proportional to the mass of the smaller of two connected body parts. This transformation limits the maximum size of a force to some reasonable value and consequently improves simulation stability. Effector values are then smoothed by averaging previous ten values. The resulting value is used as the physical torque applied to a joint of the robot. This modification eliminates sudden large forces and also improves stability of the simulation.

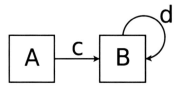
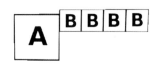

(a) Robot genotype. Recursive limits of nodes **A** and **B** are 1 and 4, respectively. Node **A** is the root node of the graph

(b) Robot phenotype created from the genotype shown on the left. Four copies of node **B** are created during the recursive traversal of the genotype. Copies of node **B** are connected using connection **d**, while nodes **A** and **B** are connected using connection **c**.

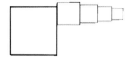
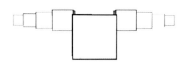

(c) Scaling factor of connection **d** has been changed from 1 to 0.8.

(d) Reflection flag for x axis of connection **c** has been enabled.

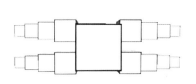
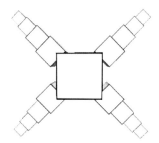

(e) Reflection flag for z axis of connection **c** has been enabled.

(f) One of three rotation angles of connection **c** has been changed from 0 to 45 degrees.

Fig. 4. Step-by-step construction of a robot. The structure of the robot genotype (a) is fixed during all steps. Robot is constructed by successive application of scaling (c), reflection for x axis (d), reflection for z axis (e) and rotation (f).

2.3 Fitness Evaluation

Fitness of each robot is evaluated in a simulated 3D physical world. Simulation proceeds in discrete simulation steps at the rate of 30 steps per simulated second. During each time step, following phases occur for each simulated robot:

1. Sensor values are updated based on the robot environment.
2. Neural signal is propagated through the distributed control system of the robot.
3. Torque is applied to each joint between body parts, according to the current value of the corresponding output neuron.
4. Simulation step is taken and the positions of all objects in the world are updated.

The simulation runs for a specified amount of time (60 seconds of simulated time) after which the fitness value of the robot is evaluated (see section 3).

Open Dynamics Engine (ODE [31]) is used to simulate the rigid body dynamics (including collision detection and the friction approximation). ODE has proved to be sufficient for the purposes of the simulation of evolving robots. However, search algorithms often exploit errors and instabilities in the physics engine to increase the fitness value of robots. Several mechanisms are needed to prevent such exploits. During the fitness evaluation, relative displacement of each two connected parts of each robot is watched and when it rises above a specified threshold (which signals that the robot is behaving unrealistically), the robot is immediately assigned zero fitness and its simulation is stopped. Individual body parts are also prohibited to reach angular or linear velocity above the specified threshold. Robot with such body parts is assigned zero fitness. Finally, an oscillation detector is used to prevent robots from moving using small oscillations (which is another way of exploiting inaccuracies in ODE). If an oscillating robot is detected, it is also discarded.

Several pre-simulation tests also need to be carried out to discover abusive robots before the simulation is started. E.g. the volume of each body part must be larger than the specified threshold as extremely small body parts cause instability in the physical engine. Such tests help detect invalid robots early, so that they do not consume computational resources during full-scale physical simulation.

2.4 Novelty Search

Instead of following the gradient of the fitness function, novelty search directs the search towards any yet unexplored parts of the behavior space. This is achieved by modifying the search algorithm to use the measure of individual's behavioral novelty instead of the fitness function. No other modifications to the underlying search algorithm are required.

Measure of individual's novelty is computed using an archive of previously discovered behaviors [23] as an average distance of the individual's behavior from k closest behaviors recorded in the archive:

$$\rho(x) = \frac{1}{k} \sum_{i=0}^{k} dist(x, \mu_i) \tag{1}$$

where μ_i is the ith-nearest neighbor of x and with respect to distance measure *dist*. This measure provides an estimate of local sparseness in the vicinity of the behavior being measured. Novelty search thus promotes individuals which are further away from already discovered behaviors.

If novelty of an individual exceeds some threshold value, individual is added to the archive. In this work the size of an archive was unlimited (previous works have successfully used limited archive [23] as well).

2.5   NeuroEvolution of Augmenting Topologies (NEAT)

NEAT is an efficient algorithm proposed originally for evolution of artificial neural networks (ANNs)[15] and recently extended for body-brain co-evolution of robots [20]. The extended NEAT is used as an underlying search method in all experiments in this work. NEAT is unique in being able to meaningfully compare and crossover individuals represented by graphs with different topologies (e.g. robots with different morphology or ANNs with different structure).

This is achieved by using *historical markings*–unique inheritable identifiers assigned to each newly created structural element (a neuron or a body part). Historical markings bring the possibility of tracing individual structure elements throughout the evolution. Each structural element is assigned a unique identifier upon its creation (either during the construction of the initial population or during mutation). Historical markings are inherited, so each node and connection can be traced back to its oldest ancestor. Historical markings are also computationally inexpensive. The genetic algorithm only needs to keep track of the value of the most recent marking assigned to a neuron or a body part. Upon each request for a new marking (e.g. when the mutation creates a new node or a new connection), the value of the most recent marking is incremented and returned.

Historical markings offer an elegant method of sensibly mating structures with different topology. Parental structures are first scanned for the presence of structure elements with matching historical markings (these are the ones, which share the common ancestor and are thus considered compatible). Since many organisms during the evolution are very close relatives, the number of matching structure elements is expected to be fairly high. An offspring is constructed by first copying the parent with a higher fitness value, followed by random exchange of internal parameters of all matching nodes and connections with another parent. This way, new offspring inherits all non-matching nodes and connections from its better parent, while all matching elements are formed by a random recombination of properties of both parents.

Historical markings are used to efficiently measure similarity of two individuals which is used in NEAT to assign individuals to species. An explicit fitness sharing is used to maintain a set of species in the population, encouraging innovation by protecting it in a separate species. Complete description of NEAT method is presented in [15, 20]. Novelty search is integrated with NEAT by simply replacing the fitness value of an individual with a measure of its behavioral novelty as described in the previous section.

3 Experiments

By searching *only* for novelty in behavior, novelty search is less prone to falling into traps set up by the objective-based fitness function in the form of local optima. However if the behavior space is too large and unconstrained, novelty search may spend most of the time exploring behaviors that are uninteresting with respect to the objective and never find any useful solutions. To test both ends of this scale in the domain of body-brain co-evolution, two experiments were performed: one non-deceptive with a large unconstrained space of behaviors (the swimming experiment) and one highly deceptive with a constrained behavioral space (the barrier avoidance experiment).

In each experiment, the performance of novelty search was compared to the performance of standard fitness-based search. Random search was used as a baseline when comparing performance in each experiment (random search was performed by assigning each individual a random fitness value).

3.1 Swimming Experiment

The objective in the swimming experiment is to evolve robots capable of moving in a water environment the largest distance from the starting position in the allocated time. The difficulty in this task is that the robot must evolve useful morphology *and* control to successfully move through water. At the start of each test, robot is placed at the origin of the coordinate system. Robot is then free to move in any direction for 60 seconds after which the position of its center of mass is recorded. Fitness function is defined as the distance between the final and starting positions of the robot. Behavior of a robot is defined as the position of the robot at the end of the test and thus consists of a vector of 3 real values. Behavior metrics is defined as Euclidean distance between the final positions of two robots. There were no barriers for the robot to avoid and no constraints on the behavior of the robot apart from the limits of the physics simulation.

3.2 Barrier Avoidance Experiment

In the barrier avoidance experiment, the robot is placed at one end of a 20m x 20m x 25m container while the target is placed at the other end of the container (see figure 5). Robot's goal is to reach the target within the time limit of 60 seconds. Task is made deceptive by putting the barrier directly in

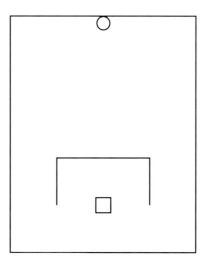

Fig. 5. Barrier Avoidance Experiment. Dimensions of the environment are 20m x 20m x 25m. The barrier is located between the robot and the target and is formed by a hollow 10m x 10m x 5m box with the bottom face left out. The target (indicated by the circle) is located at the top of the container at the initial distance of 20m from robot's initial position (indicated by the square).

front of the robot, so it obstructs the direct path to the target. Moreover, the shape of the barrier only makes it possible to reach the target if the robot first moves *away* from it, which makes this task highly deceptive. Fitness function for a robot with final position x is defined as

$$f(x) = \max(0, d(p_{start}, p_{target}) - d(x, p_{target})) \qquad (2)$$

where $d(p_{start}, p_{target})$ is the constant distance from the starting position of the robot to the target (in this case 20m) and $d(x, p_{target})$ is the distance from the final position of the robot to the target. Behavior of the robot is defined as robot's final position (same as in the swimming experiment) and behavior metrics is defined as the Euclidean distance between final positions of two robots. Target position is fixed for all experiments and robot has no information about the position of the target. As in the swimming experiment, barrier avoidance experiment is performed in a simulated water environment.

3.3   Parameter Settings

The number of individuals in a population is set to 300, with 150 generations per run. In all experiments, initial generation was initialized with uniform robots consisting of a single box with no neurons except for inputs/outputs. All parameters of the underlying search algorithm (hierarchical NEAT[20]) were set to the same values in both the standard fitness-based configuration and in the novelty-based configuration. The configuration of hierarchical

NEAT was the same as in our previous works [20]. Parameter k for computing the novelty of an individual was set to 15 and the novelty threshold for adding an individual to the archive was set to 0.1. Archive size was not limited. Each configuration was tested independently 25 times in the barrier avoidance experiment and 30 times in the swimming experiment. All significance levels were computed using Student's t-test.

4 Results

In the swimming experiment, both novelty search and fitness-based search were able to consistently find effective solutions (see figure 6a). Average maximum distance of the robot from the starting position after 150 generations was 108.32m for fitness-based search, 96.23m for novelty search and 14.46m for random search. Fitness-based search thus outperformed novelty search by 12.47% ($p < 0.05$), confirming that unconstrained behavior space without deceptive fitness function is favourable to the standard fitness-based approach.

In barrier avoidance experiment, fitness-based search never successfully found a way out of the barrier, reaching an average maximum fitness of 4.56m (see figure 6b). Performance of the fitness-based search was in this case only marginally better than random search which achieved maximum average fitness of 3.68m (although the difference was statistically significant; $p < 10^{-8}$). Novelty search was the only method that consistently escaped the trap and reached an average maximum fitness of 13.16m, significantly outperforming both fitness-based and random search ($p < 10^{-10}$).

4.1 Analysis of Behavioral Diversity

To provide better insight into how individual search methods explore the space of behaviors, the coverage of behavior space was computed for each experiment. The coverage was computed as a total number of cells from a 1m x 1m x 1m grid that contained the final position of at least one robot. This statistic was computed after each generation using all behaviors seen since the start of the run. Resulting values were averaged over 25 runs (see figure 7).

Comparison of the amount of behavior space covered by individual methods shows that novelty search was exploring the behavior space faster than the fitness based search in both experiments ($p < 10^{-10}$). In swimming experiment the results demonstrate that, although novelty search achieved on average lower fitness values than fitness-based search, it was still able to efficiently explore the behavior space. The difference in how the two methods explore the behavior space can best be seen in the set of typical runs shown in figures 8a and 8b. While novelty search thoroughly and evenly explores the search space progressing in an ever-expanding sphere (figure 8a), fitness-based search tends to exhaustively search a small number of promising directions (8b).

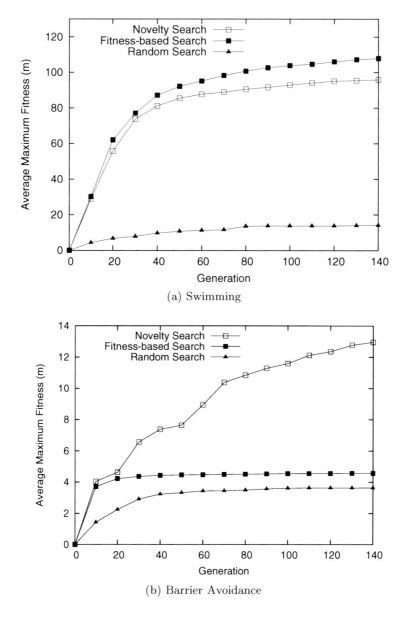

(a) Swimming

(b) Barrier Avoidance

Fig. 6. Comparison of Maximum Fitness Achieved by Novelty Search and Fitness-Based Search. The average maximum fitness is shown for the swimming experiment (6a) and for the barrier avoidance experiment (6b) averaged over 25 runs. Results show that novelty search significantly outperforms fitness-based search in constrained deceptive task (barrier avoidance), while reaching similar performance in unconstrained non-deceptive task (swimming).

In the barrier avoidance experiment, coverage of the behavior space for fitness based search never exceeded $500m^3$, which corresponds to the volume of the interior space of the barrier. A typical distribution of behaviors for such run is shown in figure 8d. While standard fitness-based approach always failed to escape the barrier, novelty search consistently explored the entire container, eventually reaching the target as well. Example of a typical run is shown in figure 8c.

4.2   Combining Novelty Search with Fitness-Based Search

To further analyze both search methods, additional experiments were performed using a combination of the novelty search and the fitness-based search. The combined search starts with novelty search and switches to fitness-based search after 20, 40, 60 or 80 generations. The motivation for switching the methods in the middle of the search is to more efficiently exploit strengths of each method in the barrier-avoidance experiment. Once novelty search overcomes the barrier (by focusing on the exploration part of the search), fitness based search may be able to quickly converge to a solution (by focusing on the exploitation).

Results of experiments with combination of Novelty Search and Fitness-based search (shown in figure 9) show, that the later the switch was made, the higher the final fitness value was achieved. The best average fitness values were reached by the novelty search (13.16m), and the combined method switching at generation 60 (12.26m) and generation 80 (13.65m). The differences between these best methods were not statistically significant ($p > 0.25$). Two other two combined search methods achieved average fitness value of 10.39m (switch after 40 generations) and 6.81m (switch after 20 generations).

Comparison of the amount of behavior space covered by individual methods confirms that novelty search explores the behavior space faster than the fitness based search. In combined experiments, later switch to fitness-based search consistently resulted in higher coverage of behavior space. Combined search methods switching at generation 20, 40, 60 and 80 reached 10.57%, 17.33%, 23.48% and 26.49% of the container, respectively. Standalone novelty search reached the highest behavior space coverage of all methods: 28.87%. All differences in behavior coverage are statistically significant ($p < 0.05$) except the difference between novelty search and combined search switching after 80 generations ($p < 0.06$).

The hypothesis that switching to fitness-based search after the barrier is overcome will improve the performance of the search was not confirmed by the experiments. Experiments with combined search switching at generations 60 and 80 indicate that switching to fitness search may provide a speedup. However, the speedup was only temporary and the difference from novelty-search was not sufficiently significant ($p > 0.1$). Moreover, behavior space analysis shows that after switching to fitness-based search, the exploration rate slowed significantly.

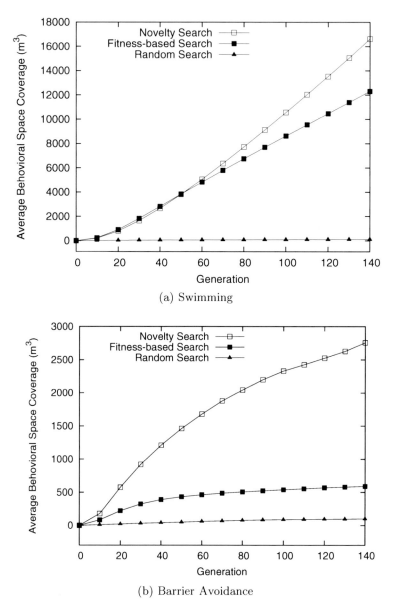

(a) Swimming

(b) Barrier Avoidance

Fig. 7. Comparison of Behavioral Diversity In Novelty Search and Fitness-Based Search. Behavioral diversity of a single evolutionary run is measured as a number of cells in 1m x 1m x 1m grid that contain at least one robot behavior discovered during the run. The cumulative behavioral diversity per generation averaged over 25 runs is shown for the swimming task (7a) and for the barrier avoidance task (7b). The main conclusion is that in both tasks, novelty search discovers significantly more behaviors than fitness-based search.

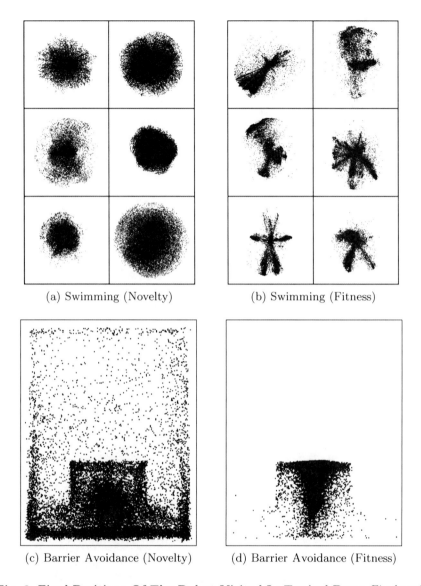

(a) Swimming (Novelty)   (b) Swimming (Fitness)

(c) Barrier Avoidance (Novelty)   (d) Barrier Avoidance (Fitness)

Fig. 8. Final Positions Of The Robot Visited In Typical Runs. Final positions of all robots discovered in selected typical runs is shown for the swimming experiment (8a, 8b) and for the barrier avoidance experiment (8c, 8d). Novelty search in both experiments explores the space of behaviors more evenly, while fitness-based search focuses on optimization of few chosen areas. For swimming experiment (8a, 8b) a 260m x 260m x 260m part of the behavior space is shown. Final positions are three-dimensional and are shown using a two-dimensional orthogonal projection.

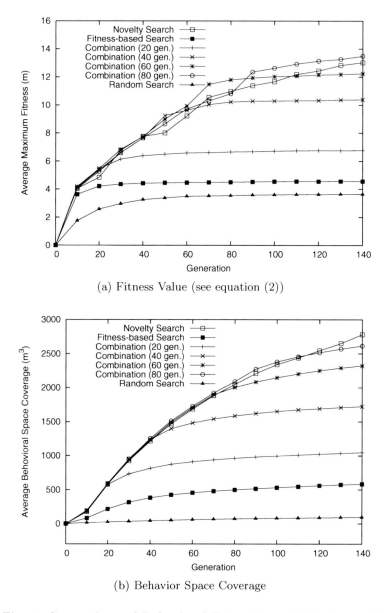

(a) Fitness Value (see equation (2))

(b) Behavior Space Coverage

Fig. 9. Comparison of Behavioral Diversity and Maximum Fitness for Different Combinations of Novelty Search and Fitness-based Search. Combined algorithm performs fitness-based search for the first fixed number of generations and then switches to using Novelty Search for the rest of the search. Figure 9a shows the average maximum fitness averaged over 25 runs and figure 9b shows the cumulative behavioral diversity per generation averaged over 25 runs.

5 Discussion

In the swimming experiment the behavior space was only constrained by the limits of the physical simulation (the speed at which a robot can move is limited by the maximum amount of torque it can exert by its joints). Such large space of possible behaviors together with absence of any obvious deceptiveness makes this problem unfavorable to the novelty search. This intuition was confirmed by the results of the swimming experiment, where novelty search was outperformed by the fitness-based search, although it was able to explore more of the behavior space (figures 6a and 7a).

However, the results of the swimming experiment do not confirm the expectation that in such unconstrained environment the novelty search will perform just as poorly as a random search. On the contrary, novelty search performed almost as well as the fitness-based search and it significantly outperformed random search. The reason is that even though novelty search does not explicitly optimize solutions towards the main objective, it still explores behavior space in a structured way (even when the behavior space is practically unlimited) progressing from simple to more complex behaviors. This is unlike the random search which often revisits the same solutions repeatedly.

In the barrier avoidance experiment, the deceptiveness of the task lies in the fact that fitness-based search leads the robot directly to the target until it reaches the barrier. At that point, in order to find better solutions, the search needs to explore behaviors that are temporarily further away from target. In the standard objective-based approach, getting further away from the target decreases the fitness value; fitness function thus has a local optimum inside the barrier where fitness-based search is likely to be trapped. This was confirmed by the results of the barrier experiment with fitness-based configuration, where individuals never fully escaped the barrier. In the same experiment, novelty search was able to explore the entire container and was capable of consistently finding the target. The effectiveness of the novelty search in this experiment can be explained by the fact that the space of all possible behaviors was constrained to the size of the container; the boundaries of the container in this case served in a sense as a guide directing the search towards the target.

6 Conclusions

In this work, we applied recently proposed novelty search algorithm to the evolution of body and brain of a simulated robot. We demonstrated advantages and disadvantages of novelty search on two tasks: swimming and barrier avoidance. Results from the swimming experiment have confirmed that novelty search does not provide an advantage for tasks where behavior space is unconstrained and fitness function is not deceptive. For such tasks, novelty search may still explore the space of possible behaviors more effectively than fitness-based search, but it takes longer to optimize solutions towards the objective.

However, the barrier avoidance experiment has shown that if the range of possible behaviors is limited (in this case by putting the robot inside a container) and fitness function is deceptive then the novelty search algorithm significantly outperforms fitness-based search.

In summary, this work confirms that the advantages of novelty search demonstrated previously in other domains can be successfully leveraged in the co-evolution of body and brain of robots. Results presented in this work demonstrated that directing the search towards behavioral novelty can help in solving deceptive problems in body-brain evolution that standard fitness-based search methods often struggle with.

Acknowledgment

The research is supported by the Charles University Grant Agency under contract no. 9710/2011.

References

1. Hornby, G.S., Lipson, H., Pollack, J.B.: Generative representations for the automated design of modular physical robots. IEEE Transactions on Robotics and Automation 19(4), 703–719 (2003)
2. Lassabe, N., Luga, H., Duthen, Y.: A new step for evolving creatures. In: IEEE-ALife 2007, Honolulu, Hawaii, April 1–5, pp. 243–251. IEEE, Los Alamitos (2007), http://www.ieee.org
3. Lipson, H., Pollack, J.B.: Automatic design and manufacture of robotic lifeforms. Nature 406(6799), 974–978 (2000)
4. Miconi, T.: In silicon no one can hear you scream: Evolving fighting creatures. In: O'Neill, M., Vanneschi, L., Gustafson, S., Esparcia Alcázar, A.I., De Falco, I., Della Cioppa, A., Tarantino, E. (eds.) EuroGP 2008. LNCS, vol. 4971, pp. 25–36. Springer, Heidelberg (2008)
5. Nolfi, S., Floreano, D.: Evolutionary Robotics. The Biology, Intelligence, and Technology of Self-organizing Machines. MIT Press, Cambridge (2nd print, 2001), (1st print, 2000)
6. Sims, K.: Evolving 3d morphology and behavior by competition. Artificial Life 1(4), 353–372 (1994)
7. Sims, K.: Evolving virtual creatures. In: SIGGRAPH 1994: Proceedings of the 21st annual conference on Computer graphics and interactive techniques, pp. 15–22. ACM Press, New York (1994)
8. Komosinski, M., Ulatowski, S.: Framsticks: Towards a simulation of a nature-like world, creatures and evolution. In: Floreano, D., Mondada, F. (eds.) ECAL 1999. LNCS, vol. 1674, pp. 261–265. Springer, Heidelberg (1999)
9. Hornby, G.S., Fujita, M., Takamura, S.: Autonomous evolution of gaits with the sony quadruped robot. In: Proceedings of the Genetic and Evolutionary Computation Conference, pp. 1297–1304. Morgan Kaufmann, San Francisco (1999)

10. Gruau, F.: Cellular encoding for interactive evolutionary robotics. In: Fourth European Conference on Artificial Life, pp. 368–377. MIT Press, Cambridge (1997)

11. Bongard, J.C., Hornby, G.S.: Guarding against premature convergence while accelerating evolutionary search. In: GECCO 2010: Proceedings of the 12th annual conference on Genetic and evolutionary computation, pp. 111–118. ACM, New York (2010)

12. Pilat, M.L., Jacob, C.: Evolution of vision capabilities in embodied virtual creatures. In: GECCO 2010: Proceedings of the 12th annual conference on Genetic and evolutionary computation, pp. 95–102. ACM, New York (2010)

13. Goldberg, D.E.: Genetic Algorithms in Search, Optimization, and Machine Learning. Addison-Wesley Professional, Reading (1989)

14. Goldberg, D.E., Richardson, J.: Genetic algorithms with sharing for multimodal function optimization. In: Proceedings of the Second International Conference on Genetic Algorithms on Genetic algorithms and their application, pp. 41–49. Lawrence Erlbaum Associates, Inc., Mahwah (1987)

15. Stanley, K.O., Miikkulainen, R.: Efficient reinforcement learning through evolving neural network topologies. In: Proceedings of the Genetic and Evolutionary Computation Conference (GECCO-2002), Morgan Kaufmann, San Francisco (2002)

16. Hornby, G.S.: Alps: the age-layered population structure for reducing the problem of premature convergence. In: GECCO 2006: Proceedings of the 8th annual conference on Genetic and evolutionary computation, pp. 815–822. ACM, New York (2006)

17. Knowles, J.D., Watson, R.A., Corne, D.W.: Reducing local optima in single-objective problems by multi-objectivization. In: Zitzler, E., Deb, K., Thiele, L., Coello Coello, C.A., Corne, D.W. (eds.) EMO 2001. LNCS, vol. 1993, pp. 269–283. Springer, Heidelberg (2001)

18. Hu, J., Goodman, E., Seo, K., Fan, Z., Rosenberg, R.: The hierarchical fair competition (hfc) framework for sustainable evolutionary algorithms, pp. 241–277. MIT Press, Cambridge (2005)

19. Mouret, J.-B., Doncieux, S.: Overcoming the bootstrap problem in evolutionary robotics using behavioral diversity. In: CEC 2009: Proceedings of the Eleventh conference on Congress on Evolutionary Computation, pp. 1161–1168. IEEE Press, Piscataway (2009)

20. Krčah, P.: Towards efficient evolutionary design of autonomous robots. In: Hornby, G.S., Sekanina, L., Haddow, P.C. (eds.) ICES 2008. LNCS, vol. 5216, pp. 153–164. Springer, Heidelberg (2008)

21. Lehman, J., Stanley, K.O.: Exploiting open endedness to solve problems through the search for novelty. In: In Proceedings of the Eleventh International Conference on Artificial Life. MIT Press, Cambridge (2008)

22. Lehman, J., Stanley, K.O.: Revising the evolutionary computation abstraction: minimal criteria novelty search. In: GECCO 2010: Proceedings of the 12th annual conference on Genetic and evolutionary computation, pp. 103–110. ACM, New York (2010)

23. Lehman, J., Stanley, K.O.: Efficiently evolving programs through the search for novelty. In: GECCO 2010: Proceedings of the 12th annual conference on Genetic and evolutionary computation, pp. 837–844. ACM, New York (2010)

24. Risi, S., Vanderbleek, S.D., Hughes, C.E., Stanley, K.O.: How novelty search escapes the deceptive trap of learning to learn. In: GECCO 2009: Proceedings of the 11th Annual conference on Genetic and evolutionary computation, pp. 153–160. ACM Press, New York (2009)

25. Doucette, J., Heywood, M.I.: Novelty-based fitness: An evaluation under the santa fe trail. In: Esparcia-Alcázar, A.I., Ekárt, A., Silva, S., Dignum, S., Uyar, A.Ş. (eds.) EuroGP 2010. LNCS, vol. 6021, pp. 50–61. Springer, Heidelberg (2010)

26. Bongard, J.: Evolving modular genetic regulatory networks. In: Proceedings of the 2002 Congress on Evolutionary Computation, CEC 2002, vol. 2, pp. 1872–1877 (2002)

27. Eggenberger, P.: Evolving morphologies of simulated 3D organisms based on differential gene expression. In: Proceedings of the Fourth European Conference on Artificial Life, pp. 205–213. MIT Press, Cambridge (1997)

28. Shim, Y.-S., Kim, C.-H.: Generating flying creatures using body-brain co-evolution. In: SCA 2003: Proceedings of the 2003 ACM SIG-GRAPH/Eurographics symposium on Computer animation, pp. 276–285. Eurographics Association, Switzerland (2003)

29. Chaumont, N., Egli, R., Adami, C.: Evolving virtual creatures and catapults. Artificial Life 13, 139–157 (2007)

30. Miconi, T., Channon, A.: An improved system for artificial creatures evolution. In: Rocha, L.M., Yaeger, L.S., Bedau, M.A., Floreano, D., Goldstone, R.L., Vespignani, A. (eds.) Artificial Life X: Proceedings of the Tenth International Conference on the Simulation and Synthesis of Living Systems, June 3-7, pp. 255–261. MIT Press, Bloomington (2006)

31. Smith, R.: Open dynamics engine manual, http://www.ode.org

From Dialogue Management to Pervasive Interaction Based Assistive Technology

Yong Lin and Fillia Makedon

Department of Computer Science and Engineering
The University of Texas at Arlington
416 Yates Street, 250 Nedderman Hall
P.O. Box 19015, Arlington, TX 76019, USA
{ylin,makedon}@uta.edu

Abstract. Dialogue management system is originated when human-computer interaction (HCI) was dominated by a single computer. With the development of sensor networks and pervasive techniques, the HCI has to adapt into pervasive environments. Pervasive interaction is a form of HCI derived under the context of pervasive computing. This chapter introduces a pervasive interaction based planning and reasoning system for individuals with cognitive impairment, for their activities of daily living. Our system is a fusion of speech prompt, speech recognition as well as events from sensor networks. The system utilizes Markov decision processes for activity planning, and partially observable Markov decision processes for action planning and executing. Multimodal and multi-observation is the characteristics of a pervasive interaction system. Experimental results demonstrate the flexible effect the reminder system works for activity planning.

1 Introduction

In traditional dialogue management systems, human-computer interactions are tightly coupled with users' speech responses. An agent speaks a sentence to a user, and waits for the responses. After the user's speech is received and recognized, the system proceeds to the next interaction. Pervasive environments break through the constraint that a user has to interact with a single computer. The sensor network can also be an attendee of an interaction. After an agent asks a question or informs an affair, the user can optionally choose to reply or simply keep silence. Without the user's responses, the agent can still discover the user's intentions or preferences by reasoning information from the sensor network. This kind of human-computer interaction forms a pervasive interaction. In this system, the communication between human and machine does not constraint to a single computer, but it involves the entire computer networks, robots and sensor networks. How to build up applications for pervasive interactions becomes an important issue in pervasive computing, artificial intelligence and HCI.

A typical example of a single computer based interaction is the human-robot interaction. A robot is often designed as an autonomous agent, with all kinds of sensors equipped and a single "mind". For the pervasive interaction, we consider a virtual agent in the environment which has the power to utilize every sensing

T. Gulrez, A.E. Hassanien (Eds.): Advances in Robotics & Virtual Reality, ISRL 26, pp. 187–200.
springerlink.com   © Springer-Verlag Berlin Heidelberg 2012

ability of the pervasive environment. It provides an integrated user-interface for the entire pervasive environment. Further discussion about robot and virtual agent can be found in [3].

In a pervasive interaction framework, an intelligent agent does not simply recognize the user's speech. Instead, all kinds of events captured by sensor networks are accepted and analyzed by the system. Human activity recognition [4] becomes an important element for event capture. On the other hand, users do not have to focus their attentions on the interaction with a single computer. Now that the pervasive environment provides services as an integrated system, the user will be released from the tightly coupled interaction, without tampering the effect of the original "conversation". As a result, the interactions will be loosely coupled with individual devices. In fact, the more loosely coupled a functional HCI application with users, the more degree of freedom users will obtain, thereafter, it will be the more friendly and easier to maintain the interactive relationship.

This chapter provides a hierarchical multimodal framework for pervasive interactions. The application background is a reminder system interacting pervasively with humans, using decision-theoretic planning. We apply this technique on the planning system that reminds individuals with cognitive impairment, to support their activities of daily living (ADL).

2   Technical Preliminary

A Markov decision process (MDP) models a synchronous interaction of a user with a fully observable environment. It can be described as a tuple $\langle S, A, T, R \rangle$, where the S is a set of states, the A is a set of actions, the $T(s, a, s')$ is the transition probability from a state s to another state s' using an action a, the $R(s, a)$ defines the reward when executing an action a in a state s. By MDP, we want to find out the policy of every state, such that the overall planning is optimal.

A POMDP models an agent's action in uncertainty world. A policy for POMDP is a function of action selection under stochastic state transitions and noisy observations. At each time step, the agent needs to make a decision about the optimal policy based on historical information. A multimodal POMDP is an extensive form of POMDP that involves multiple types of observation sources. Although different types of observation sources share the observation space, their observation probabilities may be different. The traditional POMDP can be seen as a special form of multimodal POMDP that has only a single type of observation source.

A multimodal POMDP can be represented as $\langle S, A, \Theta, O, T, \omega, \Omega, R \rangle$, where the S is a finite set of states. In each time step, the agent lies in some state $s \in S$. The A is a set of actions. After taking an action $a \in A$, the agent goes into a new state s'. The Θ is the set of multimodal observation types, and the O defines the types of observations. The T is the set of transition probabilities. The conditional probability function $T(s, a, s') = p(s'|s, a)$ presents the probability that the agent lies in s', after taking action a in state s. The agent makes an observation to gather information. The observation result is a pair (θ, o), where $\theta \in \Theta$ is the observation type, and

$o \in O$ is the observation. This is modeled as a conditional probability $\Omega(s,a,\theta,o) = p(\theta,o|s,a)$. The ω is the probability an observation belonging to an observation type, $\omega(\theta) = p(\theta|event)$. In a system that is modeled as multimodal POMDP, we have the following conditions for different observation types,

(i) $\forall(s,a,o), 0 < \Sigma_{\theta \in \Theta} \Omega(s,a,\theta,o) \leq 2$;
(ii) $\forall(s,a,\theta), \Sigma_{o \in O} \Omega(s,a,\theta,o) = 1$.

When belief state is taken into consideration, the states of the multimodal POMDP are changed to belief states. The original partially observable POMDP model changes to a fully observable MDP model, denoted as $\langle B, A, \Theta, O, \tau, \omega, R, b_0 \rangle$, where B is the set of belief states, i.e. belief space. The $\tau(b,a,b') = p(b'|b,a)$ is the probability the agent changes from b to b' after taking action a. The $R(b,a) = \Sigma_s R(s,a)b(s)$ is the reward for belief state b. The b_0 is an initial belief state.

A POMDP framework is used to control an agent. The utility is a real-valued payoff to determine the action of an agent in each time step, denoted as $R(s,a)$, which is a function of the state s and the action a. The optimal action selection becomes a problem to find a sequence of actions $a_{1..t}$, to maximize the expected sum of rewards $E\left(\Sigma_t \gamma^t R(s_t,a_t)\right)$. In this process, what we concern is the controlling effect, achieved from the relative relationship of the values. After we use a discount factor γ, the relative relationship remains unchanged, but the values can guarantee to converge. If states are not fully observable, the goal becomes maximizing the expected reward of each belief. The nth horizon value function can be built from previous value V_{n-1}, using a *backup* operator H, i.e. $V = HV'$. The value function is formulated as the following Bellman equation.

$$V(b) = \max_{a \in A}[R(b,a) + \gamma \sum_{b' \in B} \tau(b,a,b')V(b')].$$

Here, b' is the next step belief state,

$$b'(s) = b_t(s')$$
$$= \eta \Omega(s',a,\theta,o)\Sigma_{s \in S}T(s,a,s')b_{t-1}(s),$$

where η is a normalizing constant.

When optimized exactly, this value function is always piece-wise linear and convex in the belief space.

For the transition probabilities of belief points, we have

$$\tau(b,a,b') = \Sigma_{\theta \in \Theta, o \in O}\left[\Omega(b,a,\theta,o)T(b,a,\theta,o,b')\right]$$
$$= \Omega(b,a,\theta,o),$$

where $T(b,a,\theta,o,b') = 1$ if $SE(b,a,\theta,o) = b'$, $T(b,a,\theta,o,b') = 0$ otherwise. SE is an algorithm specific belief state estimation function.

We can find out $\Omega(b,a,\theta,o)$ by the following processes. First we compute b', where $b'(s) = b(s') = \eta\Sigma_{s \in S}b(s)T(s,a,s')$, where η is a normalizing factor. By $\Omega(b,a,\theta,o,s) = \eta b'(s)B(\theta,o,s)$, we can finally obtain the transition probability for $\tau(b,a,b')$.

3 Pervasive Human-Computer Interaction

3.1 Sensor Coverage in Functional Areas

Different sensors can be deployed in the functional areas of a pervasive environment. A simple but effective deployment is to put sensors in each functional area, such that the field of interest is detected. As a result, we build up an activity recognition system for the pervasive environment. The detection of activities becomes an identification of sensor tag as well as time stamp. The activity for time t becomes a posterior probability for the sensor with maximum instant change, between the measurement in time t and the measurement in time $t-1$. For time t, let y_t be the activity, x_t be the label of sensor, the probability for the activity is given by $p(y_t|x_t,y_{t-1})$.

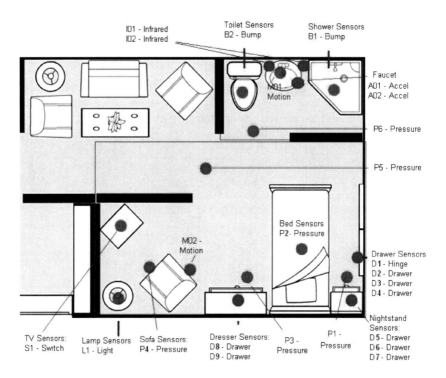

Fig. 1. Sensor Deployment in A Pervasive Environment

Difficulties of activity recognition come from concurrent activities and noises from sensor readings. When participants watch TV, they may go to the kitchen and do some washes. During the wash activity, they may return back to answer a call. These concurrent activities may result in a confusion of the activity recognition system. Existing research about the activities falls into several categories. One is the Bayesian inference, to recognize the activities [7]. Another is the classification techniques and vision-based analysis. These are often utilized to recognize human's affective state, physiology and behavior. Further research involves identifying the human's intentions for the activities, using hidden Markov models [5, 4] or conditional random fields [13]. If there are several persons in a home, the identification of activities about a specific person becomes even harder. We have to rely on face recognition or speaker recognition to differentiate the persons.

3.2 Event System and Multi-observation

Activity recognition creates events about users' activities. Every event is a segment of information about the user's activity, including activity name, sensor label, sensor type, time stamp. The sensor types for activity recognition can be accelerometer, infrared, light, pressure, video and audio. The benefit of sensor based activity recognition relies on the fact that, events are automatically detected by sensor networks, and users do not have to explicitly inform their intentions. Therefore, this is a passive interaction between humans and pervasive environments.

On the other hand, users' intentions can also be captured by speech interaction, which is a more direct interaction than the recognition of activities. From the perspective of users, it seems like an intelligent assistant lives and interacts with them. This makes the pervasive environment more friendly, especially for the elders that are living alone, the speech interaction is a necessary component. A defect of the speech interaction is that it has to rely on an explicit communication with the user. Dialogue events come from the speech interaction. In the pervasive interaction, we can simply put it as an audio event, with a sensor-label of microphone.

The combination of sensor events and dialogue events provides a friendly and integrated interface. People do not have to respond to prompts from a dialogue management agent. The events from sensor networks make up the missing information. If people would like to interact with the agent, they can also choose to "tell" the system about their activities and intentions directly. Figure 2 describes a scenario of the loose coupling pervasive interaction. When an agent prompts a user to take

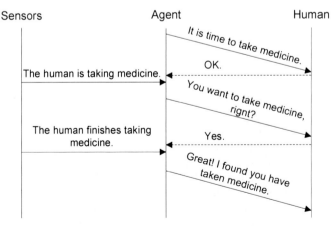

Fig. 2. A Scenario for Loose Coupling Pervasive Interaction (Remind the User to Take Medicine): dashed-lines indicate optional speech response

medicine, the user can choose to reply or not. As soon as the user takes medicine, the agent will learn the states of the user's activities by analyzing events from sensor networks. On the contrary, if it is an autonomous robot, the human's responds (marked as dashed-line in the figure) normally should not be omitted. Since the interaction is tightly coupled with the robot, the human has to make explicit respond to every scenario of the dialogue. The difference of a pervasive interaction from a single-agent dialogue is, the pervasive interaction adapts to the environments exactly, rather than adapting to the powerful function of a single agent. Thereafter, a pervasive interaction can provide a more friendly interface to users. It makes the pervasive environment to be an integrated intelligent system.

We can put the events of a pervasive interaction system into different categories, with every category belonging to a set of observations, and every observation comes from a set of noisy sensors. The application system has to identify, analyze, correlate and reason for the observations. We summarize the features of pervasive interactions as follows:

Adaptation Pervasive interaction makes users adapt to the environments perfectly;
Redundance Pervasive environments provide redundancy interfaces to sense and track the users' activities and intentions;
Cross-Reference Intelligent systems using pervasive interactions need to have the ability to learn the information by cross-reference of multiple interfaces.

4 Hierarchical Multimodal Markov Stochastic Domains

A reminder is a planning system to prompt and remind the ADL of individuals with cognitive impairments. The system requires a multimodal partially observable framework that concerns both the events from speech recognition, and the events from sensor networks. We organize the system in two levels: MDPs based activity planning, and multimodal POMDPs based action planning. An action control component is used to manipulate the reminding tasks. Figure 3 provides an overview of the multimodal planning agent's working scheme. We introduce the components separately in the following parts.

4.1 Activity Planning by MDPs

The activity planning is used to control the states of human activities. Every activity is represented as a state of the MDP model. We have the followings reasons to adopt MDP rather than POMDP for the activity planning:

(i) The uncertainty of multimodal observations is considered in a lower action planning level, instead of the level of activity planning;

(ii) Since every termination state of action planning can be determined, we can distinct the states in activity planning.

Figure 4 is a *Task State Navigation* (TSN) graph to demonstrate some example activities for the reminder system. A TSN graph contains a set of grids, with each grid represents a state of the task, labeled by the state ID. Neighboring

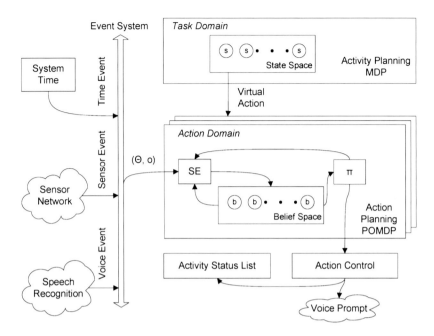

Fig. 3. Hierarchical Multimodal Framework for Reminder Planning System

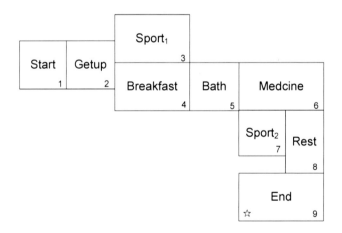

Fig. 4. Activities for Reminding Tasks

grids with ordinary line indicate there is a transitional relationship among these states. The reminding tasks start from the activity *Getup*, and end in *Sport₂* or *Rest*, whose grids are labeled with a star. For most of the states, it is certain to transmit to the next state, $T(s,a,s') = 1$. As discussed above, the transitions from *Getup* and *Medicine* to the next states are both according to the transition probabilities. We have, $T(Getup, a, Sport) = 0.5$, $T(Getup, a, Breakfast) = 0.5$, $T(Medicine, a, Sport) = 0.5$ and $T(Medicine, a, Rest) = 0.5$. The reward for state *End* is 10, and the rewards for other states are all 0.

We have only one action for the transition of different states. This is an abstract action, to indicate the end of an activity, and the start of next activity. In fact, the abstract action a for a state s is determined by the action planning subtask, which is related with the activity s. When the action planning subtask reaches the *End* state, it implies the abstract action is executed.

4.2 Action Planning by Multimodal POMDPs

Action Planning Subtask

For each activity of the reminding task, the system needs to undertake a subtask, to guide the user to execute the activity. It is an abstract action that is unrelated with the actual prompting speech of the system. The prompts are determined in the subtasks of each activity. We put the execution of different subtasks to a unifying form. These subtasks accept partial observations from sensor events as well as speech recognition events. It forms two sets of noisy observations, the sensor observations and the dialogue observations. Thus the system structure becomes a multimodal POMDP.

The TSN graph for the multimodal POMDP is presented in Figure 5. For each activity, there are six states in the action domain. For the initial state $TimeToActivity_i$, there are three possible transitions, including a state $Activity_{1:i-1}$ for previous activities, a state $Activity_{i+k+1:n}$ for future activities, and a state $Activity_{i+j}$ for current activity. If the system assumes the user is taking activity $Activity_{i+j}$, it may accept the observation from previous activities, or future activities. Due to the noises of sensors and speech recognition, these observations may be correct or not. Thus, the transitions from the state $Activity_{i+j}$ to $Activity_{1:i-1}$ and to $Activity_{i+k+1:n}$ are both small probabilities. The transitions from the state $Activity_{i+j}$ to $FinishActivity_{i+j}$

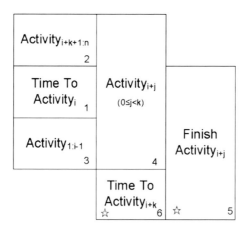

Fig. 5. The Action Domain TSN Graph for Every Activity

and to $TimeToActivity_{i+k}$ have higher probabilities. The state to which it transmits is related with the current observation.

Since both $TimeToActivity_{i+k}$ and $FinishActivity_{i+j}$ are termination states in the action domain, when the system reaches them, it will transmit to the next activity in the activity domain MDP. However, in the state $TimeToActivity_{i+k}$, we cannot determine whether or not the activity is finished. Only in $FinishActivity_{i+j}$, will the system record a status of $Activity_{i+j}$.

There are three activity states, $\{unfinished, executing, finished\}$. We establish a list to record the status of every activity in a day. This will be helpful for the system to find out unfinished activities, for future data analysis, or for reminding the user directly.

Multimodal POMDP Framework

Details of different activities are defined in a POMDP framework. Each state is related with two actions, *Prompt* and *Wait*. The *Prompt* indicates the planning system needs to provide prompt to the user. The actual content of a *Prompt* is determined by an action control, which we will introduce in next part. The *Wait* means the system takes an action of waiting.

The number of observation types (denoted as $|\Theta|$) is 2, representing dialogue events and sensor events. There are $|S|$ different observations in observation space, i.e. $|O| = |S|$. The dialogue events and sensor events are mapped into observations of different types. In fact, even if the system takes the action of *Wait*, it still has a transition probability to next states. The states to which the system transmits can be determined by the observations. Normally, the observation probabilities of dialogue events are lower than that of the sensor events, and it will decrease with the increase of age for the elderly.

The rewards are assigned as follows. For the termination states,

$$R(TimeToActivity_{i+k}, \cdot) = 1,$$

and

$$R(FinishActivity_{i+j}, \cdot) = 2.$$

In multimodal POMDP, the $\omega(\cdot)$ is a Bayesian statistic for an event belonging to an observation set. We have $\omega(x) = p(x|event)$, where x is an observation set. For other states, $R(\cdot, Wait) = 0$, and $R(\cdot, Prompt) = -0.1$.

4.3   Action Control for Reminding Tasks

The action control helps to determine atomic actions of the planning system. Since the reminder system has complex task structure and actions, the atomic actions are not correlated strictly with the actions in multimodal POMDP subtasks. Thus, we can incorporate the benefit of POMDP framework, and constrain the state space to the core part of controlling tasks. We separate the atomic actions into two categories.

The first class atomic actions are related with the actions of multimodal POMDP subtasks. Let the atomic action be $\mathfrak{a}$, we have

$$\mathfrak{a} = f(s_{activity}, s_{action}, a_{action}),$$

where $f(\cdot,\cdot,\cdot)$ is the combination function, to merge different states and actions into the text of speech. Considering the concrete actions, we have

$$\mathfrak{a} = \begin{cases} Wait, & a_{action} = Wait \\ f(s_{activity}, s_{action}, Prompt), & a_{action} = Prompt. \end{cases}$$

Thereafter, the actions in multimodal POMDP are translated to atomic actions in the reminder system, represented as the text of speech. Here is an example of the combination. Suppose $s_{activity} = Medicine$, $s_{action} = TimeToActivity_i$, and $a_{action} = Prompt$, the atomic action will be a prompt like "It is time to take medicine".

The second class atomic actions are related with the states of multimodal POMDP subtasks. Table 1 provides the correlations of the states and the second class atomic actions. These atomic actions are used to update the status types for the activities.

Table 1. Correlations for States and the Second Class Atomic Actions

ID	State	Atomic Action
1	$TimeToActivity_i$	none
2	$Activity_{1:i-1}$	update status for the activity identified from the event
3	$Activity_{i+k+1:n}$	none
4	$Activity_{i+j}$	none
5	$TimeToActivity_{i+k}$	$S(Activity_{i+k}) = unfinished$
6	$FinishActivity_{i+j}$	$S(Activity_{i:i+j-1}) = finished$

5   Experimental Evaluation

The experiments are taken in a simulation environment. Seven activities are used as an example of the daily activities (Figure 4). We have discussed other settings in previous sections. We want to find out effects of the reminding tasks.

Figure 6 demonstrates learning curve of the average reward in each time step, for the multimodal POMDP. The average reward converges to 1.3.

We made a statistic of the actions in multimodal POMDP subtasks (Figure 7). The activity state 1 represents the state $TimeToActivity_i$, and the activity state 4 represents $Activity_{i+j}$. Both states are important for the reminding subtask. The time for executing an activity is often longer than preparing the activity. This effect is reflected in the action assignments of the planning tasks. We have two actions in every state, *Prompt* and *Wait*. In the state $TimeToActivity_i$, the actions of *Prompt* are about 2/3 of all actions, and *Wait* actions occupy the rest 1/3. This is reversed for the state $Activity_{i+j}$. During the activity, the reminder system will choose *Wait* about 2/3 of all actions, only 1/3 of them are *Prompt*. This indicates our multimodal POMDP planning for the reminder system provides reasonable prompts for the user, because the system does not have to prompt many times during an activity.

Figure 8 provides a further statistic of the *Prompt* actions for different activities. For the activity state 1, $TimeToActivity_i$, there is not a single time the system does not prompt for the user. For about 70% of the activities, the system only prompts once. This percent decreases when the number of prompts increases. Only very few

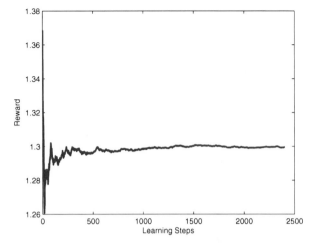

Fig. 6. Learning Curve for the Multimodal POMDP

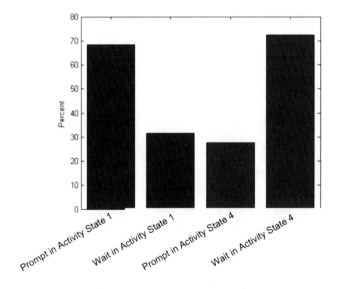

Fig. 7. Analysis of the Actions in Multimodal POMDP Subtasks

occasions the system prompt more than 4 times. This is comparable with the traditional constant of prompting tasks in a reminder system. The multimodal POMDP controls the system to adapt to multi-observation by which the system interacts with the user. The actions of the planning processes are more like a probability distribution, rather than a constant. For the state 4, $Activity_{i+j}$, there exists about 15% activities, in which the system goes to the next activity directly, without *Prompt*. For about 39% activities, the system prompts once for the user, and about 43% activities twice. When there is no prompt action, the system will perform *Wait* in every time step.

The distribution of prompts is determined by the observations accepted from the multimodal readings. The user can choose to respond the system by speech or just keep silence. Without speech responses, the user's action is determined by sensor readings. This is the reason why the number of prompts in the reminder system is not constant. These results indicate the planning system helps us build a flexible mechanism for the reminding tasks, and it is a suitable design for the pervasive interaction based reminder systems.

6 Related Work

A plan management technology are used to model an individual's daily plans in [11, 10]. The authors consider reasoning in execution of the plans, so as to make flexible and adaptive decisions for reminding tasks. A decision-theoretic planning based on POMDP is proposed to assist people with dementia to wash hands [1]. A

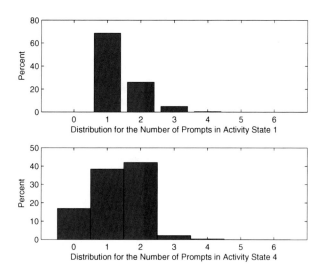

Fig. 8. Statistic for the Prompt Action in the Reminder System

time-state aggregated POMDP method is proposed for the planning of human-robot interaction [2].

The development of multimodal and human-friendly HCI is a hot topic for assistive technology [6]. In this chapter, we further discuss framework of the planning system for pervasive interaction.

Multi-task management is a challenging topic in Markov stochastic domains. It has been considered as a dynamically merging problem for MDPs [12]. Another approach is the hierarchical POMDP [9, 8]. In these works, different actions are organized in the hierarchical structure of POMDP for the interaction. The designers implement the models as hierarchical decomposition of abstract actions, until every action becomes atomic. Our system adopts the idea of hierarchical design. The task domain is modeled as MDP, whereas the activity domain is modeled as POMDP.

Another related work tries to automatically construct HPOMDP, through data-mining techniques on collected data [14]. It assumes that a collection of data may contain many repetitive patterns. If an algorithm is provided some preliminary about some given patterns, then it can take a data-mining on more data, to discover interesting pattern, such as a permutated of then original pattern. The authors want to make use of a Hidden Markov model to analyze this information, to help to construct the model of HPOMDP. Designing a learning approach to build up a Markov automata belongs to a stand-alone topic. Although the approach is interesting, we have to build up well about the pre-requisite conditions. Generally, researchers often rely on a simple Bayesian analysis to model POMDP.

7   Conclusions

Designing an integrated, friendly and loose coupling user interface is important to pervasive environments. The pervasive interaction considers not only dialogue events, but also sensor events. The observations from both of these events are used to build up the belief space of a planning system. The reminder is a typical system for pervasive interaction. It is used to remind people with cognitive impairment, about the activities of their daily livings. We model the reminding tasks in MDPs. We normalize the reminding task of every activity so that it conforms to a multimodal POMDP model. Solving the hierarchical multimodal Markov stochastic domains becomes a solution of the pervasive interaction based reminder system.

Future works about the reminder system involve the design of more detail activities and the evaluation of the hierarchical multimodal framework in real apartment pervasive environments.

References

1. Boger, J., Poupart, P., Hoey, J., Boutilier, C., Fernie, G., Mihailidis, A.: A decision-theoretic approach to task assistance for persons with dementia. In: Proceedings of IJ-CAI, pp. 1293–1299 (2005)
2. Broz, F., Nourbakhsh, I.R., Simmons, R.G.: Planning for human-robot interaction using time-state aggregated POMDPs. In: Proceedings of AAAI, pp. 1339–1344 (2008)

3. Holz, T., Dragone, M., O'Hare, G.M.P.: Where robots and virtual agents meet. I. J. Social Robotics 1(1), 83–93 (2009)
4. Kawanaka, D., Okatani, T., Deguchi, K.: HHMM based recognition of human activity. Transactions Institute Elec. Info. and Comm. Eng. E89-D(7), 2180–2185 (2006)
5. Kelley, R., Tavakkoli, A., King, C., Nicolescu, M., Nicolescu, M., Bebis, G.: Understanding human intentions via hidden markov models in autonomous mobile robots. In: Proc. of the 3rd ACM/IEEE International Conference on Human Robot Interaction, pp. 367–374 (2008)
6. Lee, H.-E., Kim, Y., Park, K.-H., Bien, Z.Z.: Development of a steward robot for human-friendly interaction. In: Control Applications (CCA), pp. 551–556 (2006)
7. Mühlenbrock, M., Brdiczka, O., Snowdon, D., Meunier, J.-L.: Learning to detect user activity and availability from a variety of sensor data. In: Proceedings of the Second IEEE International Conference on Pervasive Computing and Communications, p. 13 (2004)
8. Natarajan, S., Tadepalli, P., Fern, A.: A relational hierarchical model for decision-theoretic assistance. In: Blockeel, H., Ramon, J., Shavlik, J., Tadepalli, P. (eds.) ILP 2007. LNCS (LNAI), vol. 4894, pp. 175–190. Springer, Heidelberg (2008)
9. Pineau, J., Roy, N., Thrun, S.: A hierarchical approach to pomdp planning and execution. In: Workshop on Hierarchy and Memory in Reinforcement Learning, ICML (2001)
10. Pollack, M.: Intelligent technology for an aging population: The use of AI to assist elders with cognitive impairment. AI Magazine 26(2), 9–24 (2005)
11. Pollack, M.E.: Planning technology for intelligent cognitive orthotics. In: Proceedings of the Sixth International Conference on Artificial Intelligence Planning Systems (AIPS), p. 322 (2002)
12. Singh, S., Cohn, D.: How to dynamically merge markov decision processes. In: Proceedings of NIPS, Denver, Colorado, United States, pp. 1057–1063 (1998)
13. Vail, D.L., Veloso, M.M., Lafferty, J.D.: Conditional random fields for activity recognition. In: Proceedings of AAMAS, Honolulu, Hawaii, pp. 1–8 (2007)
14. Michael Youngblood, G., Heierman III, E.O., Cook, D.J., Holder, L.B.: Automated hpomdp construction through data-mining techniques in the intelligent environment domain. In: FLAIRS Conference, pp. 194–200 (2005)

De-SIGN: Robust Gesture Recognition in Conceptual Design, Sensor Analysis and Synthesis

Manolya Kavakli and Ali Boyali

Virtual and Interactive Simulations of Reality (VISOR), Research Group
Department of Computing, Faculty of Science,
Macquarie University Sydney, NSW 2109, Australia
{ali.boyali,manolya.kavakli}@mq.edu.au

Abstract. Designing in Virtual Reality systems may bring significant advantages for the preliminary exploration of the design concept in 3D. In this chapter, our purpose is to provide a design platform in VR, integrating data gloves and the sensor jacket that consists of piezo-resistive sensor threads in a sensor network. Unlike the common gesture recognition approaches, that require the assistance of expensive devices such as cameras or Precision Position Tracker (PPT) devices, our sensor network eliminates both the need for additional devices and the limitation of mobility. We developed a Gesture Recognition System (De-SIGN) in various iterations. De-SIGN decodes design gestures. In this chapter, we present the system architecture for De-SIGN, its sensor analysis and synthesis method (SenSe) and the Sparse Representation-based Classification (SRC) algorithm we have developed for gesture signals, and discussed the system's performance providing the recognition rates. The gesture recognition algorithm presented here is highly accurate regardless of the signal acquisition method used and gives excellent results even for high dimensional signals and large gesture dictionaries. Our findings state that gestures can be recognized with over 99% accuracy rate using the Sparse Representation-based Classification (SRC) algorithm for user-independent gesture dictionaries and 100% for user-dependent.

1   Introduction

This chapter examines the development of a VR system (De-SIGN) in which a designer can define the contour of a sketch by controlling a pointer using a pair of data gloves and can interact with the design product by using a sensor jacket in VR space. The sensor jacket and data gloves incorporate 3D position sensors so that drawing primitives entered are recreated in real time on a head mounted display worn by the user. Thus, the VR system provides a " 3D sketch pad ". The interface recognizes hand-gestures of the designer, pass commands to a 3D modelling package via a motion recognition system, produce the 3D model of the sketch on-the-fly, and generate it on

T. Gulrez, A.E. Hassanien (Eds.): Advances in Robotics & Virtual Reality, ISRL 26, pp. 201–225.
springerlink.com   © Springer-Verlag Berlin Heidelberg 2012

a head mounted display. Therefore, the project requires the integration of existing technologies in motion tracking, 3D modelling and VR, in order to design and implement a novel interface to support the design process. The implementation of 3D Sketchpad requires both technical and conceptual developments to allow natural interactions with the virtual environments. In this chapter, our focus is the technical developments.

The remainder of this chapter is organized as follows. In Section 2 We first review the motion capture technologies and discuss the strengths and weaknesses of the piezo-electrical sensing garment technology to be used in gesture recognition. We will give a sensor network system architecture for the gesture recognition system, De-SIGN for designing in VR space in Section 3. Then, we will present a novel gesture and posture recognition algorithm (SRC) we have developed for the wearable sensing garment having piezo-resistive properties, and demonstrate its accuracy providing the test results in Section 4. The chapter ends with conclusion and future work remarks.

2   De-SIGN and Motion Capture

2.1   De-SIGN System Overview

In this chapter, our purpose is to explore the development of a design platform (De-SIGN) in VR, integrating data gloves and a sensor jacket in a sensor network. Our methodology is based on the previously developed two gesture recognition systems: Designing In virtual Reality (DesIRe) [1] (Fig. 1) and DRiving for disabled (DRive) [2]. DesIRe investigated the possibility of real time dynamic interaction with a 3D object using a 5DT data glove equipped with an LED. DRive allows a quadriplegic person to control a car interface taking input from 2 LEDs and shoulder movements only. Both systems require a precision position tracking (PPT) device which is costly and causes numerous limitations for the designer. Our approach introduces a new perspective to designing in Virtual Reality combining the sensor jacket with the data gloves, allowing the position tracking of the hands with reference to the body, eliminating the need for an additional expensive optical tracking device, hence improving system's mobility and usability. Our motivation is based on the previous work of Tognetti et al.[3] which describes the basics of a sensor shirt that has also been used by Gulrez [4] for re-mapping the residual mobility of spinal cord injured patients to control a robotic wheelchair.

We planned to develop our gesture recognition system using a Sensor Jacket that operates on piezo-electrical sensors. The Sensor Jacket was produced at the Electronic Engineering Department of University Pisa, Italy [3, 5]. Since this is not a commercial product, it does not have any associated software to develop an interface for gesture recognition. Therefore, a comprehensive analysis for the sensor jacket system was required. First, we

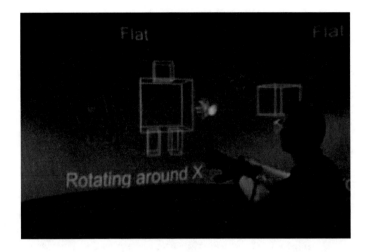

Fig. 1. Designing in virtual reality (DesIRe)[1]

conducted a set of experiments to interface the sensor jacket with a neural network proposed for gesture recognition and investigated the sensors responsible for a set of movements to be able to generate a sensor model that can be used to map the raw sensor readings into a number of gestures.

Our initial studies asserted that using only basic mathematical functions, it is not possible to create an appropriate mathematical model suitable for gesture recognition, mainly because of high nonlinearity of the signals [6]. However, our attempt for the development of an initial mathematical model and the training of an Artificial Neural Network (ANN) gave us insights for the development of a novel algorithm. We found that although the results produced by the algorithms and the ANN we have initially developed were not very precise, it still provided general information about the gestures and served as a good starting point for the further algorithm development. Following this, we developed a robust gesture and posture recognition algorithm based on an emerging research field in signal processing Compressed Sensing (CS) and Sparse Representation (SR), for the sensor jacket. The gesture recognition algorithm presented here is highly accurate regardless of the signal acquisition method used and gives excellent results even for high dimensional signals and large gesture dictionaries. Our findings state that gestures can be recognized with 100% user-dependent gesture recognition accuracy and over 99% user-independent gesture recognition accuracy rate using the Sparse Representation-based Classification (SRC) algorithm. We tested the algorithm using another gesture set collected by Infra-red (IR) camera of Wiimote of Nintendo gaming console. The system misclassified only 2 gestures out of 300 test samples, corresponding to 99.33% user-independent gesture recognition accuracy for 20 gestures in the gesture dictionary.

2.2 Motion Capture in Conceptual Design

The importance of freehand sketching in visual discoveries was readily accepted in the early 90s [7, 8, 9]. However, approaches to sketch recognition have generally treated the sketch as an image (e.g. FABEL project, [10]), and discarded information about the underlying cognitive processes and actions. Approaches that used properties of drawing production (e.g., pressure of pen, speed of drawing action, etc.) for recognition purposes include those of [11], [12] with the ROCOCO SketchPAD, and [13] with the COCKTAIL NAPKIN. Such systems are still rarely used in the idea generation in design, since they are not integrated with the demands of visual cognition. DARPA supported a system to capture knowledge via sketching called sKEA (Sketching for Knowledge Capture) [14]. sKEA uses semantics, spatial reasoning and arrows to express relationships, and analogical comparison to explore similarities and differences between sketched concepts in its limited knowledge base. The National Science Foundation (NSF) supported the Space Pen [15], which is a web annotation system to comment on virtual 3D models using gesture recognition. However, none of these systems is complex enough to support visual recognition.

Recent developments in gesture recognition led to a number of important innovations: Mulder [16] reviewed human movement behaviour in the context of hand centred input for gesture recognition. Pederson [17] proposed Magic Touch as a natural user interface that consists of an office environment containing tagged artifacts and wearable wireless tag readers placed on the user's hands. Sacco et al. [18] developed VRSHOE in which the designer can draw or modify, directly on the model and the style lines previously created in CAD, can fly-through the environment and directly interact with the model using immersive interface devices. Liu et al. [19] have started working on drawing lines in footwear design taking advantage of the VR. Bowman and Bilinghurst [20] attempted to develop a 3D design sketchpad for architects, however, the 3D interface utilizing pull-down menus did not respond to the expectations of architects.

2.3 Motion Capture Technologies

A large variety of input devices are available to interact with the objects located in a VR space, such as Head Mounted Displays, Data Gloves, Motion Capture Suits, etc. These devices give an opportunity to see, feel and control the VR space. Gesture recognition and motion tracking are the key themes in VR and the researchers have developed various techniques for this purpose. Some researchers have used video tracking of certain features of human body such as the skin colour for recognition and tracking of the hands [21]. However, this requires defining certain skin colours to be recognized and can lead to errors, if the skin has been deformed or has an unexpected colour for any reason. It is also required that the hand should be clearly seen by the cameras at all times. Therefore, in video tracking, the light in the room, background, foreground and user's clothes should all be selected carefully and the users are not allowed to move in a way that they can partly or fully cover their hands anyhow.

Another option is wearing data gloves and detecting the position of the hands using a position tracking device. The user either puts on some LEDs on the data gloves as in [1, 2] or some markers on themselves in order to be seen by the tracking device which limits the mobility [22]. The disadvantage of these approaches is the cost of motion tracking devices. Data gloves have already been used intensively in VR systems. Yanbin et al. in [23] proposed a solution for the lack of hands-on experience in traditional midwife training, where they modelled a pregnant woman and a baby and simulated the birth process using the data gloves as a virtual hand. Minoh et al. in [24] drew attention to an important issue that arises while using data gloves without any feedback whilst grasping objects. In their system, when the virtual hand replaces the data glove, its posture is adjusted so that it grasps the virtual props without misalignment.

Among the wide range of tools used in human-computer interaction (HCI), Motion Capture Suits constitute one of the most complex input devices. The functioning of Motion Capture Suits is based on various types of sensors: e.g., gyroscopic, magnetic, and opto-electrical. To perform gesture recognition, one must be able to transform the body-movement to a format recognisable by a computer. In a motion capture suit, this transformation is performed with the aid of special sensors that convert the analogue signal (the movement itself) into electrical and then into digital form. Such sensors may include gyroscopic sensors, light emitting-collecting sensors, mechano-electrical sensors and others [26]. Sensor Jacket can be used to aid quadriplegic people to control a wheelchair using the redundant muscles in their body [25] or to assist people with disabilities to drive a car [2].

There are three types of motion capture technologies we compared to use in this project:

- *Optical Motion Capture.* Optical systems utilize data captured from image sensors to triangulate the 3D position of a subject between one or more cameras calibrated to provide overlapping projections. Data acquisition is traditionally implemented using special markers attached to a user; however, more recent systems are able to generate accurate data by tracking surface features identified dynamically for each particular subject. These systems produce data with 3 degrees of freedom for each marker, and rotational information must be inferred from the relative orientation of three or more markers; for instance shoulder, elbow and wrist markers providing the angle of the elbow. Usually it is divided in two main subclasses:
  - Image processing using two infrared cameras [27]
  - Tracking of articulated pose and motion with a markerized grid suit [28]
- *Inertial Motion Capture.* Inertial Motion Capture technology is based on miniature inertial sensors, biomechanical models and sensor fusion algorithms. The motion data of the inertial sensors is often transmitted

wirelessly to a computer, where the motion is recorded or viewed. Performance of such mechanisms is based on the measures of changes in gyroscopic forces (acceleration and inclination). These rotations are translated to a skeleton in the software. Inertial mocap systems capture the full six degrees of freedom body motion of a human in real-time.

- *Mechanical Motion Capture.* Mechanical motion capture systems directly track body joint angles and are often referred to as exo-skeleton motion capture systems, due to the way the sensors are attached to the body. Performers attaches the skeletal-like structure to their body and as they move so do the articulated mechanical parts, measuring the performer's relative motion. Mechanical motion capture systems are real-time, relatively low-cost, free-of-occlusion, and wireless systems that have unlimited capture volume. Typically, they are rigid structures of jointed, straight metal or plastic rods linked together with potentiometers that articulate at the joints of the body.

All technologies and devices described above have their own positive and negative characteristics. Each of them found its application in its unique area. Qualities of these commercial systems employing different technologies are summarized in Table 1.

2.4   Motion Capture Systems for De-SIGN

Optical motion capture system

In recent years, with the rapid profileration of Micro Electro-Mechanical Systems (MEMS), the sensor technologies have become an indispensable part of our daily life. Mobile computing and communication products are getting smarter and pervasive, containing MEMs and tiny cameras. In our recent application, we investigated building an inexpensive optical motion sensing system by using Nintendo Wiimote [29].

Wiimote is the remote control of Nintendo game console that allows the user to manipulate objects in a virtual environment Fig. 2.

There are several Matlab toolboxes available for computer vision signal processing. Using the IR camera feature of Wiimote we developed 6 DOF (Degrees of Freedom) motion tracking system (MaTRACK version 1) [29] to wield a virtual object by hand and head tracking Fig. 3 In the single camera studies, Matlab camera calibration toolbox [30] is used to compute camera intrinsic parameters. We also defined Kalman and Butterworth filter parameters for optical tracking, pose and orientation estimation using a single Wiimote.

We utilized Matlab for engineering computations, however all of the algorithms designed in Matlab were deployed in Vizard Virtual Reality Engine for virtual animations. Following successful deployment of MaTRACK version 1, we developed a multi-camera motion tracking system (MaTRACK

Table 1. Positive and negative characteristics of various motion capture technologies

Motion Capture Tech.	Positive	Negative
Optical	-Rather precise - Not restrictive -Special costume not always required	Require special environment; clear line of view; special cameras; and calibration; -May require special costume
Inertial	-Precise -Sufficient data for modelling -Not restrictive	-Rather expensive -Fragile -Calibration required -Accumulated error (Drift)
mechanical	-Very precise -Almost no latency -Relatively inexpensive	-Heavy and bulky -Serious restriction on mobility -Require time to put on and calibrate

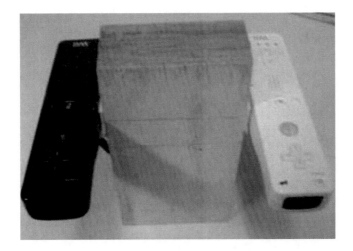

Fig. 2. Wiimote stereo system

version 2) using the Wiimotes to capture the position of the tracked IR precisely. Multi-view setups are also necessary to overcome the shortcomings of single camera systems, such as clear line of view and estimation of location of missing markers, when they are not seen by other cameras. For these purposes, open source Matlab epipolar geometry toolbox [31] including fundamental matrix computation algorithms is available.

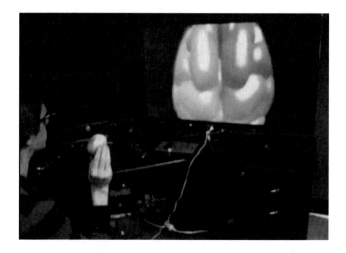

Fig. 3. Manipulation a virtual brain model in VR Enverionment

The point locations in the stereo image pairs are related to fundamental matrix in the epipolar geometry. The epipolar geometry is independent from external scene parameters and depends on the intrinsic camera parameters [32]. In Fig. 4 square marks are the point locations reported by left Wiimote in the stereo rig, wheras the lines across these points are the computed epipolar lines from the point locations reported by the right one.

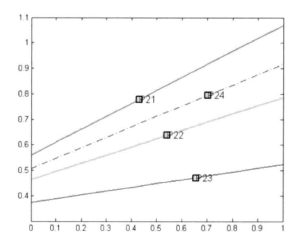

Fig. 4. Point locations seen by left Wiimote and Epipolar lines computed from the image on the right Wiimote

The use of additional cameras provides tracking system with the ability to locate the missing markers. Stereo and multi-view geometry also provide the ability to computedepth in VR space from 2D image coordinates by using well-known triangulation methods. In our stereo-vision studies we reach less than one mm accuracy in the range of one meter from the camera coordinate systems.

Piezo-Electrical sensor jacket

In this section a broader and deeper description of the piezo-electrical motion capture technology is discussed, as well as the difficulties and technical problems faced during the sensor analysis phase. Although the piezo-electrical Sensor Jacket [3, 5] document contains descriptions of all basic physical dependencies, there is no software provided with the sensor jacket hardware to support mapping the signals to gesture and posture domain. Tognetti et al [5] define posture as a geometrical model of body kinematics. According to the definition, when the sensor jacket is worn by the user and a posture is realized, the sensor network produces electrical signals strictly related to the posture. The non-linear signal behaviour of the network is modelled by the linear combination of exponential functions. The construct (a mapping function F) between the sensor space and kinematic configuration readings, and corresponding measurements by using a goniometer are stored in a database, and sensor readings are mapped using multi-variate piece-wise interpolation and function inversions. The proposed method in [3] is considered to be time-consuming since a high number of matrix inversions are necessary.

In this chapter we intend to give clear descriptions of the physical processes taking part in generating the output data and studies on signal processing for gesture and posture recognition. Piezo-electrical effect or just piezo effect is a phenomenon of changes in the material's electrical qualities when it is stretched or under pressure. Operation of the piezo-electric sensors is based on the fact that the impedance of the material changes when the deformation forces are applied to it.

The sensor jacket is realized smearing a mixture of graphite solution on an elastic fabric substrate [3]. Wearable sensing garments produced by this technology have some advantages and disadvantages over other motion capture technologies. These are as follows:

Positive qualities:

- It is wearable and has low weight
- It does not accumulate static errors
- It is not obtrusive
- It fits any body proportions
- It is relatively inexpensive and can be produced in laboratory conditions
- It does not require any special environment and clear line of view

Negative qualities:

- It is not wireless, however by adding a simple signal communication circuit, the captured data can be transmitted over a wireless protocol
- The time response of electrical characteristics of the sensor is nonlinear and depends on the deformation speed. Moreover, after it stretches, time elapsed for returning the steady state position in terms of its electrical characteristics, even though the threads return to their original unstretched lengths
- Hence the same gesture, performed with different speeds, gives different outputs which are not only different in peaks and valleys but also in the wave forms

The sensor jacket brings considerable advantages as an input device for our conceptual design studies, therefore, we have implemented various signal processing methods to prepare the signals acquired from its sensor threads for gesture and posture recognition algorithms.

In the classification of signals, if there is no analytical solution available, systems are treated as a black box and for this approach Artificial Neural Networks (ANN) are highly successful. We intended to train an ANN for gesture recognition and conducted experiments with the trained ANN. The input data for training the ANN were collected using the data gloves and the HMD. These input set was not sufficient for training the ANN thus the ANN gave poor results in the gesture recognition studies.

In the following sections we present a brief description for the sensor fusion and network structure and then present a novel method in gesture and posture recognition based on a hot topic in signal classification Compressed Sampling (CS) and Sparse Representation (SR).

3   The Sensor Network System Architecture, Hardware and Software Components

The aim of this chapter is constructing sensor fusion in the context of research project (DP0988088 funded by Australian Research Council (ARC) and to develop a system that allows multiple designers to work simultaneously on the same design remotely using upper body gestures. A pair of data gloves, a sensor jacket and a Head Mounted Display (HMD) are planned to be used for gesture recognition.

De-SIGN system comprises 3 main elements: a sensor jacket (sensor shirt), 5DT data gloves, a head mounted display and software platforms, a gesture recognition system and a CAD system based on Vizard VR Toolkit (Fig. 5).

The **Sensor Network System for De-SIGN** consists of a sensor data acquisition unit, a data processor, a network communication unit (NCU), a feed forward artificial neural network (to be trained by back propagation) and a visualisation unit (Fig. 6). The data acquisition unit, the data processor and the transmitter half of the NCU are coded in C++. We have designed

Fig. 5. De-SIGN I/O system

the NCU receiver, the artificial neural network and the visualisation unit in Python using Vizard Virtual Reality Toolkit.

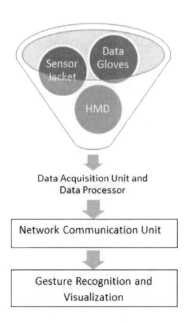

Fig. 6. Sensor network system architecture

3.1   Data Acquisition Unit (DAU)

The DAU consists of a pair of data gloves, a sensor shirt and an HMD. The gesture recognition is to be done using the information gathered out of the sensor values of data gloves and the sensor shirt, and HMD will be used to view the 3D design from different perspectives at all times.

3.2   The Sensor Jacket

General characteristics of the Sensor Jacket is described in section 2. The sensor jacket has 52 sensor threads smeared on a flexible fabric. These threads have piezo-resistive property whose resistance changes under deformation. In order to measure varying resistance, sensor threads are fed with a constant current source and voltage drops are measured for each thread.

A number of sensor threads are responsible of each region. These produce signals when wrists, elbows and shoulders are in motion. Being symmetric on both right and left sides, 6 sensors are used for front shoulder, 8 for the back shoulder and 12 for the limb, resulting at 26 sensors for each side. We use the sensor jacket to determine position of the arms and hands over the active threads related to these sections.

We captured test signals for a set of gestures using a NI (National Instruments) data acquisition card. The NI-DAQ card has a development and signal processing software platform which allows the user to read individual channels connected to the sensor jacket. The sensor readings acquired by using NI-DAQ system depicts nonlinear behavior. The observed nonlinearities in our first trials depend on the size of person and settling time of the threads when they return to their undeformed lengths.

3.3   Data Gloves

5DT data gloves are Lycra gloves that are equipped with five embedded fiber optic sensors and one tilt sensor. Thus, the gloves measure finger flexure as well as pitch, roll and yaw angles of tilt sensor. Finger flexure is measured on the upper joint on each finger. The voltage level, delivered by the gloves, varies as a finger is bent or straightened. Therefore, the shape of the fingers can be detected accurately in any time instant. The tilt sensors provide orientation of the hands. If additional Degrees of Freedom (DOF) are required to incorporate the upper-limb movement of the users to the system, we need to incorporate additional means of motion capture, as this was the case in this project.

3.4   Head Mounted Display

The Head Mounted Display is an Icuiti VR920 HMD which is suitable specifically for virtual reality and high-end gaming applications, for viewing 3D design from various view-points. There is a 3D gyroscope system encased in Icuiti HMD. It measures the orientations, pitch, yaw and roll accurately. The HMD system comes with a Software Development Kit (SDK) which allows the user to collect information from the sensors.

3.5 Data Processor (DP)

A Data Processing software unit is designed to process incoming data. Raw signals are read and the noise on the signals is eliminated using appropriate filters to prevent any distortion on each signal. In order to animate virtual characters without jitter and reduce the number of inputs to the Neural Network (NN) proposed for gesture recognition, the raw sensor signals are filtered and Euler angles for each joint (shoulder, elbow, wrist, fingers and neck) are computed in this unit. As we were not provided any software and model for the sensor jacket, at the very beginning of our studies we only utilized the data captured by using the HMD and the data gloves. In this case, the readings from sensor threads on the sensor jacket for the gesture sets were assumed to be output values, whereas the orientation parameters provided by the HMD and the data gloves were inputs for training data.

3.6 Network Communication Unit (NCU)

Our project requires a system on which multiple remote designers can perform design on the same object simultaneously. There is only one data acquisition card to which sensor jacket is connected. Therefore a network structure is necessary which enables the researchers to connect to the system using their user IDs from any PC. The NCU takes the IP addresses as inputs and send the processed data to the visualization server over a network protocol. We designed a transmitter which facilitates the user connections in C++ and Python. A UDP protocol was employed to increase the communication speed as TCP protocol incurs additional time for handshaking processes.

3.7 Visualisation Unit

The modifications by any designer performing object manipulation using De-SIGN would be rendered on the 3D screen. The visualization unit was designed to enable the designer visualize the modification real-time. The ANN interprets the actions and displays them over this unit. We produced a limited number of animations for the first phase using the data gloves and the HMD.

We integrated the HMD, data gloves and sensor jacket for gesture recognition to be used in a CAD environment. Vizard VR Toolkit were exploited to receive the input from both the sensor jacket and data gloves. Incoming data were then processed to animate the upper limb of animated characters in real-time using the skeleton models that Vizard provides. As a result, the gestures were imitated by virtual characters in the VR space.

3.8   Sensor Network Graphical User Interface (GUI)

The system usability is a great concern for De-SIGN. Instead of sub-optimized user interfaces, the system was optimized by designing a Graphical User Interface (GUI) (Fig. 7) for the usability of the sensor network. The code was written in C++. The GUI was developed in QT which is a software development platform provided by Nokia. The GUI for the sensor network allows the developer to define the devices, target computer's IP address, filtering coefficients, and data logging preferences.

Fig. 7. A sight into the sensor network transmitter GUI

4   Gesture Recognition

Gesture classification and recognition is a problem of pattern recognition for time series. There are abundant classification and matching methods available specific to application and many general methods for pattern recognition. As the sensor jacket is a fairly new non-commercial technology and there is no specific software that maps the sensor readings into gesture domain. Therefore a mathematical model is required for this kind of function construction. We have spent a great deal of time analyzing the sensor readings to build such an approximate mapping function.

We found that using only basic mathematical functions, it is not possible to create an appropriate mathematical model suitable for gesture recognition, mainly because of high nonlinearity of the signals without additional precise measurement devices. The researchers who developed the sensor jacket also proposed an approximating mathematical models by modeling the behavior of the system using exponential functions. They used goniometers and collected the sensor readings for their gesture sets from the sensor jacket as well as goniometers. State space models and linear combination of exponential functions to approximate the response of the sensors are utilized for multivariate approximation of the interval postures in the study [5].

The first phase of our gesture recognition studies was based on the development of an Artificial Neural Network structure to map the sensor signals into gesture domain by training the network. Along with our initial studies on the model construction for the sensors, the first phase provided general information about the signal structure and served as a good starting point for the further algorithm development. Following this, we developed a robust gesture and posture recognition algorithm based on Compressed Sensing (CS) and Sparse Representation (SR) for the sensor jacket.

Compressed Sensing (CS) and Sparse Representation (SR) are new research fields which allow signals to be recovered from a few number of samples much below the well-known Nyquist sampling rate by random/non-adaptive measurements as long as the signal is sufficiently sparse in measurement domain [33-37].

Mathematically, given a sufficiently k-sparse signal $x \in R^n$ whose members consists of a few nonzero k-elements and zeros in a measurement domain with an orthonormal basis Ψ (such as Fourier, direct cosine transformation or wavelet bases), the whole content of x can be recovered by sampling via a random matrix $U \in R^{nxm}$ which satisfies the Restricted Isometry Property (RIP) by preserving the lengths of k-sparse elements with the condition that $m << n$. The resulting equation $y = U\Psi x$ is then solved by the linear programming method ℓ_1 minimization.

Wright et al. successfully applied the underlying principle of CS and SR, namely discriminative nature of sparsity to face recognition problem [40] and developed a robust face recognition algorithm which they call Sparse Representation based Classification (SRC), working within a high confidence interval even with the faces under varying pose and illumination conditions. Following their studies, the approach is applied in biometrics to iris recognition successfully handling wide variety of error on iris images [41].

The SRC algorithm uses a dictionary matrix consisting of training samples. In the algorithm. first, training samples of k classes and the test image are converted to a column vector and projected on a lower dimensional space using a generated random measurement matrix. Then training vectors are stacked into a matrix to construct the dictionary. The resulting equation is solved to recover the sparse signal x by ℓ_1 minimization methods after the columns of the dictionary and test vector are normalized. The SRC algorithm assumes the test image vector lies in the linear span of training samples (Eq. 1) associated with the same class of object where the signal $x = [0, 0, 0, 0, \alpha_1, \alpha_2, \dots \alpha_n, 0, 0, 0, 0]$.

$$y = [\alpha_{i,1}\nu_{i,1} + \alpha_{i,2}\nu_{i,2} + \dots + \alpha_{i,n}\nu_{i,n}] \qquad (1)$$

The pseudo code for the algorithm is as follows:

- Construct the dictionary matrix $A = [\nu_{i,1}, \nu_{i,2}, \ldots, \nu_{i,k}] \in R^{m x n}$ for k classes by reducing the dimension using a random matrix having RIP and converting the samples, and test image vector to one dimensional vectors ν_i and y
- Normalize the columns vector of reduced $\tilde{A}$ and the test image vector $\tilde{y}$,
- Solve the ℓ_1 minimization problem $\hat{x} = argmin_x \|x\|_1$ subject to $\tilde{A}x = \tilde{y}$
- Compute the residuals $r_i(y) = \|\tilde{y} - \tilde{A}\delta_i\|_2$ where δ_i is a selection operator for $\hat{x}$ corresponding the ith class span in $\tilde{A}$
- Identify y by finding the minimum of $r_i(y)$

The CS is first used in the study [38] with a complementary algorithm Affinity Propagation (AP) proposed by Frey& Dueck [39] which clusters the data by message passing between the points. In this study, the gestural data consists of accelerometer sensor readings are separately saved in the matrices corresponding to three axes R_x, R_y and R_z. Then, the gesture recognition problem is defined for each axis as

$$\bar{y}_x = \bar{R}_x x + \epsilon_x \tag{2}$$

where $\bar{y}_x$ is the randomly sampled x component of the gesture to be recognized and $\bar{R}_x$ is the classes of the remaining gesture traces after narrowing the AP results.

The recognition problem is converted to a CS problem by introducing a pre-processing operator to meet the incoherency and sparsity requirements.

$$Q_x = Orth(\bar{R}_x^T)^T \tag{3}$$

$$W_x = Q_x \bar{R}_x^\dagger \tag{4}$$

Where Q_x is the orthonormal basis for $\bar{R}_x$ and $\bar{R}_x^\dagger$ is the pseudo-inverse of $\bar{R}_x$ thus the gesture recognition problem has a new form of

$$h_x = W_x \bar{y}_x = Q_x x + \epsilon_x' \tag{5}$$

$$\hat{x}_{eq} = \hat{x}_x^2 + \hat{x}_y^2 + \hat{x}_z^2 \tag{6}$$

Equation (5) is solved for all axis components to identify the gesture class by computing (6) and taking the minimum of $\hat{x}_{eq}$.

The algorithm gives higher accuracy rates than the traditional methods when the candidate gestures are narrowed by measuring the similarities between the test gesture and candidate gesture exemplars specified by the AP method. If the AP method is not used the rate of recognition accuracy dramatically decreases for our three gesture sets.

We adopt the SRC algorithm for gesture recognition and convert the recognition problem to the CS problem type without introducing additional operator.

Three gesture sets are used to test the proposed SRC based gesture recognition algorithm. The starting point to this study is the fact that every gestural signal is sparse in Discrete Cosine Transformation (DCT) domain for our test gestures. In the Fig. 8, a sensor reading from the single thread of the sensor jacket and its recovered form are given. The recovered signal by sampling from sparse domain using a random measurement matrix and solving ℓ_1 minimization problem and the original signal are nearly identical in the Fig. 8.

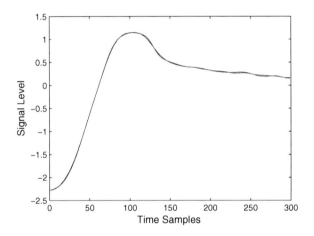

Fig. 8. Original and recovered signal

The sensor jacket has 52 individual piezzo-resistive sensor strips which are located from wrist to shoulders on the right and left side of the jacket. The data is acquired by National Instrument Data Acquisition Unit. There are three gestures to be recognized performed by the sensor. These are; moving the arm from relaxed position to front at the shoulder level, from relaxed position to side at the shoulder level and from side to front at the shoulder level.

A sample gesture data is given in Fig. 9. As seen in the figure, there are active and inactive sensors, and the acquired data is noisy.

Each sensor reading is sparse in Discrete Cosine Transform domain. Before using sensor readings we apply a light Gaussian smoother to the readings to eliminate jitter on the data. All sensor readings are normalized, so as to make them have a zero mean and a variance of one. All inactive sensor signals are eliminated defining an absolute total variation threshold. The remaining sensor readings are concatenated in a vector for an individual gesture. A

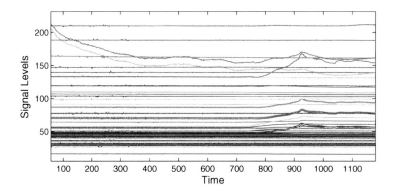

Fig. 9. Active and inactive sensor readings of the Sensor Jacket

gesture dictionary is constituted from three gesture classes by stacking the gesture traces as columns of the dictionary. The rest of the solution is to design a measurement matrix and apply ℓ_1 minimization.

CS theory states that if the signal is sparse in any domain, signals can be recovered with an overwhelming probability by random measurement matrices having The Restricted Isometry Property (RIP) condition. Random Gaussian or Achlioptas' matrix [42] can be used for both dimensionality reduction and measurements, since the values having RIP properties preserve distances in the embedding space. We use Achlioptas' matrix since $2/3^{rds}$ of the matrix is sparse, making it easier to construct than a Gaussian matrix thus saving computation time.

Achlioptas' matrix is defined as

$$U_{ij}\sqrt{3} = \begin{cases} +1 \ withprobability \ \frac{1}{6} \\ 0 \ \ \ withprobability \ \frac{2}{3} \\ -1 \ withprobability \ \frac{1}{6} \end{cases} \tag{7}$$

The pseudo codes for the gesture recognition algorithm are as follows.

- Normalize the data, apply a Gaussian smoother and reshape all the gestures and the test gesture, and take DCT of each
- Find the longest length of gestures (lh_{max}) and make the other gestures of the same length by zero padding, and stack the training gestures into training matrix $A_G \in R^{lh_{max}xn}$

$$A_G = \begin{pmatrix} G_{1,1} & G_{1,2} & \cdots & G_{1,k} \\ G_{2,1} & G_{2,2} & \cdots & G_{2,k} \\ \vdots & \vdots & \ddots & \vdots \\ 0_{lh_{max}-n1,1} & 0_{lh_{max}-n2,1} & \cdots & 0_{lh_{max}-nk,k} \end{pmatrix} \tag{8}$$

- Take m measurement from both the test gesture vector and the training matrix with the designed random measurement matrix U^{mxn} to form the reduced equations $(\tilde{y}) = (y)x$ where $x = [0, 0, 0, 0, \alpha_1, \alpha_2, \ldots \alpha_n, 0, 0, 0, 0]$ consists of a few nonzero coefficients corresponding to gesture class.
- Solve the ℓ_1 minimization problem $\hat{x} = argmin_x \|x\|_1$ subject to $\tilde{A}x = \tilde{y}$
- Compute the residuals $r_i(y) = \|\tilde{y} - \tilde{A}\delta_i\|_2$ where δ_i is a selection operator for $\hat{x}$ corresponding the ith class span in $\tilde{A}$
- Identify y by finding the minimum of $r_i(y)$

In the sensor jacket gesture recognition application there are three gesture classes consist of 10 gesture traces for the first experiment. We randomly choose 6 gesture traces from each class for the dictionary matrix and the remaining gesture traces in total 12 traces are used to test the recognition algorithm. Only the active signals are used in the algorithm. The noise on the active signals is eliminated using a light Gaussian smoother. The SRC based algorithm gives 100% recognition accuracy for the gestures performed by the sensor jacket. We conducted the second experiment by designing a new test gesture set that consists of 5 different gesture classes. There are 15 gesture traces are captured, 10 for dictionary and 5 for testing purpose. A Wi-imote mouse is employed to determine the beginning and end of the gestures in experiments in data collection step. The system gives 100% recognition accuracy once again for 25 test gestures.

We tested the SRC based gesture recognition algorithm further with two different gesture sets composed by using different signal domains. The first gesture set that contains 23 gesture types is composed from the gesture sets used in the studies [38, 43] Fig. 10.

Gestures are captured by using the IR camera of Wiimote in two dimension x and y. Each gesture has two dimensions and different lengths. Before the gesture recognition process each gesture is smoothed using a Gaussian smoother and then normalized by making the mean of gestures zero and variance one. 15 gestures, 10 for the gesture dictionary and 5 for testing are captured from 3 subjects. In total, we have 1035 gesture traces stored in xml format. These gesture files are then converted to .mat files which are then used as input for the SCR gesture recognition algorithm.

Our system (SenSe) gives 100% user-independent recognition accuracy for each person with a full dictionary matrix which consists of 10 gestures for each gesture class including 230 columns in total. To be able to provide a user independent gesture recognition system, the dictionary is built by arbitrarily chosen gestures from each subject with the same number of gestures from each class. Then, we used the remaining gestures for testing purposes. The system misclassified only 2 gestures out of 300 test samples, corresponding to 99.33% recognition accuracy for 20 gestures. 3 gestures were removed from the database, since these 3 classes of gestures were performed in a very different manner for all subjects.

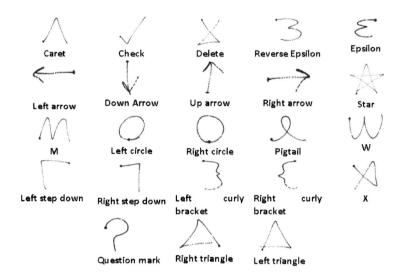

Fig. 10. 23 Gestures combined from the studies [38, 43]

5 Conclusion, Discussion and Future Works

In this chapter, we presented a system architecture for a gesture recognition system (De-SIGN) for designing in VR space. We first provided a brief overview of existing motion capture technologies and devices and then discussed the strengths and weaknesses of the piezo-electrical sensing garment technology that was used in De-SIGN. We developed an algorithm (SenSe) for sensor analysis and synthesis and then conducted a set of experiments to test its accuracy in gesture recognition. We presented the gesture and posture recognition algorithm (SRC) based on CS and SR we have developed for the wearable sensing garment having piezo-resistive properties. The SRC algorithm we presented in this section is highly accurate, regardless of the signal acquisition method, and it demonstrates high performance even for high dimensional signals and large gesture dictionaries. Our findings state that gestures can be recognized with over 99% accuracy rate for user-independent gesture recognition studies using the SRC algorithm. The algorithm gives 100% recognition accuracy for user dependent gesture recognition with rich gesture sets. In order to test the gesture.

We test our system also with the gesture sets of $1 gesture recognition study [43]. The set consist of 16 different gestures collected from 11 subjects. Each gesture is repeated 10 times in 3 speed profiles fast, slow and medium speeds by each subject. The gesture dictionary is built by choosing two random gestures belong to any speed profile from 5 subject's gesture folder so that every gesture class consists of 10 gesture. The test gestures are chosen from the remaining subject's folder at random. The SRC gesture recognition algorithm misclassified only 2 gestures out of 80 test gestures with a corresponding accuracy 97.5% even the gesture dictionary is composed of randomly chosen gestures from various speed profiles of 5 subjects. The accuracy level for the experiments stays at the same level when the number of representative gestures in the dictionary decreased to 3.

The application of the SRC algorithm is an original contribution to gesture recognition and brings following novelties:

- Multi-dimensional gesture signals can be represented as a one dimensional vector reshaping them as in the study face recognition implemented by Wright et al in [40] and have high recognition accuracy
- No prior clustering algorithm is necessary unlike other CS based pattern recognition algorithm proposed in [38] for gesture recognition, however, pre-classification algorithms can be used to reduce the computation time
- There is no upper and lower bound for the number of classes and the number of gesture traces in the dictionary
- It can be applied to any gesture signal acquisition domain.
- The algorithm achieves high accuracy for rich gesture databases event the presence of very similar gestures in the database
- The feature selection for CS measurements is random

While we try to develop a 3D Sketchpad, an appropriate GUI is the primary concern for improving its usability, as there are many parameters to configure, such as the parameters of designed system SRC which may give rise to confusion for any person who is not familiar with the system codes. A good design practice also requires allowing only expert access to sensitive code in a complex system, while letting users to get their work done with the system easily.

Our system employs devices which are used to capture data in several domains. We design and deploy several filtering approaches in our studies. The Wiimote, data glove and the sensor jacket signals are highly noisy due to the high sampling rates. The noisy structure of the signals amplified through the computations for visualization and cause unwanted jitter on the manipulated objects and animation characters. Following figures demonstrate the noisy raw signals acquired by the sensor jacket (Fig. 11-a) and the Wiimote (Fig. 11-b), as well as filtered signals. We deployed a couple of filtering methods such as low-pass Butterwoth, Kalman, Moving Averages, as well as Gaussian smoothers for our applications as required.

Further research and experiments about the most common gestures performed by novice and expert designers will improve the usability and efficiency of the system. In order to perform further analysis a larger group of experimental data will be required. The range of experiments should be broader and the amount of repetitions should be increased. Future research and development is necessary in order to assure an ideal design platform based on gesture recognition. The most important step for this purpose is developing an accurate (sensor domain to body kinematics) mapping model for the sensor jacket and integrating it into the transmitter side of the system.

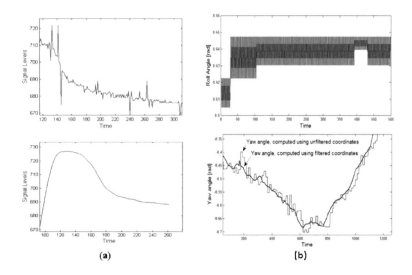

Fig. 11. Noise levels on the signals [38, 43]

An important improvement for the CS based gesture recognition is to devise a partial gesture recognition algorithm which may provide a streaming recognition and visualization. This may bring about a solution to one of the principle problems in gesture recognition, gesture spotting.

Acknowledgments. This project has been sponsored by the Australian Research Council (ARC) Discovery Grant scheme (DP0988088). Special thanks to Mine Cicek (MEng), Alexei Novoselov (MIT, BEng), Dr Tauseef Gulrez, Meredith Taylor, Iwan Kartiko and John Porte of the Department of Computing, Macquarie University, for their assistance in specific parts of this project.

References

1. Kavakli, M., Taylor, M., Trapeznikov, A.: Designing in virtual reality (DesIRe): A gesture based interface. In: Proceedings of 2nd International Conference on Digital Interactive Media in Entertainment and Arts, vol. 274, pp. 131–136 (2007)
2. Kavakli, M.: Gesture recognition in virtual reality, special issue on: Immersive Virtual, Mixed, or Augmented Reality Art. The International Journal of Arts and Technology 1(2), 215–229 (2008)
3. Tognetti, A., Lorussi, F., Bartalesi, R., Quaglini, S., Tesconi, M., Zupone, G., De Rossi, D.: Wearable kinesthetic system for capturing and classifying upper limb gesture in post-stroke rehabilitation. J. Neuro Engineering and Rehabilitation 2, 1–16 (2008)

4. Gulrez, T.: Body machine interface: Remapping residual mobility of spinal cord injured patients to control assistive robotic devices. PhD Thesis, Macquarie University, Sydney (2008)
5. Tognetti, A., Bartalesi, R., Lorussi, F., De Rossi, D.: Body segment position reconstruction and posture classification by smart textiles. Transactions of the Institute of Measurement and Control 29(3-4), 215–253 (2007)
6. Novaselov, A.: Wearable sensor analysis for gesture recognition. Dissertation, Macquarie University, Sydney (2009)
7. Fish, J., Scrivener, S.: Amplifying the mind's eye: Sketching and visual cognition, vol. 23 (1), pp. 117–126. The MIT Press, Leonardo (1990)
8. Goldschmidt, G.: The dialectics of sketching. Creativity Research Journal 1(2), 123–143 (1991)
9. Goel, V.: Sketch of Thoughts. The MIT Press, Cambridge (1995)
10. Schaaf, J.W., Nowak, M., Voß, A.: Using Gestalten to Retrieve Cases. In: Haton, J.-P., Manago, M., Keane, M.A. (eds.) EWCBR 1994. LNCS, vol. 984, pp. 136–150. Springer, Heidelberg (1995)
11. Negroponte, N.: Recent advances in sketch recognition. In: Proceedings of the National Computer Conference and Exposition, AFIPS 1973, pp. 663–675. ACM Press, New York (1973)
12. Scrivener, S.A.R., Harris, D., Clark, S.M., Rockoff, T., Smyth, M.: Designing at a distance via real-time designer-to-designer interaction. Design Studies 14(3), 261–282 (1993)
13. Gross, M.D.: Recognizing and interpreting diagrams in design. In: Proceedings of 2nd ACM Workshop on Advanced Visual Interfaces AV 1994, pp. 124–125 (1994)
14. Forbus, K.D., Lockwood, K., Klenk, M., Tomai, E., Usher, J.: Open-domain sketch understanding: The nuSketch approach. In: Proceedings of AAAI Fall Symposium on Making Pen-based Interaction Intelligent and Natural, pp. 58–63 (2004)
15. Jung, T., Gross, M.D., Do, E.Y.: Annotating and sketching on 3D web models. In: Proceedings of the 7th International Conference on Intelligent User Interfaces, pp. 95–102 (2004)
16. Mulder, A.: Hand Gestures for HCI, Hand Centered Studies of Human Movement Project. Technical Report 96-1. School of Kinesiology, Simon Fraser University, Vancouvar (1996)
17. Pederson, T.: Human hands as a link between physical and virtual. In: Proceedings of Conference on Designing Augmented Reality Systems (DARE 2000), pp. 153–154 (2000)
18. Sacco, M., Vigano, G., Paris, I.: Virtual reality and CAD/CAM systems applied to custom shoe manufacture on a mass market basis. In: Proceedings of International Conference on Mass Customisation (2002)
19. Liu, X., Hinds, B., McCartney, J., Dodds, G.: Design and visualisation of footwear in a VE. The Virtual Engineering Center, Queens University, Belfast (2003)
20. Bowman, D., Billinghurst, M.: Special issue on 3D interaction in virtual and mixed realities: Guest editors' introduction. Virtual Reality 6(9), 105–106 (2002)
21. Montero, J.A., Sucar, L.A.: A Decision-theoretic video conference system based on gesture recognition. In: Proceedings of 7th IEEE International Conference on Automatic Face and Gesture Recognition FG 2006, pp. 387–392 (2006)

22. Aleotti, J., Caselli, S.: A Low-cost humanoid robot with human gestures imitation capabilities. In: Proceedings of The 16th IEEE International Symposium on Robot and Human interactive Communication, pp. 631–636 (2007)
23. Yanbin, P., Hanwu, H., Jinfang, L., Dali, Z.: Data-glove based interactive training system for virtual delivery operation. In: Second Workshop on Digital Media and its Application in Museum & Heritages, pp. 383–388 (2007)
24. Minoh, M., Obara, H., Funtatomi, T., Toyoura, M., Kakusho, K.: Direct manipulation of 3D virtual objects by actors for recording live video content. In: Proceedings of the Second International Conference on Informatics Research for Development of Knowledge Society Infrastructure, pp. 11–18 (2007)
25. Gulrez, T., Kavakli, M.: Precision position tracking in virtual reality environments using sensor networks. In: IEEE International Symposium on Industrial Electronics (ISIE 2007), pp. 1–6 (2007)
26. Bowman, D.A., Kruijff, E., LaViola, J.J., Poupyrev, I.: 3D User Interfaces: Theory and Practice. Addison Wesley Professional, Reading (2005)
27. Torige, A., Kono, T.: Human-interface by recognition of human gesture with image processing-recognition of gesture to specify moving direction. In: Proceedings of IEEE international Workshop on Robot and Human Communication, pp. 105–110 (1992)
28. Karlekar, J., Le, S.N., Fang, A.C.: Tracking of articulated pose and motion with a markerized grid suit. In: 19th International Conference on Pattern Recognition (ICPR 2008), pp. 1–4 (2008)
29. Boyali, A., Kavakli, M., Twamley, J.: TReal-time six degree of freedom pose estimation using infrared light sources and wiimote IR camera with 3D TV demonstration. In: Proceedings of MobiQuitous 2010: 7th International ICST Conference on Mobile and Ubiquitous Systems (2010)
30. Bouguet, J.Y.: Matlab camera calibration toolbox, http://www.vision.caltech.edu/bouguetj/calib_doc/
31. Mariottini, G.L., Prattichizzo, D.: Matlab epipolar geometry toolbox (2006), http://egt.dii.unisi.it/ (access August 2010)
32. Hartley, R., Zisserman, A.: Multiple view geometry in computer vision. Cambridge University Press, Cambridge (2003)
33. Candés, E., Romberg, J., Tao, T.: Robust uncertainty principles: Exact signal reconstruction from highly incomplete frequency information. IEEE Trans. Inform. Theory 52(2), 489–509 (2006)
34. Candés, E., Tao, T.: Near optimal signal recovery from random projections: Universal encoding strategies? IEEE Trans. Inform. 52(12), 5406–5425 (2006)
35. Donoho, D.: Compressed sensing. IEEE Trans. Inform. 52(4), 1289–1306 (2006)
36. Candés, E., Wakin, M.B.: An Introduction to compressive sampling. IEEE Signal Processing Magazine 21, 21–30 (2008)
37. Baraniuk, R.G.: Compressive sensing [lecture notes]. IEEE Signal Processing Magazine 24(4), 118–121 (2007)
38. Akl, A., Valaee, S.: Accelerometer-based gesture recognition via dynamic-time warping, affinity propagation, & compressive sensing. In: IEEE International Conference on Acoustics Speech and Signal Processing (ICASSP), pp. 2270–2273 (2010)
39. Brendan, J.F., Delbert, D.: Clustering by passing messages between data points. Science 315(5814), 972–976 (2007)
40. Wright, J., Yang, A., Ganesh, A., Sastry, S., Ma, Y.: Robust face recognition via sparse representation. IEEE Transactions on Pattern Analysis and Machine Intelligence (PAMI) 31(2), 210–227 (2009)

41. Pillai, K.J., Patel, V.M., Chellappa, R., Ratha, N.K.: Secure and robust iris recognition using random projections and sparse representations. IEEE Transactions on Pattern Analysis and Machine Intelligence, 1–1 (2011)
42. Dimitris, A.: Database-friendly Random Projections. Journal of Computer and System Sciences - Special issue on PODS 66-4 (2001)
43. Wobbrock, J.O., Wilson, A.D., Li, Y.: Gestures without libraries, toolkits or training: A $1 recognizer for user interface prototypes. In: Proceedings of the ACM Symposium on User Interface Software and Technology (UIST 2007), pp. 159–168 (2007)

Image Segmentation of Cross-Country Scenes Captured in IR Spectrum

Artem Lenskiy

School of Electrical, Electronics & Communication Engineering,
Korea University of Technology and Education,
1800 Chungjeol, Byeongcheon, Dongnam Cheonan, 330-708, Korea
a.a.lensky@gmail.com

Abstract. Computer vision becomes a major source of information for autonomous navigation for different types of vehicles including military robots and mars/lunar robot rovers. Nevertheless, the attention is mostly focused on texture descriptors with fixed scales. Besides, the majority of work is done in analyzing images captured in visual spectrum. In this chapter we elaborate on the problem of segmenting cross-country scene images in IR spectrum using salient features. Salient features are robust to variations in scale, brightness and angle of view. As for salient features we choose Speeded-Up Robust Features (SURF). We provide a comparison of two SURF implementations. SURF features are extracted from input imaged and for each of features terrain class membership values are calculated. The values are obtained by means of multi-layer perceptron. The features class membership values and their spatial positions are then applied to estimate class membership values for all pixels in the image. The values are used to assign a terrain class to each pixel in the image. To decrease the effect of segmentation blinking and speed up segmentation, we are tracking camera position and predict positions of features.

Keywords: autonomous ground vehicle, visual navigation, texture segmentation.

1 Introduction

A navigation system designed for autonomous driving in outdoor, non-urban environment is more sophisticated than a system designed for traversing in urban environment with plenty straight lines. A variety of navigation systems have been proposed for off-road navigation. Such systems process data obtained from varies sensors including laser-range finders, color and grayscale cameras. Each type of sensors provides its own advantages and disadvantages.

In this chapter we elaborate on the problem of segmenting cross-country scene images using texture information. Our approach takes advantage of salient features that are robust to variations in scale, brightness and angle of view. The discussed texture segmentation algorithm can be easily incorporated into navigation system with laser range scanners and other sensors.

T. Gulrez, A.E. Hassanien (Eds.): Advances in Robotics & Virtual Reality, ISRL 26, pp. 227–247.
springerlink.com

Moreover, using salient features the computer vision system using the discussed algorithm can be easily extended to object recognition simply by adding new type of objects during the learning process.

The rest of the chapter is organized as follows. Section 2 gives an overview of current achievements in the field of navigation of unnamed ground vehicles in cross-country environment. At the end of section 2, a block diagram of the proposed segmentation system is given. Section 3 compares two popular implementations of SURF algorithms. Section 4 describes feature extraction stage. Texture model construction is described in section 5. The segmentation procedure is proposed in section 6. In section 7 we elaborate on 3D reconstruction and tracking of matched SURF features. Experimental results and conclusions are given in section 8 and 9 respectively.

2   Literature Overview

The majority of systems for unmanned ground vehicles utilize various type of sensors including, cameras and laser-range scanners. The data obtained by laser-range scanners is applied for precise 3D reconstruction of neighbor environment. From the reconstructed 3D scene it is possible to distinguish flat regions, tree trunks and tree crowns [1]. Moreover, there is no difference for lase-range scanners at what time of day it operates. On the other hand one disadvantage of laser scanners consists in inability to distinguish some types of terrain. For instance it is impossible to distinguish gravel, mud and asphalt as they have similar point distributions. It is also difficult to recognize either a reconstructed object is a tall patch of grass, a rock or a low shrub. Another possible disadvantage of 3D laser-range scanners is their high price.

Another source of information that is commonly used for terrain segmentation and road detection is color. Color carries information that is useful for distinguishing gravel and mud, or grass and rocks. The recognition accuracy in this case depends on surrounding illumination. To minimize the influence of variations in illumination, a representative training data set should be collected, that covers all expected environmental conditions. Additionally, the classifier should be able to adequately represent the variability of perceived color within each single class. Manduchi et al. [2, 3] proposed a classifier that uses Gaussian Mixture Models to estimate density functions for each class. The motivation to use a number of Gaussians comes from the fact that, for the same terrain type the color distributions under direct and diffuse light often correspond to different Gaussians modes. To improve color based terrain type classification Jansen et al. went even further [4]. Using Greedy expectation maximization algorithm they firstly clustered training images into environmental states. Then before an input image is segmented its environmental state is determined. Color distributions vary from one environmental state to another even within the same terrain class. Overall, they were able to classify sky, foliage, grass, sand and gravel with the lowest probability for correct classification of 0.80. Nevertheless, such classes as grass, trees and bushes are still indistinguishable using color information, moreover color information is not available at night.

Texture is another characteristic that can be employed for terrain segmentation. Texture features extracted from grayscale or color images allow us to separate between grass, trees, mud, gravel, and asphalt. A number of computer vision algorithms have been successfully applied for terrain segmentation using various texture features. To classify texture features into six terrain classes, Sung et al. [5] applied a two-layer perceptron with 18 and 12 neurons in first and second hidden layers respectively. The feature vector was composed of the mean and energy values computed for selected sub-bands of two-level Daubechies wavelet transform, resulting in 8 dimensional vector. These values were calculated for each of three color channels. Thus, each texture feature contained 24 components. The experiments were conducted in the following manner. First, 100 random images from the stored video frames were selected and used to extract training patches. Then, among them ten were selected for testing purposes. The wavelet mean and energy were calculated for fixed 16x16 pixel sub-blocks. Considering a resolution of input images of 720x480 pixels, the sub-block of 16x16 pixels is too small to capture texture characteristics, especially at larger scales, which therefore leads to a poor texture scale invariance. They achieved the average segmentation rate of 75.1%. Considering that color information was not explicitly used and only texture features were taken into account, the segmentation rate is promising, although there is still room for improvement. Firstly, color components are correlated, secondly texture patches of fixed size ignores image context.

Castano et al. [6] experimented with two types of texture features. The first type of features is based on Gabor transform, particularly the authors used Gabor transform with 3 scales and 4 orientations as described in [7]. The second type is based on the histogram approach, where amplitudes of complex Gabor transforms are partitioned into windows and histogram are built from Gabor features for each window. The classifier for the first type of features modeled the probability distribution function using mixtures of Gaussian and performed a Maximum Likelihood classification. The second classifier represents local statistics by marginal histograms. Comparable performances were reached with both models. Particularly, in the case when half of the hand segmented images were used for training and the other half for testing, the classification performance on the cross-country scene images was 70% for mixtures of Gaussian and 66% for histogram based classifiers. Visual analysis of the presented segmentation results suggests that the wrong classification happens due to the short range of scale independence of Gabor features, and rotational invariance, that make texture less distinguishable.

To be able to apply texture features for navigation at night, cameras that are capable of capturing image in IR spectrum should be used. The obtained IR images are useful at daytime as well. The terrain appearance in IR spectrum is less affected by shadows and illumination changes. Furthermore,

the percentage of reflection in visible light range (0.4–0.7 µm) for grass and
trees (birch, pine and fir) is indistinguishable. On the other hand there is
substantial difference in reflectance in infrared range. By segmenting the surrounding environment the robot is able to avoid trees and adjust the speed
for traversal. Kang et al. [8] proposed to use texture features to segment
urban environment from a moving vehicle. The futures were extracted from
multiband images including color and near-infrared bands. Prior texture segmentation they reconstructed 3D scene using structure from motion module.
The reconstructed scene was divided into four depth levels. Depending on the
depth level the appropriate mask size is selected. Overall they used 33 masks
for each band constituting 132 dimensional feature vector. For each mask a
product is calculated by multiplying pixels in the mask together according to
their patterns. The obtained features are named higher-order local autocor-
relation(HLAC) features [9, 10]. The authors presented segmentation results
for four urban scenes, with best segmentation recognition rates of 84.4%,
76.1%, 84.5% and 76.1%. We tend to believe that features extracted (approximately 1000 per image) from any reasonable number of training images
(< 100) will very sparsely distributed in such high dimensional feature space.
Therefore, any classifier will not be reliable with comparably, to the number
of feature dimensions, small training set.

The best segmentation accuracy is achieved then different source of information are combined and used in the segmentation process. An interesting
work has been done by Blas et al. [11, 12]. They combined color and texture
descriptors for online, unsupervised cross-country road segmentation. The
27-dimensional descriptors were clustered to define 16 textons. Then each
pixel is classified as belonging to one of them. During the segmentation process, a histogram for a 32x32 window is estimated. The histogram represents
number of occurred textons. The obtained histograms are clustered into 8
clusters. A histogram from each window is then assigned to one of the clusters. Unfortunately, the authors did not present quantitative segmentation
results for cross-country terrain segmentation due to probably unsupervised
nature of the algorithm. Lenskiy et al. [13] also applied texton based approach
for terrain recognition, however instead of features extracted from fixed size
windows, they employed SIFT features [14]. SIFT features uses scale-space
extrema detection to automatically detect scales i.e. window size and location
of interest points.

Rasmussen [16] provided a comparison of color, texture, distance features
measured by the laser range scanner, and combination of them for the purpose
of cross-country scene segmentation. The segmentation was the worst when
texture features were used alone. In the case when 25% of all features were
used for training, only 52.3% of the whole feature set was correctly classified.
One explanation of this poor quality of segmentation is related to the feature
extraction approach. The feature vector consisted of 48 values representing
responses of the Gabor filter bank. Specifically, it consisted of 2 phases with
3 wavelengths and 8 equally-spaced orientations. The 48-dimensional vector

appears to have enough dimensions to accommodate a wide variety of textures, however as we mentioned above it is still high, considering that training set consisted of 17120 features. Besides the feature dimensionality, the size of texture patches also influenced the segmentation quality. The size of the patch was set to a relatively small constant value equal to 15x15, which led to poor scale invariance. Furthermore, features locations were calculated on the grid without considering an image content. Another reason of the problematic segmentation accuracy is in the low classifiers capacity. As a classifier, the author used a neural network with only one hidden layer with 20 neurons. A one layer feed-forward neural network is not capable of partitioning concave clusters, while terrain texture features are very irregular.

In this chapter we elaborate on an early proposed texture segmentation algorithm that takes advantages of salient features [15, 17]. We present an analysis of the speeded up robust features (SURF)[19] for the purpose of terrain texture segmentation in IR images. For the reasons described above instead of 64 dimensional SURF we apply 36 dimensional version of the SURF [18]. Generally, salient features are detected in two steps: (1) features are localized and (2) descriptors are constructed. Features are searched at different image scales, i.e. size of image patches is not fixed that allows overcoming the problem mentioned above with fixed windows. In the next step robust to changes in brightness and angle of view descriptor is generated. The extracted features are applied to construct a texture model, which is used in the terrain segmentation system (fig. 1). These features are extracted from hand segmented images. After the texture model is constructed the system processes input video frames and returns segmented images, which further can be used for navigating the AGV.

3   Comparison of SURF Implementations

There are two sources of difficulties related to the real-world texture segmentation. The first one is a high variety of transformations affecting texture appearance. Even the same patch of texture, significantly changes it appearance under projective transformations including scale variations. Another source of complexity comes from high variety of textures within a class. For instance, the class corresponding to trees contains different textures of firs, pines etc.

One solution to the first problem is to find texture features invariant under projective transformation and brightness changes. A number of salient features have been proposed to satisfy such conditions. However, they have been applied mostly for object recognition and 3D reconstruction. Lenskiy and Lee provided a number of experiments on different variations of SURF features and found that 36 dimensional SURF [19] features with no-rotational invariance (U-SURF) achieve the best segmentation accuracy on day time images [18]. In this chapter we compare two SURF implementations and choose the one that the best suits our goals.

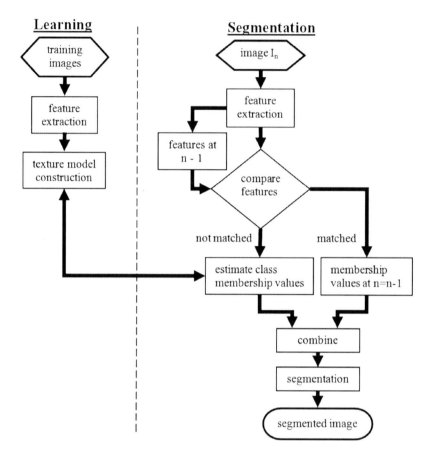

Fig. 1. The segmentation system block scheme

The SURF algorithm consists of three stages. In the first stage, interest points and their scales are localized in scale-space by finding the maxima of the determinant of the Fast-Hessian matrix. In the second stage, orientation is determined for each of features prior to computing the feature descriptor. We use rotation dependent SURF, thus we omit this step. Finally a descriptor based on sampled Haar responses is extracted.

The SURF algorithm has become a standard for salient features extraction. The SURF algorithm has been implemented for various hardware/software platforms such as Android, iPhone, Linux and others. A comparison of some open source implementations is given by Gossow et. al [20]. They found that some implementations produce up to 33% lower repeatability and up to 44% lower maximum recall than original implementation. We compare an open source implementation of the SURF named OpenSURF [20] and the original SURF [19] implementation. For this purpose we took five image pairs from

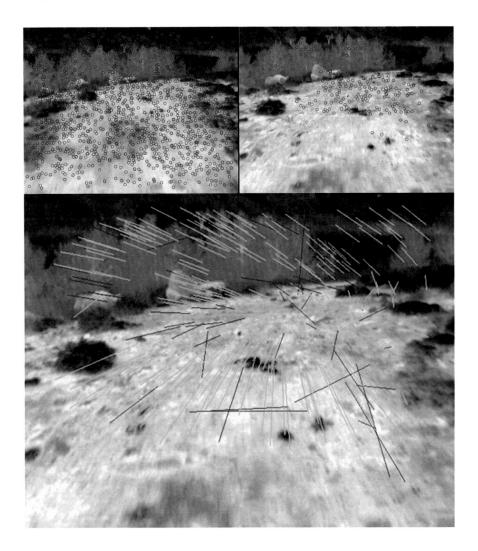

Fig. 2. Top left image show detected SURF features, top right show matched features in the second image, and in the bottom image cyan lines connects corresponding points and red lines are outliers.

a recorded video. Images in the pair are taken at different but close times, showing almost the same scene. For instance, if first frame is number N in the video sequence then the second is $N+50$. We extract SURF features from each image and then match them. Figure 3 shows three pair of images. Locations of detected SURF features in the top-left image, and matched features in the top-right image are marked with blue circles. The parameters for both

SURF implementations are set identical. The comparison results are shown in table 1. First column shows what implementation of the SURF algorithm is used to extract features. Second column in the table shows the number of detected features in each image in the analyzed pairs. Third column show the number of matched features. As it can be seen the number of matched features is significantly less than the number of detected features. Changing the parameters in the matching algorithm we can customize the number of matches. The matching algorithm calculates Euclidian distance between two features and if the distance is less than a predefined threshold, these two features are considered as a match. Finally using Random Sampling Consensus (RANSAC) algorithm [21] we estimated fundamental matrix. The fundamental matrix estimation process is described in section 7. Fig. 2 shows selected by RANSAC algorithm corresponding points (cyan lines). The number of inliers that satisfies the estimated fundamental matrix is shown in forth column. Last column in the table show the ratio between the number of inliers and the average number of features detected in both images in the pair. It can be seen from the table 1 that the original SURF implementation constantly shows higher number of inliers in the relation to the total number of detected features. Thus, in our further experiments we will use the original implementation rather than OpenSURF implementation.

Table 1. Comparison of two SURF implementations.

Implementation	Detected features I_1/I_2	Matched features	Inliers	Ratio
OpenSurf	1021/971	332	220	0.22
OpenSurf	1188/1171	255	168	0.14
OpenSurf	975/961	132	82	0.08
OpenSurf	753/676	181	103	0.15
OpenSurf	1021/971	332	217	0.22
OriginalSURF	1043/982	355	260	0.26
OriginalSURF	1307/1275	262	205	0.16
OriginalSURF	1096/1031	148	99	0.09
OriginalSURF	737/673	155	108	0.16
OriginalSURF	1043/982	355	266	0.27

Figure 3 shows some examples of features detected by original and open SURF algorithms.

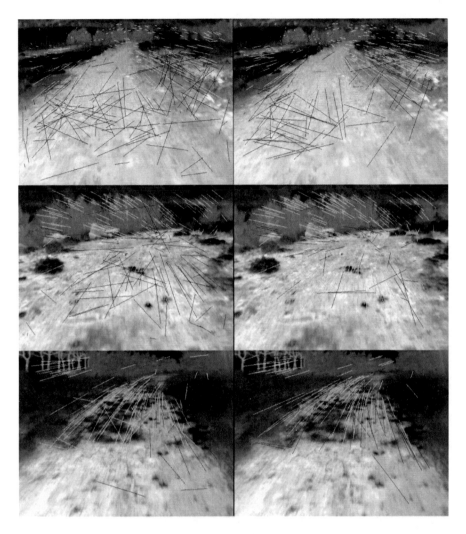

Fig. 3. Detected SURF features and their correspondences shown with cyan lines. Images in the left column are processed with OpenSURF and in the right with original SURF implementations.

4   Features Extraction

The U-SURF consists of two steps. Firstly, using a blob detector, interest points are detected. Then, around each interest point a region descriptor is calculated. The number of extracted features is highly dependent on the image content and on the blob response threshold. The large number of points will slow down the segmentation process and more irregular regions will generate higher quantity of features leading to non-uniform feature distribution.

On the other hand the small number will lead to a low spatial segmentation resolution. To maintain the number of detected features around a limited quantity, we detect interest points at a low blob response threshold and then among them we select the strongest and uniformly distributed ones. Specifically, an image is partitioned into a grid of squares. Strengths of points from each square are compared and the feature with the highest strength is selected, while others are omitted (fig. 4(b)). Selected features are sorted into class arrays according to their labels obtained from hand segmented maps (fig. 4(a)). We considered three terrain classes: a) grass and small shrubs, b) trees and c) road including gravel and soil. Among 3506 video frames, we selected 70 frames for training purposes and 10 for testing. Overall 45361 features were extracted from the training images.

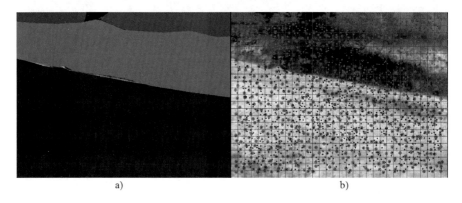

a)   b)

Fig. 4. a) Hand segmented image, b) Detected features, selected features are circled. Blue, green and dark red features represent trees, grass and road correspondingly.

To overcome the mentioned above problem of high interclass variability we train a classifier to distinguish one class of features from others. Lenskiy and Lee [18] analyzed a number of classifiers for the purpose of SURF features classification. A classifier based on multilayer perceptron(MLP) was found to be the most suitable in terms of classification accuracy, memory and processing time requirements. Although, a MLP is able to generalize and classify complex domains of data, some data preprocessing must be performed. The necessity comes from the fact that terrains of different classes are often mixing up making it impossible to separate one terrain region from the others even by human. Therefore, a hand segmented region often contains fragments of a few types. To exclude features representing such fragments as well as non-informative features we omit those of them which reside far apart from the features of its own class [18]. After excluding such features we are left with 41887 features, specifically 9080, 5852 and 26955 features in grass, trees and road classes correspondingly.

For the purpose of analyzing how good terrain classes are separated in the SURF features space we divided the whole training set into two subsets. The first subset contains features surrounded by at least three/five neighbors of the same class. We call this subset a dense subset. The second subset contains all the remaining features. The number of features when at least three neighbors are of the same class is equal to 16705 (40%), the number of remaining features is 25182 (60%). When at least five neighbors are of the same class, there are 16494 (40%) features in the dense subset and 25393 (60%) features in the non-dense subsets. To check how well features from different terrain classes are separated we calculated interclass and intraclass variability as follows:

$$\nu(x,y) = \frac{1}{N_1 N_2} \sum_{i=1}^{N_1} \sum_{j=1}^{N_2} \sqrt{\sum_{k=1}^{36} (x_{i,k} - y_{j,k})^2} \tag{1}$$

where x, y are feature sets, N_1 and N_2 are number of features in feature sets x and y respectively.

Tables 2-4 show estimated variability for the whole training set, for the non-dense and dense subsets respectively. Values on the main diagonal represent interclass variability, and the remaining values represent variability between corresponding classes. As it can be observed in cases of grass and road features, the values of interclass variability are smaller than values for intraclass variability. However, in the case of features from the class associated with trees, the interclass variability is not smaller than the variability calculated between trees and road classes even for features from the dense-subset. This although does not necessary mean that the class is not separable at all, it may means that the class features are not normally distributed. To check if this is the case we visualize features distribution. For this purpose we apply principal component analysis(PCA) to each of two subsets and selecte among 36 components two main components. Plotting two main components of each subset we are able to see features distributions in dimensionally reduced space (fig 5). First of all even in two dimensional space, a solid structure is visible, which supports the assumption that terrains are separable in SURF feature space. Secondly, as it can be seen from the plots(fig 5 b,d), features from the tree class are split up into two clusters. Therefore a nonlinear classifier such as MLP is needed to separate classes.

Table 2. Inter- and intraclass variations for the whole training feature set.

y \ x	grass	tress	road
grass	0.79	0.85	0.77
tress	0.85	0.83	0.79
road	0.77	0.79	0.69

Table 3. Inter- and intraclass variations for the non-dense feature subset.

x y	grass	tress	road
grass	0.78	0.83	0.77
tress	0.83	0.82	0.79
road	0.77	0.79	0.74

Table 4. Inter- and intraclass variations for the dense feature subset.

x y	grass	tress	road
grass	0.81	0.97	0.83
tress	0.97	0.87	0.84
road	0.83	0.83	0.65

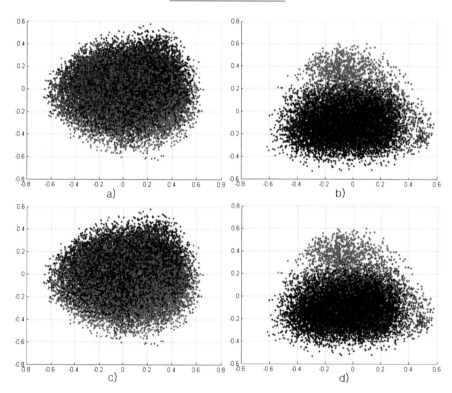

Fig. 5. Dimensionally reduced feature spaces. a) and c) show non-dense feature subset, where for each feature among 5 (a) or 3 (c) its neighbors, one or more features belong to a different class. b) and d) show dense features subset, where among each feature all 5 (b) or 3 (d) neighbors belong to the same class.

5   Texture Model Construction

As it was mentioned, the number of features in each class is large. To generalize and transform the data into a compact form we applied multi-layer perceptron (MLP). This way features are transformed into synaptic weights of the MLP, the quantity of synaptic weights is considerably less than the number of features multiplied by the dimension of a SURF descriptor. Therefore, significant memory reduction and reduction of processing time can be achieved.

Moreover, it has been proven that a neural network with one hidden layer with sigmoid activation functions is capable of approximating any function to arbitrarily accuracy [22]. However, there is no precise answer on the number of neurons necessary for approximation. It was numerically shown [23] that a MLP with two hidden layers often achieve better approximation and requires fewer weights. Fewer weights mean less likeliness to get caught in local minima, while solving the optimization problem that is required in the training process. Thus, we experimented with a MLP with two hidden layers.

The training process is associated with solving a nonlinear optimization problem that should be solved in the training process. Due to the fact that the error function that should be minimized is not convex, the training procedure is very likely to end-up in a local minimum. The error function in most general form is written as follows:

$$\varepsilon(w) = \sum_x \left(f(w, x) - class(x) \right)^2 \longrightarrow min \qquad (2)$$

where $f(w, x)$ is the output of the MLP, x is a training set consisted of feature descriptors, and $class(x)$ desired class label obtained by hand segmentation. $class(x)$ returns three dimensional vector, where only one component set to one, and the others are zeros.

To overcome the problem associated with the training process a number of minimization algorithms have been suggested. We experimented with 'resilient propagation' (RPROP) and the $Levenberg - Marquardt$ (LM) training algorithms [18]. The former one requires less memory, but needs a great number of iterations to converge. The latter one converges in less iteration, and usually finds a better solution, however it requires a great amount of memory. We experimented with the MLP of the following architecture: 36-40-20-3, where 36 is the length of a SURF descriptor, 40 and 20 are number or neurons in the first and the second hidden layers, and 3 is the number of classes. The decision function of the MLP with two hidden layer can be expressed as follows:

$$f(d) = \sum_{i=0}^{3} w_{m,i}^{(3)} \, \sigma \left(\sum_{j=0}^{20} w_{i,j}^{(2)} \, \sigma \left(\sum_{k=0}^{40} w_{j,k}^{(1)} \, d_k \right) \right) \qquad (3)$$

where σ is a sigmoid activation function, $w^{(m)}$ are interlayer weight matrices. The total number of interlayer weighs is equal to $N_w = (36 + 1) \times 40 + (40 + 1) \times 20 + (20 + 1) \times 3 = 2363$ which is considerably less than the total number of feature components $N_c = 41887 \times 36 = 1507932$. The computational complexity does not depend on the size of the training set, yet only depends on the number of neurons.

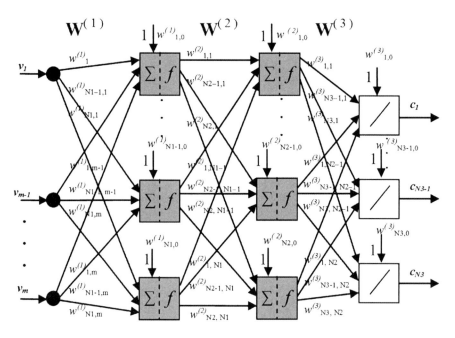

Fig. 6. The structure of a MLP with two hidden layers. $m = 3$, $N_1 = 40$, $N_2 = 20$, $N_3 = 3$

6   Segmentation Procedure

The segmentation procedure consists of two steps. First SURF features are extracted from the input image. Descriptors of each extracted feature are processed with a classifier, resulting in a 3-dimensional vector p, where 3 is the number of terrain classes. The vector's components represent how likely the descriptor belongs to the corresponding class. These values along with features' spatial positions $l_k = (x_k, y_k)$ are utilized by the segmentation algorithm. The algorithm is summarized in the following list of steps:

Algorithm 1 (Segmentation algorithm)
  Input: Texture features $f_j = \{l_j, d_j\}$, $j = 1 \ldots N$ where N is the number of features and d is a feature descriptor.
Pixels coordinates $q_i = (x_i, y_i)$, $i = 1 \ldots$ width $\times$ height
1. Select indexes of extracted features located within the radius r around pixel $q_i = (x_i, y_i)$:

$$T(l) = \{j | \|q_1 - l_j\| < r, j = 1 \ldots N\} \tag{4}$$

2. Estimate the membership value $P_{1\ldots3}(q_i)$ of a pixel q_i to belong to each of the three classes:

$$P_{1\ldots m}(q_i) = \frac{1}{\#(T(c))} \sum_{k=1}^{\#(T(c))} f(d_k) \cdot \frac{1}{\sigma\sqrt{2\pi}} e^{-\frac{\|q_i - l_k\|^2}{2\sigma}} \tag{5}$$

3. Choose the class with maximum probability:

$$class = ind(q_i) = arg\left(max\left(P_{1\ldots m}(q_i)\right)\right) \tag{6}$$

Repeat steps 1-3 for all the pixels.
  Output: Segmented image

If the maximum of estimated **p** is lower than a predefined threshold, then the pixel's class is mark as unknown.

7 Feature Tracking

To reduce the calculation time and minimize segmentation blinking caused by classifying the same region to different classes from frame to frame, we match features in the current frame with features extracted in the previous frame. If a corresponding point is found, the membership values calculated in the previous frame are transferred to corresponding points in the current frame. The correspondence is found by calculating Euclidian distance between feature descriptors in images $I_{(i-1)}$ and I_i. To optimize the calculation time, the searching area in I_i is restricted by a circle defined by a feature coordinates in $I_{(i-1)}$ and a radius r. The feature with a minimum Euclidian distance in the 36-dimensional space is considered as a match if the distance is less than a predefined threshold. Based on the found matches a fundamental matrix F is estimated. A fundamental matrix F is the unique 3×3 rank 2 homogeneous matrix which satisfies:

$$m'^T F m = 0 \tag{7}$$

for all corresponding points (m, m'). Another definition of the fundamental matrix is $l' = Fm$ for the epipolar line l'. Since m' belong to l', $m'^T l' = 0 \rightarrow m'^T Fm = 0$. We only need 8 points to solve equation (7) and find F. However, this solution would not be robust due to the fact that the feature matching results are not always correct and my contain outliers. To filter them out and find fundamental matrix the RANSAC algorithm is applied [21]. Generally it contains the following steps:

1. A random subset of N points correspondence is selected and then equation (7) is solved using singular value decomposition.
2. The total number of inliers in the whole set of point correspondence is determined. A point is considered inliers if the distance to its epipolar line is less than ε.
3. Step 1 and 2 is repeated until the largest inliers is found for a particular F.

As soon as a fundamental matrix is estimated an essential matrix E is calculated in the following way:

$$E = W^T F W, \qquad (8)$$

where W is a matrix of intrinsic parameters. Then E is decomposed [24] into $P = A[I|0]$, $P' = A[R|t]$. P and P' are calibrated perspective projection matrices (PPM). Using reconstructed PPMs and feature coordinates we can triangulate them and obtain 3D coordinates M (fig. 7).

As soon as a metric 3D reconstruction of a current camera position and orientation is provided the prediction of the next position and orientation can be performed [25]. Then, predicted PPM is applied to previously reconstructed 3D points M to obtain a projection onto an image plane. To predict the cameras position and orientation we use two separate trackers. For tracking the camera position we use the regular Kalman filter [26] and to predict camera orientation we applied Extended Kalman Filter, where orientation is represented using quaternion notation [27].

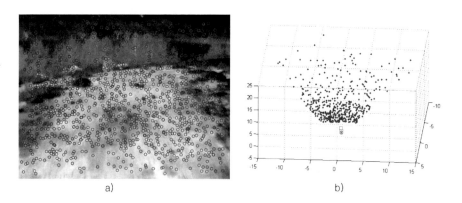

a)   b)

Fig. 7. a) Detected SURF features in one image in the pair, b) Reconstructed points from a pair of frames, green squares represent positions of cameras, and the blue circle show predicted position of the camera

8   Experimental Results

To test the segmentation algorithm we run experiments on two image sets:
the training set and the testing set. The training image set consists of 70
images, and the testing set contains 10 images. We experimented with MLP
trained by RPROP and LM algorithms. Both networks have the same above
described architecture. The mean, standard deviation, minimum and maximum segmentation error rates are shown in table 5.

Table 5. Error rates using different classifiers and data sets

Classifier	mean	std	min	max
NN (training set)	5.03	1.66	2.89	10.7
MLP 40-20 LM(training set)	15.25	5.26	7.27	29.31
MLP 40-20 RPROP(training set)	18.90	6.42	8.95	37.30
NN (test set)	16.67	9.17	9.86	32.1
MLP 40-20 LM (test set)	19.1	10.39	12.5	37.52
MLP 40-20 RPROP(test set)	21.51	9.72	13.14	38.29

Fig. 8. The first and second rows shows segmentation results obtained for test and
training images respectively, when nearest neighbor classifier was applied.

The lowest error rate is achieved with the nearest neighbor classifier (NN). The NN classifier looks for a nearest feature and assigns the nearest feature's class to the input feature. The distance to a nearest feature is used to calculate the strength associated with the assigned class [18]. Because in the case of NN classifier all training directly used in the classification process, the segmentation error is the lowest compared to the MLP classifier. On the other hand MLP converts the training set into a compact set of interlayer weights that considerably reduce memory demands and decrease calculation time. The MLP trained by the LM algorithm shows comparably low segmentation error rate followed by the MLP trained by RPROP. In [17] a network with 40 neurons in each hidden layer was trained with the LM algorithm. Here we reduced the number of neurons in the second hidden layer to 20. We noticed that each time the network is retrained different segmentation accuracy is obtained. Moreover the increase of number of neurons up to 60 neurons in each hidden layer (experiments not shown) did not lead to better quality of segmentation. One reason is that the network gets stuck in local minima as more variable/weights makes the optimization problem more

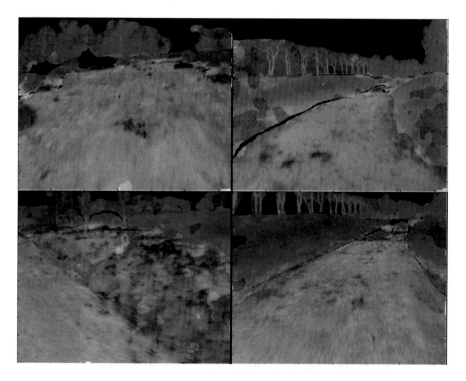

Fig. 9. The first and second rows shows segmentation results obtained for test and training images respectively, when MLP classifier trained by LM algorithm was applied.

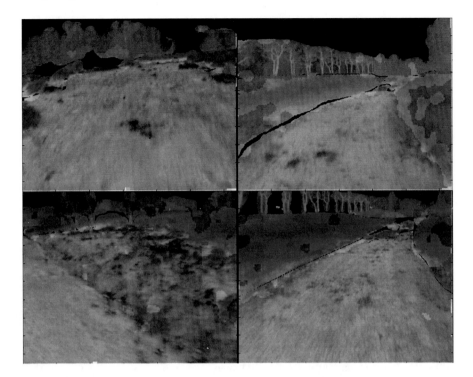

Fig. 10. The first and second rows shows segmentation results obtained for test and training images respectively, when MLP classifier trained by RP algorithm was applied.

complex. Another reason of worse classification especially on the test data is poorer generalization of a network with large number of weights. In this case the network memorizes so-called noisy data. The segmentation results obtained with the NN and MLP trained by the LM and RPROP algorithms are shown in figures 8, 9 and 10.

The lowest error rate is achieved with the nearest neighbor classifier (NN). MLP trained by LM algorithm shows comparably low segmentation error rate followed by MLP trained by RPROP. In [17] a network with 40 neurons in each hidden layer was used. Here we reduced the number of neurons in the second hidden layer to 20. We noticed that each time the network is retrained different segmentation accuracy is obtained.

9   Conclusions

We compared original and open SURF implementations. We found that upright version of original implementation gives slightly higher ratio between the number of extracted features and the number of correctly detected corresponding points than the ratio calculated for the OpenSURF implementation.

The features extracted from the training image set were analyzed with the subject to find the quality of separability of terrain classes represented in the 36-dimensinal SURF features space. We demonstrated that SURF features indeed cares meaningful for terrain segmentation information. Then, we applied segmentation algorithm that extracts 36 dimensional SURF features and estimates class membership values for each pixel. We considered three terrain classes: grass, bushes and trees. The proposed segmentation approach effectively supplements those vision systems that use salient features for 3D reconstruction and object recognition. We applied MLP for class membership estimation. The segmentation system based on the network estimator shows slightly higher segmentation error rate compared to the estimator based on nearest neighbor algorithm. Nevertheless, the number of interconnecting weights is significantly lower than the number of training samples, which leads to considerable memory and computational savings.

References

1. Lalonde, J.-F., et al.: Natural terrain classification using three-dimensional ladar data for ground robot mobility. Journal of Field Robotics 23, 839–861 (2006)
2. Manduchi, R.: Learning Outdoor Color Classification. IEEE Transactions On Pattern Analysis And Machine Intelligence 28(11), 1713–1723 (2006)
3. Manduchi, R., et al.: Obstacle detection and terrain classifcation for autonomous off-road navigation. Autonomous Robot 18, 81–102 (2005)
4. Jansen, P., et al.: Colour based Off-Road Environment and Terrain Type Classification. In: 8th International IEEE Conference on Intelligent Transportation Systems (2005)
5. Sung, G.-Y., Kwak, D.-M., Lyou, J.: Neural Network Based Terrain Classification Using Wavelet Features. Journal of Intelligent and Robotic Systems 6 (2010)
6. Castano, R., Manduchi, R., Fox, J.: Classification experiments on real-world texture. In: Third Workshop on Empirical Evaluation Methods in Computer Vision, Kauai, Hawaii (2001)
7. Lee, T.S.: Image Representation Using 2D Gabor Wavelets. IEEE Transactions On Pattern Analysis And Machine Intelligence 18 (1996)
8. Kang, Y., et al.: Texture Segmentation of Road Environment Scene Using SfM Module and HLAC Features. IPSJ Transaction on Computer Vision and Applications 1, 220–230 (2009)
9. Kurita, T., Otsu, N.: Texture Classification by Higher Order Local Autocorrelation Features. In: Asian Conference on Computer Vision (1993)
10. Otsu, N., Kurita, T.: A New Scheme for Practical Flexible and Intelligent Vision Systems. In: IAPR Workshop on Computer Vision (1988)
11. Blas, M.R., et al.: Fast color/texture segmentation for outdoor robots. In: International Conference on Intelligent Robots and Systems, Nice, France (2008)
12. Konolige, K., et al.: Mapping, Navigation, and Learning for Off-Road Traversal. Journal of Field Robotics 26(1), 88–113 (2009)

13. Lensky, A.A., et al.: Terrain images segmentation and recognition using SIFT feature histogram. In: 3rd Workshop on military robots, Adgency for Defence Development, KAIST (2008)

14. Lowe, D.G.: Object recognition from local scale-invariant features. In: Proceedings of the International Conference on Computer Vision, vol. 1, pp. 1150–1157 (1999)

15. Lenskiy, A.A., Lee, J.-S.: Rugged terrain segmentation based on salient features. In: International Conference on Control, Automation and Systems, Gyeonggi-do, Korea (2010)

16. Rasmussen, C.: Combining Laser Range, Color, and Texture Cues for Autonomous Road Following. In: International Conference on Robotics & Automation, Washington, DC (2002)

17. Lenskiy, A.A., Lee, J.-S.: Terrain images segmentation in infra-red spectrum for autonomous robot navigation. In: IFOST 2010, Ulsan, Korea (2010)

18. Lenskiy, A.A., Lee, J.-S.: Machine learning algorithms for visual navigation of unmanned ground vehicles. In: Igelnik, B. (ed.) Computational Modeling and Simulation of Intellect: Current State and Future Perspectives, IGI Global (2011)

19. Bay, H., et al.: Speeded-Up Robust Features (SURF). Computer Vision and Image Understanding 110(3), 346–359 (2008)

20. Gossow, D., Paulus, D., Decker, P.: An Evaluation of Open Source SURF Implementations. In: RoboCup 2010: Robot Soccer World Cup XIV (2010)

21. Fischler, M.A., Bolles, R.C.: Random Sample Consensus: A Paradigm for Model Fitting with Applications to Image Analysis and Automated Cartography. Comm. Of the ACM 24, 381–395 (1981)

22. Cybenko, G.: Approximation by superposition of a sigmoidal function. Math. Control. Signals, Syst. 2, 303–314 (1989)

23. Chester, D.L.: Why Two Hidden Layers are Better than One. In: Proc. Int. Joint Conf. Neural Networks, Washington, DC (1990)

24. Hartley, R., Zisseman, A.: Multiple View Geometry in Computer Vision, 2nd edn. Cambridge University Press, Cambridge (2003)

25. Lensky, A.A., Lee, J.S.: Tracking feature points in static environment. In: Conference on Ground Weapons, Daejon, South Korea (2009)

26. Brown, R.G., Hwang, P.Y.C.: Introduction to Random Signals and Applied Kalman Filtering with MATLAB Exercises and Solutions, vol. 496. John Wiley & Sons, Chichester (1997)

27. Goddard, J.S.: Pose and Motion Estimation From Vision Using Dual Quaternion-based extended Kalman Filtering, The University of Tennessee, Knoxville (1997)

Part III

3D Virtual Reality Environments

Virtual Reality as a Surrogate Sensory Environment

Theodore W. Hall, Mojtaba Navvab, Eric Maslowski, and Sean Petty

University of Michigan, Ann Arbor, Michigan, USA
twhall@umich.edu, moji@umich.edu, emaslows@umich.edu, sepetty@umich.edu

This chapter examines certain aspects of virtual reality systems that contribute to their utility as surrogate sensory environments. These systems aim to provide users with sensory stimuli that simulate other worlds. The fidelity of the simulation depends on the input data, the software, the display hardware, and the physical environment that houses it all. Robust high-fidelity general-purpose simulation requires a collaborative effort of modelers, artists, programmers, and system administrators. Such collaboration depends on standards for modeling and data representation, but these standards lag behind the leading-edge capabilities of processors and algorithms. We illustrate this through a review of the evolution of a few of the leading standards and case studies of projects that adhered to them to a greater or lesser extent. Multi-modal simulation often requires multiple representations of elements to accommodate the various algorithms that apply to each mode – for example, alternative geometries for visualization, auralization, and collision detection. Tools and algorithms to assist in the extraction of these representations from common base data will expand the pool of high-quality multi-modal simulations. In the final analysis, the output stimuli depend on aspects of the display hardware and its physical setting that might not be adequately accounted for by idealistic algorithms. It is important to measure these actual stimuli in order to validate and fine-tune the simulation system.

1 Introduction

Virtual Reality (VR) aims to elicit sensory responses to simulated environments. Although the sensory responses of interest are most often human, VR may be directed as well toward animals or machines. In any simulation, it is important to determine at the outset: the critical stimuli; the VR system's capabilities to produce them; and the expected range of sensory and behavioral responses to those stimuli.

VR simulations frequently serve as surrogates for inaccessible physical environments. The physical environment might be too dangerous or too expensive to construct or visit in person. It might be a distant target of remote

T. Gulrez, A.E. Hassanien (Eds.): Advances in Robotics & Virtual Reality, ISRL 26, pp. 251–273.
springerlink.com

sensing and manipulation in real time (or as close to real time as the speed of light will allow), or it might be a historical situation that no longer exists, or a hypothetical situation that does not yet exist. These simulations put a premium on realism: they aim to present a natural interface and elicit the same sensations as the physical environment, as near as possible without endangering the user. The motivation may be to inform the design of the environment itself, or to develop and exercise systems and procedures to operate there.

VR simulations also serve as tangible sensible metaphors for structures and phenomena that normally evade the senses due to scale or abstraction, such as the inner workings of an atom, the evolution of a galaxy, or a bioinformatics graph relating genes, diseases, and symptoms. Here the emphasis is on symbolism, clarity, consistency, and strategy in allocating the dimensions of stimulus to the parameters of the structure or phenomenon. The motivation is to develop a cognitive model as a basis for exploration and discovery.

Whatever the motivation, the means is sensory stimulation. Progress in virtual reality is measured in terms of the system's capabilities to produce content-rich multi-modal stimuli in real time in response to intuitive (or at least easily mastered) user actions. This chapter examines progress in virtual reality along several fronts:

- Section 2 reviews the technical evolution of system capabilities, user expectations, and standardization.
- Section 3 discusses the stimulus and sensory pathways with an emphasis on measurement, calibration, and validation.
- Section 4 concludes with summary observations regarding the current state of virtual reality as a source of sensory stimuli and offers thoughts about the path forward.

2 Technical Evolution

Exponential increases in computing power, along with other innovations in display system hardware and software, have enabled dramatic improvements in the quality of simulation. Newer systems are capable of more modes of sensory stimulus (such as vision, hearing, and touch) with greater detail at higher frame rates. More physical phenomena can be computed in real time (rather than precomputed and "baked in"), allowing for higher fidelity interaction.

Nevertheless, the quality of a simulation depends not only on the hardware and software, but also on the data – in particular, on the artistry and craftsmanship invested in creating scene assets. This is well-known in the entertainment and gaming industries,[1] but sometimes not fully appreciated in academic and research settings where "art" might be seen as irrelevant or even antithetical to "science." A skilled content creator will tailor scene assets to the capabilities of the display system – for example, in regard to polygon count, textures, lighting, and levels of detail – to achieve an optimum balance of frame quality and frame rate.

2.1 Standards for Environmental Representation

A robust simulation system must support certain standards for environmental representation. It is standardization that allows a broad clientele of content creators to prepare models, using their tools of choice, with reasonable expectation that they can be imported and displayed by the simulation software and hardware. Though standards usually lag behind the ultimate capabilities of leading-edge systems, their evolution echoes the progress of common capabilities and expectations.

The Virtual Reality Modeling Language (VRML) is representative of simulation capabilities that emerged in the early 1990s. VRML version 1 was practically a clone of Silicon Graphics' OpenInventor version 2.0 file format, first published in 1994 [1]. VRML version 2, also known as VRML97, was adopted as International Standard ISO/IEC 14772 in 1997 [2]. The standard specifies a file format, a content model, and algorithms for its interpretation. The model is a directed acyclic graph that includes nodes for geometry, color, texture, and light, as well as sound. However, it provides for only one texture per shape and one pair of texture coordinates per vertex. Strict adherence to the "ideal VRML implementation" actually *prohibits* calculation of shadows and indirect lighting – it specifies equations that do not accommodate such effects. The emphasis is on real-time navigation and manipulation of a virtual world at a frame rate sufficient for animation, rather than high-quality rendering per frame. In 1997 there was little expectation that systems could compute shadows and indirect lighting in real time. Moreover, the techniques for doing so remain experimental and not ready for standardization. Today, the VRML97 lighting equations may be best understood as a *minimal* implementation, rather than an *ideal* one. The VRML97 sound specification is similar in its lack of provision for occlusion, absorption, reverberation, and similar dynamic effects. The model does not encode enough material properties to allow for higher-quality rendering of either light or sound.

Early applications of virtual reality exhibit the rather cartoon-like rendering accommodated by the VRML97 standard and the systems of the era. The lack of bump maps and normal maps results in flat lighting of rough textures, appearing more like wallpaper rather than truly rough surfaces. The lack of

---

[1] As a case in point, consider `http://www.creativeheads.net/` – a job site "for the video game, animation, VFX, and software/technology industries." As of 3 Feb. 2011, a search for "programmer" matched 45 jobs, "software" matched 127, "artist" matched 177, and "art" matched 184.

shadows and indirect lighting impairs the sense of emplacement. Users' emotional responses often focus more on the simulation technology rather than the simulated environment. Still, the simulations are adequate for recognition, manipulation, and navigation through scenes. Simple two-dimensional projections allow for sight-line and visibility analysis, and three-dimensional stereographic projections provide a sense of immersion and scale. Some examples follow in Sect. 2.3.

The VRML97 standard was amended in 2002 to add geospatial and NURBS support, but the shape, material, lighting, and sound specifications remain unchanged [3].

The Extensible 3D (X3D) standard, "the successor to the Virtual Reality Modeling Language," was adopted as ISO/IEC 19775 in 2008 [4]. It preserves much of the structure of VRML97 but recasts it in an XML schema and extends it in several significant areas. In particular, it provides for multi-texturing and multiple vertex attributes, and adds components for humanoid animation, programmable shaders, 3D textures (for volume rendering), rigid body physics, and particle systems. The programmable shaders allow simulations to override the default lighting algorithm with bindings to the GLSL, HLSL, and Cg shading languages. The sound model, however, remains essentially unchanged from the VRML97 standard.

The COLLADA^{TM} (COLLAborative Design Activity) format is also an XML schema. It originated from Sony Computer Entertainment but has since been adopted as an open standard by the Khronos Group, a not-for-profit industry consortium. Version 1.0 was released in 2004, and version 1.5.0 in 2008 [5, 6]. Its aim is to provide a universal import and export format among the variety of applications that may contribute to a digital-content-creation pipeline, including architectural and engineering modeling, sculpting, scene composition, and animation. It provides for multi-texturing, animation, programmable shaders, rigid body physics, and kinematics, but says nothing about sound.

There are many other "standard" file formats – far too many to enumerate here.[2] These are almost entirely focused on the properties necessary to render a visual stimulus: geometry, color, texture, light, and mechanisms for animating them.

Meanwhile, systems for high-quality real-time auralization (sound rendering) must rely on other formats. For example, Noisternig et al. [7] describe an open-source auralization framework that assigns an acoustic material to each polygon, defined by octave band absorption coefficients (from 32 Hz to 16 kHz) and a scattering coefficient. Such properties are absent from all of the leading model exchange formats. The framework incorporates the Virtual Choreographer software for model loading, which defines its own custom format, comparable to but distinct from VRML97 and X3D [8]. Even so,

---

[2] As of 21 Feb. 2011, version 4.40 of the 3D Object Converter, available at http://web.axelero.hu/karpo/, claimed to support 543 formats.

the auralization relies on a separate material file to define the octave-band absorption coefficients.

Haptics and force-feedback are closely related to real-time physics. Rigid-body physics simulation adds requirements for properties of mass, density, force, velocity, static and dynamic friction, restitution, and non-hierarchical joints and constraints between bodies that don't adhere to the scene-graph paradigm. Cloth and soft-body simulation adds bending and stretching stiffness and tensile strength or tearing factors. Fluid simulation adds viscosity and surface tension. None of the aforementioned standards accommodate all of these material and attachment properties. VRML97 provides for basic collision detection between the navigation avatar and rigid scene geometry. X3D and COLLADA add general rigid bodies and joints, but lack cloth, soft bodies, and fluids. The NVIDIA PhysX^{TM}software development kit supports all of these concepts, but relies on a custom XML format for conveying them [9]. Content-creation software must export graphics and physics aspects in separate files, leaving it to the VR simulation system to separately import and reunite the aspects in coherent entities.

Finer-grained tactilization might be partially achieved by re-purposing or overloading graphic properties – for example, using visual texture maps to convey tactile grain – but this is not yet well-developed in the virtual-reality context. A temperature attribute is conceptually simple, but none of the aforementioned standards include it. The concept of tactilization is most often applied in the context of making visible images accessible to the visually impaired – as with braille displays and printers. This often aims at substituting touch for sight in an abstract manner, rather than an accurate simulation of touch as such [10].

Olfaction for virtual reality (involving the sense of smell) is still in its infancy. The literature focuses primarily on the hardware and software for detection (input) and delivery (output) of aromas and odors, but says very little about how to encode and convey olfactory data in a generalized way. Whereas standards for recording, encoding, and reproducing visual and auditory stimuli have been well established for many years, no such standard exists for olfaction [11, 12].

Another complicating factor is that different aspects of a multi-modal real-time simulation may require alternative representations of physical entities. For example, auralization may require fewer, larger polygons than visualization, because the classical laws of geometrical acoustics consider reflecting surfaces to be much larger than the wavelength of the sound [13]. (For frequencies between 32 Hz and 16 kHz, the wavelength ranges from 10.7 m down to 21.4 mm.) Meanwhile, real-time physics collisions typically require *convex* shapes, but neither visualization nor auralization impose that restriction.

Ideally, a standard will soon emerge that accommodates all of the properties necessary for plausible real-time visualization, auralization, and tactilization, animated by properties of physics and kinematics. The algorithms and hardware systems for these sensory modes are already well developed and widely used. Other sensory simulation modes, such as smell and taste, are still experimental and not ready for standardization. Since the XML-based standards (such as X3D and COLLADA) are explicitly intended to be *extensible*, they seem to offer the best path forward.

2.2   Content Creation and Display

A content creator for a multi-modal VR simulation is often faced with the task of creating multiple representations of scene objects. For example, visible geometry of arbitrary complexity might need to be approximated by a small number of invisible convex primitives for real-time physics. Similarly, auralization might require the creation of invisible simplified geometry for acoustic boundaries. Populating a scene with even a small number of individualized virtual people becomes very time-consuming if each one involves a unique exercise in modeling, rigging, and animating. The specialized demands on time, skill, and software tools impede the creation of these simulations. Projects often originate from groups with application-specific expertise but limited in-house resources for modeling. A shared interdisciplinary VR facility may possess the knowledge, skill, and software, but lack the personnel to hand-craft every project for every client. Anything that can be done to streamline the creation of multi-modal scene assets will increase the pool of these assets and the overall quality of simulation. We have been addressing this need along several fronts.

For improved visualization, including real-time shadows and indirect lighting, we are developing an advanced rendering engine that implements algorithms published by Kaplanyan and Dachsbacher [14], and others developed in-house, taking advantage of modern graphics hardware with GLSL shader programs. Aside from providing a more interactive and immersive experience, this also spares the content creators from having to pre-compute and "bake in" static lighting effects with their textures.

As a first pass for automated physics content, our system provides an easy way to wrap graphic entities in invisible axis-aligned bounding boxes of a specified density. However, this is often a poor approximation of the visible form and yields unnatural behavior. Simple convex hulls are often not much better. For the next phase, we are investigating procedures for automatic convex decomposition of non-convex entities, based in part on algorithms by Gottschalk, Lin, and Manocha [15], and by Lien and Amato [16].

To populate scenes with realistic humans, we have developed a universal character system that allows the content creator to rapidly derive individuals from weighted averages of a few predefined prototypes. By maintaining a consistent skeleton and mesh topography but interpolating the vertex offsets according to a few general classifications, we are able to produce a large variety of body forms with minimal custom modeling.

2.3   Example Projects

A few examples may serve to illustrate the points in the preceding sections. The following projects are selected from the authors' experiences in the University of Michigan Virtual Reality Laboratory (VRL) and 3D Lab (UM3D).[3] They show a progression in visual fidelity and overall realism from early projects to current developments.

We continue to develop an in-house virtual-reality system, *Jugular*, to address issues related to content creation, exchange, and high-fidelity real-time multi-modal display. The system integrates various open-source and custom modules for file loading, graphics, sound, animation, and physics. It supports monocular and binocular perspective projection in anaglyph, split-screen (passive), and active stereo, on the desktop, the stereo wall, or the four-screen immersive enclosure.

Acrophobia Treatment

Between 1996 and 1998 we worked with the Department of Psychiatry at the University of Michigan Medical School to develop an elevator simulation for the treatment of acrophobia (fear of heights). This was part of a clinical trial to compare the efficacy of treatment in virtual and real worlds [17]. Subjects diagnosed with acrophobia would be randomly assigned for progressive desensitization in either the virtual or the real elevator. As shown in Fig. 1, modeling of the virtual elevator aimed for a good degree of realism, but explicitly omitted certain features that were deemed too difficult to achieve at the time, such as window reflection, tinting, and dust (which would provide visual cues that a glass barrier is present), and animated people or animals outside (which would provide a sense of height and scale). Conversely, to control for the influence and limitations of VR equipment, subjects in the real elevator would be asked to wear mock VR shutter glasses, and subjects in both environments would be instructed to avoid touching the walls or operating the buttons themselves. Despite the limitations, the VR treatment was effective for at least some subjects. The project left open the questions of which stimuli were most significant, and the extent to which that varies between patients.

Detroit Metro Airport, McNamara Terminal Project

In 1999 we developed a VRML97 model for a new state-of-the-art passenger terminal – the McNamara Terminal – being designed for Detroit Metropolitan Wayne County Airport (DTW). The project included 97 gates, airfield

---

[3] The Virtual Reality Lab (VRL) resided in the College of Engineering from 1993 to 2008. Its successor, the 3D Lab (UM3D) resides in the Digital Media Commons, a division of the Library. As such, it is a service unit for the entire University of Michigan community that provides 3D scanning, 3D printing & rapid prototyping, motion capture, and advanced visualization including virtual reality.

Fig. 1. Virtual elevator for the treatment of acrophobia, 1998

connections via aprons and taxiways, a large parking structure with 11,500 spaces, a multi-level system of access roads, and a power plant. A major design issue was the visibility of the far gates and taxiway from the control tower. The VRML97 model included interactive controls to adjust several design parameters in real time, including the length of the terminal building and number of gates, as well as the height and placement of the tower. (Although the tower already existed and there were no plans to replace it, the model allowed hypothetical variations in its design.)

With neither textures, shadows, nor indirect lighting, the rendering lacked realism. Nevertheless, it was useful as a visualization and design tool especially when viewed full-scale in an immersive stereoscopic VR enclosure, as shown in Fig. 2. Major design decisions derived from insights gained from this model [18, 19].

Fig. 2. Visibility analysis for the new McNamara Terminal at Detroit Metro Airport, 1999. Photograph by Douglas G. Ashley for Northwest Airlines

Virtual Disaster Simulators

In 2004 and 2009 we worked with the Department of Emergency Medicine and the Clinical Simulation Center at the University of Michigan Medical School to develop and operate two renditions of virtual disaster simulators. These aimed for a level of realism and sensory immersion sufficient to elicit emotional responses from "first responders" arriving on the scene to assist the injured [20, 21]. Both simulations included surround-sound effects with localized prerecorded voices attached to characters in the scenes. First responders interacted with each of these in an immersive stereoscopic VR enclosure, as shown in Figs. 3–4.

Fig. 3. Virtual Disaster Simulator version 1, 2004

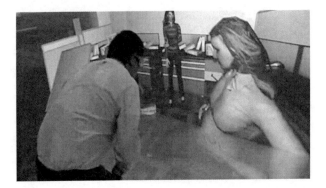

Fig. 4. Virtual Disaster Simulator version 2, 2009

The scenario for the first exercise in 2004 was an explosion at the University's football stadium, which seats over 100,000 spectators. We developed a VRML97 model that included shapes for the stadium, surrounding structures, and a few humans, and applied photographic textures throughout. We modeled humans in VRML97-style as hierarchies of rigid limbs connected at rotation joints. To simulate hands-on treatment, we placed a physical human patient simulator (HPS) in the VR enclosure, and projected victim images onto it from predetermined viewpoints.

For the second exercise in 2009 the scenario was an explosion in an office building. To avoid the limitations of VRML97, we developed this model in a proprietary format adopted from the game industry that supports multiple textures per surface (including bump maps and normal maps as well as diffuse maps) and animated skin-mesh human models on skeletal rigs. Since the emphasis in this exercise was triage rather than treatment, sight and sound were more important than touch. The human models were much higher fidelity, including accurate renditions of injuries such as burns and compound fractures, and the HPS was deemed unnecessary. Medical residents were tasked with locating the victims and rapidly assessing their condition from sight and sound. A live examiner provided vital signs (e.g., pulse) and scripted answers per victim in response to specific questions.

Figure 5 contrasts the simply-textured stick-figure characters of the 2004 model with the multi-textured[4] mesh characters of the 2009 model. The second exercise achieved a considerably greater degree of realism by departing from VRML97 and adopting modern game graphic techniques. Unfortunately, this relies in part on proprietary licensed software unavailable to most content creators outside the Lab, which limits its applicability.

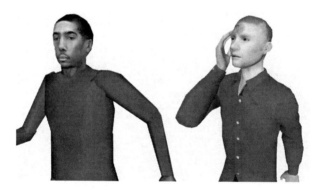

Fig. 5. Example character models for the Virtual Disaster Simulator: (**a**) VRML97 stick figures in version 1, 2004; (**b**) animated meshes in version 2, 2009

We hope that open standards, such as X3D, COLLADA, or a successor will evolve to satisfy the model definition and exchange requirements. Beyond that, the simulation software requires sophisticated algorithms to produce high-quality renderings of the model at a frame rate sufficient for animation in a multi-screen stereoscopic head-tracked display environment.

---

[4] Including bumps, normals, and diffuse colors.

Advanced Renderer

Figure 6 shows a variation of the Crytek Sponza scene [14] (to which we have added a moving curtain in the courtyard) loaded from a VRML97 file and rendered with a single directional light source (a "sun") according to the VRML97 algorithm. Polygons facing toward the sun are bright, regardless of occluding objects. Polygons facing away from the sun are lit by a uniform ambient light component. There is neither shadow nor indirect illumination.

Fig. 6. The Sponza scene in the VRML renderer

Fig. 7. The Sponza scene in the advanced renderer, including dynamic real-time shadows and indirect lighting

In 2010 we rewrote our rendering system to implement algorithms for real-time shadows and indirect lighting, adapted from Kaplanyan and Dachsbacher [14]. It employs GLSL shader programs to take full advantage of modern graphics hardware. Figure 7 shows the same view rendered by the new algorithm. As the curtain moves in the scene: the flag poles cast moving shadows on the curtain; the curtain casts moving shadows on the walls and floor; and the curtain casts moving indirect illumination on the walls.

Of course, high-quality rendering codes have been widely available for many years. For convincing virtual reality, the challenge is to achieve these effects at a frame rate sufficient to animate moving lights, occlusions, and reflecting surfaces, on multiple synchronized stereoscopic screens, in real time. The improved visual fidelity is vital for psychological immersion in the simulation.

Universal Character System

Figure 8 shows samples from the "Universal Character System" that we have developed to populate virtual scenes with a variety of life-like animated human figures, without the need to model each individual from scratch. The system comprises a skeletal rig and associated mesh, and a set of six prototypical body forms: obese, under-weight, aged, male, female, and muscular. The system interpolates the mesh vertex XYZ coordinates from weighted averages of the six prototype forms. The mesh topology, its texture UV coordinates, and its connection to the skeleton, all remain consistent. This saves considerable time in generating individuals, and also facilitates mixing and matching of characters with libraries of skin type, clothing, and skeletal motion data. This does not remove the need for a skilled artist, but this allows the artist to focus on sculpting the fine details.

Again in reference to Fig. 8: when the visual stimulus achieves a certain level of realism and familiarity, once can imagine the added sense of immersion that olfaction might provide.

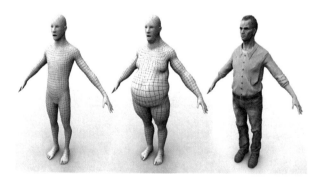

Fig. 8. The Universal Character System: (**a**) base mesh; (**b**) obese prototype; (**c**) a character derived from weighted averages of slightly obese, middle-aged, and male, with final details overlaid in an art package

3 Stimulus and Response

There is an old adage in computing: "garbage in, garbage out." One cannot expect reasonable output from unreasonable input. However, the inverse is not true. That is to say, *not* garbage in (data and algorithms) does not guarantee *not* garbage out. Beyond the input, the latter requires that the various display systems – video, audio, haptic, tactile, olfactory – be properly calibrated. There are limits to what these systems can produce, which are typically well inside the limits of what a healthy person senses in the real world. For example, the full intensity of sunlight is unattainable in a VR simulation with current common hardware. However, human perceptual response is a step removed from raw sensory stimulus. It involves higher-level neural processes that acclimate to ambient conditions. In creating perceptions, ratios are nearly as important as absolute values.

There are at least two good reasons for measuring the physical stimuli provided by VR display systems:

- To determine whether the output is faithful to the input. This is especially important for design simulations that aim to inform design decisions – for example, regarding architectural lighting intensity, material color, acoustic reverberation, speech intelligibility, and so on.
- To gauge the VR system itself as a testbed for human perception and sensory response to stimuli.

To that end, we have invested considerable time and attention to measuring the stimuli delivered by our virtual reality enclosure. The following subsections describe the physical setup and the measurements we have taken.

3.1 Display and Measurement Hardware and Configuration

Figure 9 shows schematic and fish-eye views of our immersive enclosure, which is similar to many CAVE-like systems. It is a cube measuring 3.048 m (10 feet) in width, depth, and height. It runs on a cluster of six workstations, with one control computer, one motion-tracking computer, and four rendering computers.

The renderers are Boxx Tech Workstations, with quad-core CPUs at 2.6 GHz, 8 GB RAM, and NVIDIA Quadro FX 5600 + GSync graphics cards. Four Christie Mirage S+4K projectors produce 3D images on the left, front, right, and floor surfaces. The walls are rear-projected whereas the floor is front-projected from a mirror in the ceiling. The resolution per surface is 1024 x 1024 pixels. The stereo mode is frame-sequential (alternating left-right) at 96 frames per second. Infrared emitters synchronize Stereo Graphics CrystalEyes® liquid crystal shutter glasses with the projectors.

A Vicon MX13 system with eight 1.3-megapixel cameras provides wireless motion-tracking of reflective markers attached to the shutter glasses and to a Logitech RumblePad game controller.

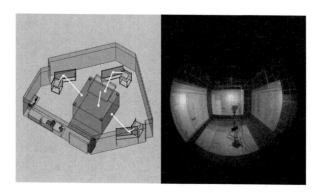

Fig. 9. The immersive VR enclosure: (**a**) schematic of the room housing the enclosure, with arrows indicating the light paths; (**b**) fish-eye view

The sound system comprises four Klipsch speakers mounted in the upper corners of the enclosure, a Klipsch subwoofer on the floor a short distance away, and an amplifier rated at 100 watts per channel. Alternatively, it also supports high-fidelity stereo headphones. It uses the motion-tracking data to account for the position of the user's head relative to the speakers as well as to the virtual sound sources.

The software is an ongoing in-house development, named *Jugular*, that integrates several open-source, proprietary, and custom-developed subsystems for graphics, sound, animation, physics, motion-tracking, data management, and networking.

Figures 10–11 show typical sensor setups for our measurements. We mounted spectrometers and microphones to a manikin head on a tripod and used a notebook computer to collect and monitor the readings. We took spectrometer readings both with and without the shutter glasses on the manikin. To account for the head-related transfer function (HRTF) in acoustics, we placed the microphones in anatomically accurate ear canals that we embedded into the manikin. We manufactured the external ears and the canals from 3D scans of an actual ear. In addition to the manikin instrumentation, we also used fish-eye photography and a 48-microphone acoustic camera (shown next to the manikin in Fig. 11) to capture the complete visible and audible field.

Figure 12 shows a small video camera mounted to a hat and aimed at the wearer's eye in order to measure the pupil diameter in response to changing light conditions. The pupil diameter is an indicator of the degree to which the eye's rods and cones are activated by the scene's luminance.

3.2 Light Measurements

Figures 13–14 are graphs of the spectral power distribution (SPD) for absolute irradiance as a function of wavelength, obtained from the spectrometers,

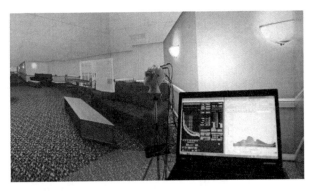

Fig. 10. Equipment setup for video and audio measurements in the VR enclosure: instrumented manikin head, tripod, and notebook computer. The background image is a projected virtual scene

Fig. 11. Equipment setup for video and audio measurements in the VR enclosure. The manikin head is fitted with spectrometers behind the shutter glasses to account for the glasses' absorption and tint, and microphones in the ears to account for the acoustic head-related transfer function. The 48-microphone acoustic camera is next to the manikin. The background image is a projected virtual scene

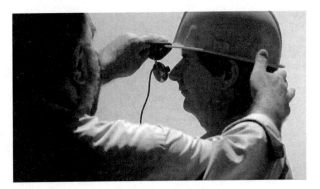

Fig. 12. A video camera attached to the hat monitors the wearer's pupil diameter in response to luminance levels in the VR enclosure

for white surfaces projected into the enclosure. In both graphs, we obtained the upper curve by aiming the spectrometer directly at the ceiling-mounted projector that shines down onto the floor, with nothing intervening between the projector and the spectrometer. We obtained the next lower curve by aiming the spectrometer at the front screen. Assuming that the front projector is equivalent to the (ceiling-mounted) floor projector, the difference between the curves represents the transmission loss through the front screen.

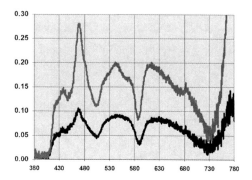

Fig. 13. Spectral power distribution (SPD) with absolute irradiance (μw/cm^2/nm) as a function of wavelength (nm) for a white surface projected in the enclosure. The top curve was measured directly from the ceiling-mounted projector (shining down onto the floor); the lower curve was measured from the front screen. Assuming the two projectors are equivalent, the difference is the transmission loss through the screen

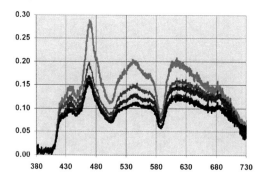

Fig. 14. Spectral power distribution (SPD) with absolute irradiance (μw/cm^2/nm) as a function of wavelength (nm) for a white surface projected in the enclosure. Similar to Fig 13, the top curve is from the projector; the lower curves include successive transmission losses through the screen, screen plus inactive shutter glasses, and screen plus active shutter glasses

In Fig. 13, the spike at the right edge of the graph, between 730 and 780 nm, is almost entirely due to the near-infrared LEDs surrounding each of the motion-tracking cameras mounted above the enclosure walls. In fact, the cameras operate at 780 nm. These LEDs also emit visible red light between 730 and 750 nm. (Above 750 nm is not visible to humans.) A comparatively tiny amount of visible red light also arises from the infrared LEDs that synchronize the shutter glasses with the projectors. The upper curve captures all of this LED light directly, whereas the lower curve captures the portion reflected from the front screen.

In Fig. 14, the domain is restricted to wavelengths between 380 and 730 nm. For the top two curves we removed the shutter glasses from the manikin. For the next lower curve we placed inactive shutter glasses (with the lenses held "clear") in front of the spectrometer, whereas for the lowest curve we placed active shutter glasses (with the lenses alternating between "clear" and "opaque"). The lowest curve is most representative of the stimulus delivered to a human user of the system.

The differences in magnitude between Figs. 13 and 14 are due to different placements of the spectrometer. For Fig. 13, we placed it close enough to the front screen to omit the influence of the left and right side screens. For Fig. 14, we placed it in the center of the enclosure, where it picked up additional light from the sides. Also, measuring through active shutter glasses is problematic due to the non-synchronized frame rates of the glasses and the spectrometer. In these measurements, the ratios across the domain of wavelengths are more significant than the absolute values.

Although the transmission losses through the screen and shutter glasses are significant, there is no significant wavelength bias. Besides projecting white surfaces, we also projected pure primary red, green, and blue surfaces onto different screens and measured the cross-talk between them. We also encoded and projected color samples from the Munsell Color Order System, and used correlated color temperature (CCT), color rendering index (CRI), and color quality scale (CQS) to evaluate actual versus intended color output [22]. These tests verify that color rendition is very good: 96 on a CQS scale of 0–100. They also provide quantitative guidance for fine-tuning.

Measurements of pupil diameter, as shown in Fig. 12, verify that the range of illumination levels in the enclosure is adequate to stimulate scotopic, mesopic, and photopic vision,[5] though the dark and light extremes of human vision are not achievable. Incidental sources of ambient light in the enclosure limit the dark end of the scotopic range, and the projectors lack the power to reach the bright end of the photopic range. Still, the system achieves a simultaneous contrast ratio of 10 : 1 and an overall luminance ratio (not simultaneous) of 100 : 1.

---

[5] Scotopic vision is the "dark" range from rod threshold to cone threshold, in which there is no sensitivity to color. Mesopic vision is the intermediate range from cone threshold to rod saturation. Photopic vision is the "bright" range above rod saturation.

3.3 Light versus Sound

Light emittance and reflectance throughout a scene are encoded as intensities of red, green, and blue (RGB) primaries on a scale of 0 to 1. For emitters these represent fractions of the projectors' peak intensity. For reflectors these represent scale factors for the reflected light as a fraction of the incident light. Ultimately, RGB intensities for each of the millions of pixels are precisely calculated according to some algorithm with reference to a common basis. Textures that are created outside the control of the algorithm (e.g., from digital photography) are nevertheless assumed to represent surface reflectance under "normal" lighting conditions (typically daylight) compatible with the algorithm. All of these influences are commingled at run-time to arrive at RGB intensities for each pixel.

The rendering process for sound is qualitatively different. Sound sources may be recorded at vastly different sensitivities – for example: a whisper versus a public announcement versus a church organ versus a jet engine. Sound is not reducible to a small number of "primary" tones or pitches the way light is, but rather is analyzed according to octave bands and center frequencies, with each band representing a doubling of frequency. When placed in a common scene, the algorithm distributes and mixes these contributions across a very small number of speakers, according to the relative proximities of the sources and their attenuation characteristics. In many enclosures, including ours, the ratio of rendered light elements (pixels) to rendered sound elements (speakers) is over a million to one. The overall intensity of the sound output is much more readily and arbitrarily adjustable than the light output. It involves merely turning a single knob on an amplifier, versus stepping through hierarchical menus on a projector. Therefore, the types of measurements and evaluations applicable to sound are different than those for light.

In particular, we are interested in the system's ability to localize the perception of a virtual sound source anywhere within the scene by balancing its loudness across just four speakers. To a certain extent, simulations can "cheat" because audible things are often also visible. Visual localization provides perceptual reinforcement for audio localization. Where a sound source is visibly occluded, it is often audibly occluded as well, and localization is more ambiguous.

Okamoto et al. [23] describe a system for real-time room auralization that uses 157 speakers arrayed across the ceiling and four walls. This system is strictly for auralization. It would be challenging to integrate this with a visualization system such that neither system occluded the other.

3.4 Sound Measurements

The rear-projection fabric screens that comprise the three walls of the VR enclosure are acoustically transparent at low frequencies. In effect, the enclosure inherits the acoustic properties of the room that houses it, shown in the left half of Fig. 9. The T60 reverberation time for this room (for a sound to decay to one thousandth its original level) is between 0.5 and 0.6 seconds. The ambient noise level with all of the equipment running is 43 dB(A).

To test the localizability of virtual sound sources rendered by the enclosure's four speakers, we configured a scene with eight virtual sources. We positioned four to correspond with the locations of the physical speakers and the other four at intermediate locations. Each of these eight sources played a short excerpt of a Mozart string quartet, chosen for its range of tones. We configured the eight sources to play one at a time, rather than simultaneously, and used both subjective and objective means to measure their apparent versus intended locations.

Figure 15 shows the results of the subjective test. We projected a visible grid on each of the surfaces with no indication of the sound source locations. We asked nine students, one by one, to use a laser pointer to locate each of the apparent sound sources on the grid, relying solely on their hearing. There was some spread in their responses for each of the eight locations – the ones aligned with the actual speakers as well as the intermediate ones. Most of the students located each of the eight virtual sources within about $\pm\pi/8$ radians of its intended location.

Figure 16 shows the results of the objective test, obtained with the 48-microphone acoustic camera system. That system interpolates sound level contours, projects them onto boundary planes, and saves the results as an animation. The time-line at the top of the figure represents the audio track recorded by the acoustic camera, with several discrete musical segments. As the system plays back its analysis, the interpolated contours move to indicate the apparent locations of the musical sources. This provides an objective measurement of the stimulus that forms the basis for the listeners' subjective perceptions of the locations.

Our greatest challenge in achieving a realistic aural perception is overcoming the 43 dB(A) background noise. The biggest contributors to this are the projectors' cooling fans. The six workstations as well as the building air handling system also contribute. To compete with this using the room speakers, whispers need to be amplified considerably to be audible at any distance from the virtual source. What is not audible at some distance is not localizable. It may be possible to add more acoustic absorption around the noisy equipment without impeding its ventilation.

Alternatively, the user can wear the headphones, which have muffs that block most of the background noise. These are essential for high-quality auralization especially at low virtual sound levels, but they impede conversation with other live participants in the room. We're also considering options to improve their integration with the 3D shutter glasses and the motion-tracking reflectors.

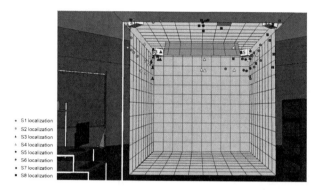

Fig. 15. Subjective localizations of virtual sound sources, indicated by a laser pointer on a grid according to each student's perception of the sound

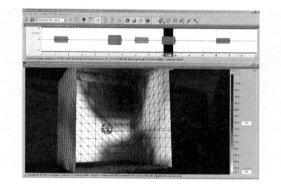

Fig. 16. Objective localizations of virtual sound sources, recorded and interpolated by a 48-microphone acoustic camera. This is one frame of an animation that moves from source to source

4 Conclusion

Virtual reality, if true to its name, should elicit realistic perceptions of virtual (unreal) worlds. There will always be observable differences between virtual and real worlds – as long as the objects of our algorithms are bigger, fewer, and less parallel than the quarks and leptons that compute the real world. The goal should be to achieve, as Samuel Taylor Coleridge phrased it, "that willing suspension of disbelief for the moment, which constitutes poetic faith." This requires a collaboration of model data, algorithms, output systems, and the ambient environment that houses it all. The overall impression is only as strong as the weakest link in that chain.

Standards for scene representation are crucial to enable a broad cliente of content creators to make use of VR systems. Creating scene assets often involves a level of modeling and artistic skill that is rare among the people who develop and maintain the software and hardware. Standards enable many people from many disciplines to play nice together to create compelling fictional worlds. These are useful not merely as entertainment, but as surrogate environments for testing designs, situations, and people's reactions to them. There is currently no one standard that encompasses all of the parameters necessary to specify a realistic multi-modal simulation (except perhaps by reliance on ambiguously specified meta data). Standards lag behind the cutting edge hardware and software technology. Those that are designed to be extensible (e.g., derived from XML), seem well poised to fill in the blanks.

Development efforts in virtual reality have focused primarily on the visual sense, followed by the aural, haptic and tactile senses. Each of these has its own representational issues. Though geometry plays a role in all of them, alternative geometric representations are often required to optimize performance in each individual mode. Tools and algorithms to assist in streamlining the extraction of these alternative representations from a common base will go a long way toward supporting multi-modal simulations for a broad clientele.

Olfaction remains an experimental mode that is difficult to discretize, digitize, encode and decode. It could contribute greatly to emotional immersion, considering the connection between smell and memory.

Display systems generally have output ranges that are quite narrow in comparison to the threshold and saturation or tolerance levels of human senses. Moreover, at the low end of the range these systems must compete with ambient noises that arise from the enclosing environment as well as from the VR system itself. Audio equipment may spill light. Video equipment certainly spills sound. Everything spills heat. The room that houses all of this equipment must be seen as an integral part of the system in order to properly isolate and absorb this noise.

Careful measurement of the actual stimuli delivered by the VR system components is an essential part of debugging and improving the system's overall performance.

References

1. Wernecke, J.: The Inventor Mentor: Programming Object-Oriented 3D Graphics with Open Inventor[TM], Release 2. Addison-Wesley, Reading Menlo-Park New-York Don-Mills Workingham Amsterdam Bonn Sydney Singapore Tokyo Madrid San-Juan Paris Seoul Milan Mexico-City Taipei (1994)
2. ISO/IEC: The Virtual Reality Modeling Language (ISO/IEC 14772-1:1997). International Organization for Standardization, Geneva (1997)
3. ISO/IEC: The Virtual Reality Modeling Language Amendment 1 – Enhanced Interoperability (ISO/IEC 14772-1:1997/Amd. 1:2002). International Organization for Standardization, Geneva (2002)
4. ISO/IEC: Extensible 3D (X3D) (ISO/IEC 19775-1:2008). International Organization for Standardization, Geneva (2008)

5. Arnaud, R., Barnes, M.C.: COLLADA^{TM}: Sailing the Gulf of 3D Digital Content Creation. AK Peters, Wellesley (2006)
6. Barnes, M., Finch, E.L. (eds.): COLLADA – Digital Asset Schema Release 1.5.0 Specification. The Khronos Group and Sony Computer Entertainment, Clearlake-Park Tokyo Foster-City London (2008)
7. Noisternig, M., Katz, B., Siltanen, S., Savioja, L.: Framework for Real-Time Auralization in Architectural Acoustics. Acta Acustica united with Acustica 94, 1000–1015 (2008)
8. Jacquemin, C.: Virtual Choreographer Reference Guide (version 1.4). LIMSI-CNRS and University Paris 11 (2007)
9. NVIDIA Corporation: PhysX^{TM}. NVIDIA Corporation, Santa Clara (2008)
10. Asakawa, C., Takagi, H., Ino, S., Ifukube, T.: Auditory and Tactile Interfaces for Representing the Visual Effects on the Web. In: Hanson, V.L., Jacko, J.A. (eds.) Proceedings of the Fifth International ACM Conference on Assistive Technologies (ASSETS 2002), pp. 65–72. Association for Computing Machinery, New York (2002)
11. Davide, F., Holmberg, M., Lundström, I.: Virtual Olfactory Interfaces: Electronic Noses and Olfactory Displays. In: Riva, G., Davide, F. (eds.) Communications Through Virtual Technology: Identity, Community, and Technology in the Internet Age, pp. 193–220. IOS Press, Amsterdam (2001)
12. Başdoğan, Ç., Loftin, R.B.: Multimodal Display Systems: Haptic, Olfactory, Gustatory, and Vestibular. In: Nicholson, D., Schmorrow, D., Cohn, J. (eds.) The PSI Handbook of Virtual Environments for Training and Education – Developments for the Military and Beyond: VE Components and Training Technologies, vol. 2, pp. 116–135. ABC-CLIO, Santa Barbara (2009)
13. Christensen, C.L.: Odeon Room Acoustics Program Version 10.1: Industrial, Auditorium and Combined Editions. Odeon A/S, Lyngby (2009)
14. Kaplanyan, A., Dachsbacher, C.: Cascaded Light Propagation Volumes for Real-Time Indirect Illumination. In: Proceedings of the 2010 ACM SIGGRAPH Symposium on Interactive 3D Graphics and Games (I3D 2010), pp. 99–107. Association for Computing Machinery, New York (2010)
15. Gottschalk, S., Lin, M.C., Manocha, D.: OBBTree: A Hierarchical Structure for Rapid Interference Detection. In: Proceedings of the 23rd Annual Conference on Computer Graphics and Interactive Techniques (SIGGRAPH 1996), pp. 171–180. Association for Computing Machinery, New York (1996)
16. Lien, J.M., Amato, N.M.: Approximate Convex Decomposition of Polyhedra. In: Lévy, B., Manocha, D. (eds.) Proceedings of the 2007 ACM Symposium on Solid and Physical Modeling (SPM 2007), pp. 121–131. Association for Computing Machinery, New York (2007)
17. Huang, M.P., Himle, J., Beier, K.P., Alessi, N.E.: Comparing Virtual and Real Worlds for Acrophobia Treatment. In: Westwood, J.D., Hoffman, H.M., Stredney, D., Weghorst, S.J. (eds.) Medicine Meets Virtual Reality: Art Science, Technology: Healthcare (R)evolution, pp. 175–179. IOS Press, Amsterdam (1998)
18. Beier, K.P.: Web-Based Virtual Reality in Design and Manufacturing Applications. In: Bertram, V. (ed.) Proceedings of the 1st International Conference on Computer Applications and Information Technology in the Maritime Industries (COMPIT 2000), Potsdam, pp. 45–55 (2000)
19. Beier, K.P.: Web-Based Virtual Reality in Design and Manufacturing Applications. Hansa International Maritime Journal 137(5), 42–47 (2000)

20. Wilkerson, W., Avstreih, D., Gruppen, L., Beier, K.P., Woolliscroft, J.: Using Immersive Simulation for Training First Responders for Mass Casualty Incidents. Academic Emergency Medicine 15(11), 1152–1159 (2008)
21. Andreatta, P.B., Maslowski, E., Petty, S., Shim, W., Marsh, M., Hall, T.W., Stern, S., Frankel, J.: Virtual Reality Triage Training Provides a Viable Solution for Disaster-Preparedness. Academic Emergency Medicine 17(8), 870–876 (2010)
22. Davis, W.L., Ohno, Y.: Evaluation of Color Difference Formulae for Color Rendering Metrics. In: Proceedings of the ISCC/CIE Expert Symposium. Inter-Society Color Council and Commission Internationale de l'Eclairage, Reston and Vienna (2006)
23. Okamoto, T., Katz, B.F.G., Noisternig, M., Iwaya, Y., Suzuki, Y.: Implementation of Real-Time Room Auralization Using a Surrounding 157 Loudspeaker Array. In: Suzuki, Y., Brungart, D., Kato, H., Iida, K., Cabrera, D., Iwaya, Y. (eds.) Proceedings of the International Workshop on the Principles and Applications of Spatial Hearing, World Scientific, Singapore (2009)

Operating High-DoF Articulated Robots Using Virtual Links and Joints

Marsette A. Vona

College of Computer and Information Science,
Northeastern University, Boston MA
vona@ccs.neu.edu

Abstract. This chapter presents the theory, implementation, and application of a novel operations system for articulated robots with large numbers (10s to 100s) of degrees-of-freedom (DoF), based on *virtual articulations* and *kinematic abstractions*. Such robots are attractive in some applications, including space exploration, due to their application flexibility. But operating them can be challenging: they are capable of many different kinds of motion, but often this requires coordination of many joints. Prior methods exist for specifying motions at both low and high-levels of detail; the new methods fill a gap in the middle by allowing the operator to be as detailed as desired. The presentation is fully general and can be directly applied across a broad class of 3D articulated robots.

1   Introduction

Normally the connectivity structure of an articulated robot's links and joints—its kinematic topology—is considered invariant.[1] This chapter explores a new kind of human interface based on virtually making such changes, by allowing virtual links and joints to be interactively connected to a virtual model of the robot. We call such virtual kinematic constructions *virtual articulations*.

We present the architecture of an implemented simulator [38] for arbitrary topology 3D articulated robots with virtual articulations, and show how it can be applied as an intuitive human interface to operate coordinated motion in robots with many joints. Experimental results are presented with the 36 degree of freedom (DoF) All Terrain Hex Limbed Extra-Terrestrial Explorer (ATHLETE) [41] and with a simulated modular mobile tower structure with 100s of joints [8].

Due to their application flexibility, high-DoF robots are increasingly common: humanoids with 20 or more DoF are now available off-the-shelf, snake-like robots with 10s of joints have long been studied, and modular/self-reconfiguring robots can have 10s to 100s of actuated joints. This flexibility is

---

[1] Robots capable of on-line reconfiguration exist, but are unusual.

T. Gulrez, A.E. Hassanien (Eds.): Advances in Robotics & Virtual Reality, ISRL 26, pp. 275–305.
springerlink.com   © Springer-Verlag Berlin Heidelberg 2012

especially attractive in space exploration, where the extreme costs of transportation from Earth are balanced by maximizing versatility in the delivered hardware. But remote operation of complex robots can be tedious; a main feature of this chapter is a set of examples where virtual articulations are used to operate inspection and manipulation tasks on the ATHLETE hardware that would have been challenging in prior interfaces. We also show an application to a simulated modular mobile tower that could be useful in space, where structures need to be assembled from compact storage. If such structures were controllably mobile and deformable they could simultaneously support manipulation, inspection, and locomotion, thus greatly reducing costs relative to launching separate robotic hardware.

Though some of the results in this chapter have been reported previously [8, 9, 36, 37], here we include for the first time the details of the multipriority system architecture which enables motion constraint with virtual articulations and scaling to 100s of joints.

Section 2 explains the idea of virtual articulations with some examples. We then review related work in Section 3, and present the system architecture in Section 4. Section 5 gives several detailed application examples, followed by conclusions in Section 6.

2 Virtual Articulations

What makes operating high-DoF robots difficult? One issue is that there are often many ways the robot could move to achieve a task (kinematic redundancy) and some motions may be preferred due to secondary goals [28]. Sometimes a human operator can quickly visualize a desired motion, but expressing it in an operations system can be tedious. The approach described in this chapter allows the operator to graphically and intuitively specify motions using virtual articulations. Task-specific coding and low-level motion sequencing is minimized. It can also be time consuming to specify motion for large systems with many simultaneously moving DoF, even when motions for subsystems are easily defined, simply due to scale. In a technique we call *structure abstraction*, virtual articulations can be applied to impose hierarchical organization, and can enable re-use of substructure motion constraints.

To see how virtual articulations can help, consider an example with ATHLETE. With 36 revolute joints in six identical limbs, ATHLETE can be applied in a broad range of tasks. Often, an intended motion is easily visualized, but expressing it could be tedious—even using custom-built direct manipulation hardware [22]. Figure 1 shows one example of how virtual articulations can help: the operator would like to tilt the field of view of an un-actuated camera by rotating the robot's body about a virtual axis in space, with the legs moving as necessary. Using virtual articulations, the intended rotation axis can be defined by attaching a virtual revolute joint between the robot body and the ground. Rotating the joint drives the intended motion, and the system automatically computes compatible motions for all the legs.

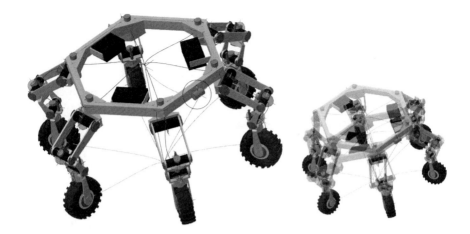

Fig. 1. Operating ATHLETE with virtual articulations.
A virtual revolute joint (circled) is interactively added by an operator to facilitate tilting the field of view of one of ATHLETE's cameras, which are rigidly mounted to the robot (the camera is not visible in the model). Camera tilt can then be commanded by rotating the virtual joint; feasible whole-robot supporting motions are automatically computed on-line. Figure 14 shows this experiment on the actual hardware. ATHLETE model courtesy of the RSVP team, NASA/JPL/Caltech.

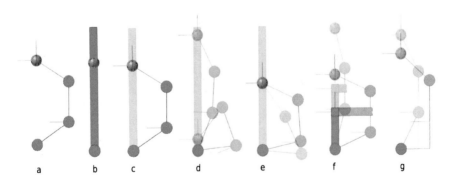

Fig. 2. Structure abstraction and kinematic constraint.
Virtual articulations can be used both for *structure abstraction* and to constrain intended motion. The actual robot mechanism in (a) is an open chain of four revolute joints. The operator is designing a piston-like behavior of the endpoints. In our system, this can be modeled as a revolute-prismatic-revolute construction (b) which virtually replaces the model of the actual mechanism (c,d), forming an abstraction. Left alone, the middle link remains underconstrained (e); the operator restricts it by adding an assembly of two virtual prismatic joints (f) below the abstraction barrier. The same constraint can also be defined using a shorthand *Cartesian-2* joint (g).

Virtual articulations can also can help organize large systems with 10s to 100s of DoF. Abstraction is well known for managing complexity in algorithms, but so far there have been few corresponding techniques in kinematics. By applying virtual articulations in a specific hierarchical framework, *structure abstraction* can be used to effectively hide the complexity of a kinematic subsystem in a way that corresponds to traditional algorithmic abstraction. Figure 2 shows a simple example where a series chain of four revolute joints is abstracted as a piston-like assembly, hiding the motion of the middle link. This assembly is later used as a subsystem in Section 5.3.

3 Research Context

A few related ideas have previously been reported. For example, Flückiger [14] shows a virtual interface where a model of the robot is provided to the operator for pose manipulation. Our approach fundamentally differs from this type of system in that we allow the operator to virtually change and augment the kinematic structure, in addition to moving it.

In [26, 27], Pratt et al introduced *virtual model control*, where virtual physical dynamic elements—often springs and dampers, but in a few cases also kinematic joints—are added to a robot model as a means to design motion control strategies. In their reports, the virtual elements are manually designed, tuned, and implemented on a case-by-case basis for each task of each mechanism. Overall, physics based approaches require more parameter tuning than the purely kinematic models we use, potentially adding additional burden on the operator as virtual elements are added/removed/reconfigured. And, while algorithms for efficiently simulating the physical dynamics of reduced coordinate models and closed kinematic chains [12] are known in the literature, currently available real-time physics engines, such as ODE [33], are mostly based on Cartesian coordinate models (called "maximal coordinate methods" in [20]) and need to be carefully tuned to efficiently and stably simulate kinematic chains, especially in the presence of chain-closing joints.

Geometric constraints have been proposed for defining and constraining motion in figure animation [2, 25, 40]. In these works typically only a few simple constraints are available, for example, point-on-plane and point-on-line. The constraints are not themselves considered to be kinematic joints, and can only be added between geometric objects situated within links of the original articulated model. In our approach the operator is free to construct arbitrary additions including both virtual links and a variety of virtual joint types.

Compared to other operator interface approaches, virtual articulations fill a gap between existing low-level methods to operate articulated robots, including forward and inverse kinematic control, and existing high-level methods such as goal-based motion planning and programming-by-demonstration.

Arguably the simplest low-level methods are traditional forward- (joint space) or inverse-kinematic (task space) operation. Without higher-level goals or constraints, both are potentially tedious in the high-DoF case. Task priority and task space augmentation approaches [5] can maintain such constraints, but do not themselves offer a convenient way to *design* them.

A few authors have previously explored the idea of constraining the kinematic motion of redundant robots by adding chain-closing virtual joints in specific cases. Ivlev and Gräser [18, 19] defined virtual chain closures to restrict the motion of redundant revolute serial-chains to avoid known obstacles in the environment. They treated only smaller examples which could be solved analytically on paper. Bruyninckx [4] also showed an example where a virtual closed chain minimizes the otherwise unconstrained twist motion about the axis of a round tool on a revolute-jointed industrial manipulator.

Another related idea is to introduce optimization metrics, such as energy minimization, joint limit avoidance, or manipulability maximization [21, 28]. These are useful but blunt instruments—there is no direct way for the operator to arbitrarily customize the intended motion. Virtual articulations are one language for such customization: the operator can constrain motion as much or as little as desired, and a large class of motions can be developed starting from a relatively small catalog of virtual joints. Traditional optimization metrics can still be added as additional priority levels alongside virtual articulations.

Turning to high-level methods, goal-based motion planning, the "piano moving" problem of achieving a target configuration among obstacles, is typically not directly applicable when the operator would also like to specify detailed aspects of the motion. Again, something more is needed if the operator needs to arbitrarily customize the motion *on the way* to a goal configuration.

Another high-level technique, programming-by-demonstration, allows more specific motion specification, but is hard to apply when the robot topology diverges from preexisting systems (including known animals). Thus it has been used with some success for humanoids (e.g. [24]), or when mimicking hardware is available, as in some of our earlier work with ATHLETE [22]. Virtual articulations are not tied to any particular topology. And the experiment in Figure 12 shows that that an integration of programming-by-demonstration with virtual articulations can have some of the advantages of both methods.

There seem to be only a few prior works related to our concept of structure abstraction. Davis [7] explored spatial geometric abstraction at a conceptual level, including an interesting but brief mention of a "kinematic device as black box." Zanganeh and Angeles [44] give a more detailed and formal development of the idea of separating topologically large kinematic structures into sub-mechanisms, but their approach does not specifically separate interface from implementation—it just demarcates sub-mechanisms. Our structure abstraction includes this capability, but also allows interface mechanisms that stand-in for underlying implementations.

4 System Architecture

Articulated systems, which we call *linkages*[2], are composed of links (rigid
bodies) interconnected by joints. The binary connectivity of joints and the
arbitrary connectivity of links naturally suggest representing the topology of
a linkage as a graph, as in Figure 3, where the links are vertices and the joints
are edges. Such graphs have been used at least since the 1960s [11].

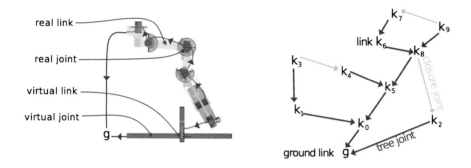

Fig. 3. Kinematic graph (*linkage*) examples.
Left: a mixed real/virtual model with superimposed linkage graph; right: a second
example. Links (rigid boides) correspond to vertices and joints correspond to edges
(such graphs are not novel). We always identify a spanning tree (dark edges) with
the remaining *closure joints* (light edges) closing kinematic cycles.

The 3D kinematic simulation system we present here—the *Mixed Real/
Virtual Simulator and Interface* (MSim)—allows the construction of arbitrary
topology linkages. Unlike some prior kinematic simulators [3, 10, 29, 43],
to support interactive construction of virtual articulations we must support
both on-line structure changes (adding, removing, reconnecting joints) and
closed kinematic chains (cycles in the linkage graph). Figure 4 shows the
architecture of our prototype implementation, which we have made freely
available in source and executable form [38].

When the linkage graph is cyclic there may be multiple paths between a
given pair of links; our system always maintains a spanning tree of the graph,
so that the relative pose of any two links can always be uniquely defined by
the path between them consisting only of *tree joints*; the remaining joints,
which complete kinematic cycles, are called *closure joints* or just *closures*.
This tree also induces a natural directivity on the edges it contains: such
edges point towards the root of the tree. (The same edge direction also gives
the forward sense of the associated joint transform.)

---

[2] Some authors use this term more specifically, e.g. only for 1-DoF systems, or for
systems with only revolute joints.

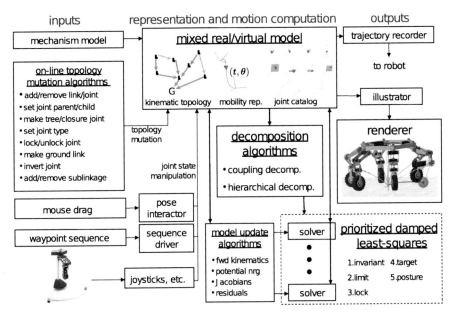

Fig. 4. Overall architecture of our implementation.

How do linkage graphs arise and evolve? Algorithms for topological mutation of kinematic graphs have not often been presented in detail, in part
because robots do not often change topology. But we must add, remove,
and reconfigure virtual articulations on-line. We have thus developed a set
of *structure mutation* primitives (table 2) for common kinds of structural
(topological) changes. These are sufficient to build any linkage structure from
scratch. We omit the algorithms here due to space constraints, see [39] for
details.

The following formal definition of the components of a linkage and their
properties will be helpful (also see Figure 3).

Definition 1. *Each joint (graph edge) j is directed to connect links p_j and
c_j the parent and child link of j. The edge direction is from c_j to p_j.*

Definition 2. *Each link k (graph vertex) has an associated coordinate frame
F_k called the link frame of k. The link frame of the child (resp. parent) link
of a joint j is called the joint's child (resp. parent) frame F_{c_j} (resp. F_{p_j}).*

Definition 3. *For each joint j there is an associated rigid-body joint transform X_j that defines the pose of the child frame relative to the parent frame.*

Definition 4. *The joint transform X_j must lie within the joint mobility
space $\mathcal{M}_j$, a subspace of the full 6-dimensional space of 3D rigid body transforms (the special Euclidean group $SE(3)$) depending on the joint's type and
limits.*

For example, a revolute joint allows only motion within a 1-dimensional subspace of $SE(3)$—rotation about a fixed axis. Limits, if specified, further restrict motion to a closed bounded interval of this subspace.

Our joint mobility representation, which is based on the exponential map $(t, \theta) \in \mathbb{R}^6$, is fully described in [39]. Here we present only the final catalog of 11 joint types available in the system (Figure 5). These include all lower-pair joints except helical; i.e. revolute, prismatic, cylindrical, spherical, and planar. In addition, some higher pairs are included for convenience: a *pin-slider* joint combines a prismatic and revolute joint with orthogonal axes; a *point-slider* is similar but replaces the revolute joint with a spherical joint; the *point-plane* joint combines a planar and a spherical joint; *Cartesian-2,3* joints are serial chains of two or three orthogonal prismatic joints; and finally a *general* joint allows full 6-DoF motion (and is equivalent to a serial chain of three orthogonal prismatic joints followed by a spherical joint).

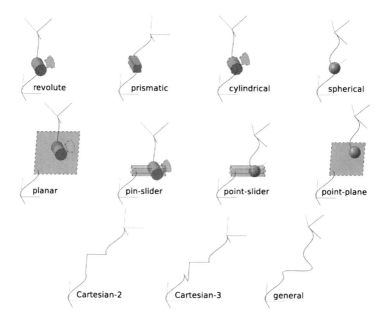

Fig. 5. 3D joints available in our implementation.
Each joint type is represented graphically in our interface system. Connections to links are shown as Bézier curves.

Definition 5. *The degrees of freedom (DoF) f_j of a joint is the dimension of the joint space: $f_j = \dim(\mathcal{M}_j)$. The degrees of invariance (DoI) is $i_j = 6 - \dim(\mathcal{M}_j)$. The state of a tree joint t is a vector $x_t \in \mathbb{R}^{f_t}$; the state of a closure joint c is always taken as its full transform $y_c \in \mathbb{R}^6$.*

We define joint types as axis-aligned subspaces of the exponential map parameterization of $SE(3)$, so a tree joint state x_t is always just a subset of the full 6-DoF (t, θ) vector. (Constant structural transforms in the linkage still allow the joint mobility directions to be posed arbitrarily.)

Definition 6. *Exactly one of the links is identified as the* ground *link* g.

Definition 7. *Every link* $k \neq g$ *has exactly one of its outgoing joints identified as its* parent joint p_k, *and* g *has no parent joint. The incoming joints of a link* k *are called* child joints *of* k.

Definition 8. *The set* T *of all parent joints of all links forms a directed spanning tree with root* g; *all edges in* T *point towards* g. *A transform* $X_{k_0 \leftarrow k_1}$ *gives the pose of any link frame* F_{k_1} *relative to any other link frame* F_{k_0} *by following the unique path from* k_0 *to* k_1 *in* T:

$$X_{k_0 \leftarrow k_1} = \left(\prod_{i \in g \leftarrow k_0} X_i \right)^{-1} \left(\prod_{i \in g \leftarrow k_1} X_i \right). \tag{1}$$

Definition 9. *The* support S_j *of a joint* j *is the minimal ordered sequence of tree joints from the parent link* p_j *to the child link* c_j. S_j *is partitioned into the* downchain $S\!\downarrow_j$ *of joints from* p_j *to the least common ancestor link*[3] LCA(p_j, c_j) *and the* upchain $S\!\uparrow_j$ *from* LCA(p_j, c_j) *to* c_j:

$$S_j = \left(S\!\downarrow_j, S\!\uparrow_j \right). \tag{2}$$

Definition 10. *If* J *is the set of all joints in a linkage with tree joints* $T \subset J$, *then the remaining joints* $C = J \setminus T$ *are chain closures.*

The mobility transform X_j for a closure joint j with support $S_j = \left(S\!\downarrow_j, S\!\uparrow_j \right)$ is computed from the mobility transforms of the supporting tree joints based on Eq. 1 (here we omit the constant joint posing transforms for brevity):

$$M_j = \left[\left(\prod_{i \in S\uparrow_j} X_i^{-1} \right) \left(\prod_{i \in S\downarrow_j} X_i \right) \right]^{-1} \tag{3}$$

with the products taken in chain order.

Let L be a linkage with joints J and tree joints $T \subseteq J$ with ordering

$$T = \left(t_0, \cdots, t_{|T|-1} \right). \tag{4}$$

---

[3] I.e. the topologically nearest link that is an ancestor of both p_j and c_j.

Definition 11. *The* mobility space $\mathcal{M}$ *of a linkage L is the product space of the individual tree joint mobility spaces, in order:* $\mathcal{M} = \prod_{0 \leq i < |T|} \mathcal{M}_{t_i}$.

Definition 12. *The number of degrees of freedom* DoF *f of L is the dimension of $\mathcal{M}$: $f = \dim(\mathcal{M})$. The number of degrees of invariance* DoI *i of L is the sum of the DoI of the closure joints in L: $i = \sum_{j \in J \setminus T}(6 - \dim(\mathcal{M}_j))$.*

Definition 13. *The* tree state *$\boldsymbol{x}$ of L is the concatenation of the states of its tree joints, in the order given above*

$$\boldsymbol{x}^T = \left(\boldsymbol{x}_{t_0}^T, \cdots, \boldsymbol{x}_{t_{|T|-1}}^T \right) \in \mathcal{M} \subset \mathbb{R}^f. \tag{5}$$

Similarly, let

$$C = \left(o_0, \cdots, o_{|J \setminus T|-1} \right) \tag{6}$$

be an ordering of the closure joints $J \setminus T$ of L.

Definition 14. *The* closure state *$\boldsymbol{y}$ of L is the concatenation of the states $\boldsymbol{y}_{o_j}$ of its closure joints in order, and the* invariant error *$\boldsymbol{e}_i$ of L is the concatenation of the invariant errors of the closure joints in order*

$$\boldsymbol{y}^T = \left(\boldsymbol{y}_{o_0}^T, \cdots, \boldsymbol{y}_{o_{|C|-1}}^T \right) \in \mathbb{R}^{6|C|} \tag{7}$$

$$\boldsymbol{e}_i^T = \left(\boldsymbol{e}_{i_{o_0}}^T, \cdots, \boldsymbol{e}_{i_{o_{|C|-1}}}^T \right) \in \mathbb{R}^i. \tag{8}$$

$\boldsymbol{e}_i$ is a projection Π_i of $\boldsymbol{y}$,

$$\boldsymbol{e}_i = \Pi_i \boldsymbol{y}, \tag{9}$$

with Π_i an $[i \times 6|C|]$ binary matrix mapping $\boldsymbol{y}$ to $\boldsymbol{e}_i$ by selecting only the DoI of each closure joint.

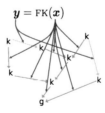

$$\boldsymbol{y} = \text{FK}(\boldsymbol{x})$$

Fig. 6. The forward kinematic mapping.
The forward kinematic mapping FK computes the *closure state* $\boldsymbol{y}$ as a function of the *tree state* $\boldsymbol{x}$. $\boldsymbol{y}$ is a concatenated vector of the closure joint $(\boldsymbol{t}, \boldsymbol{\theta})$ state vectors; $\boldsymbol{x}$ is a concatenated vector of the DoF of the tree joints.

Definition 15. *The* forward kinematic mapping *(Figure 6)* FK *of L computes* $\boldsymbol{y}$ *given* $\boldsymbol{x}$

$$\text{FK} : \mathcal{M} \to \mathbb{R}^{6|C|} \quad \text{FK}(\boldsymbol{x}) = \boldsymbol{y}, \tag{10}$$

and is computable by composing appropriate instantiations of Eq. 3.

The differentiability of the underlying operations implies that FK is itself differentiable; an algorithm to compute derivatives of FK is given in [39].

4.1   Hierarchical Linkages

Many real-world robots contain repeated kinematic sub-structure, e.g. multiple copies of identical legs. When possible, breaking a model up into decoupled components can speed up motion computations (Section 4.1). Further, subdivision is important in using virtual articulations in an operator interface for high-DoF mechanisms—in many cases the operator can specify motions independently for smaller parts of the model, and thus break up the job of motion specification. Taking this a step further, virtual articulations combined with model hierarchy enable our technique of *structure abstraction*, where a complex kinematic subsystem is abstracted as a simpler virtual mechanism capturing the intended motion.

The mixed real/virtual interface supports these kinds of subdivisions by structuring the overall model into a hierarchy (tree) of properly nested *sublinkages*. Figure 7 shows an example; details are given in [39] and [37]. We allow sub-linkages to be *driving* (considered rigid with respect to their parent), *driven* (parent considered rigid), or *simultaneous* (grouping only). *Crossing joints* span the boundary between a parent and child sub-linkage.

Sub-linkages and virtual articulations can be used to implement a version of abstraction in the domain of kinematics.

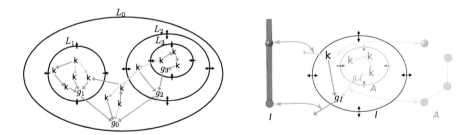

Fig. 7. Hierarchical linkages and structure abstraction.
Left: sub-linkage L_1 is driving, L_2 is simultaneous, and L_3 is driven. Tree joints are dark colored arrows; closure joints are light. Right: an example of structure abstraction. The "implementation" A, a series chain of four revolute joints, is virtually replaced by a simpler piston-like "interface" I. Also see Figure 2.

Definition 16. Structure abstraction *is achieved by (a) encapsulating a connected part A of a linkage L s.t. A becomes demarcated as a simultaneous sub-linkage of of L, and (b) substituting a virtual linkage I for A in L, with I simultaneous in L and A a driven sub-linkage of I. A is the* abstracted sub-linkage *(the "implementation") and I is its* interface linkage.

Figure 7 shows an example. The concept parallels traditional abstraction in computing: I should be simpler than A, but capture all of the behavior of A that would be relevant to the surrounding mechanism.

Making A a driven sub-linkage of I is not the only option. For example, keeping A simultaneous could also make sense. However, our choice to make A driven defines a strong form of abstraction: motion of L, with I substituting for A, is *independent* of the actual motion of A. This permits A to be effectively decoupled, a simplification that can be helpful both (a) to an operator designing a motion for a high-DoF robot, and (b) to the system as it computes motions for a large linkage.

This power has a trade-off: there is no built-in constraint to ensure that A can reach every configuration to which it may be driven by I.

Definition 17. *I is a* proper abstraction *of A if all of the reachable configurations of I drive reachable configurations of A.*

It would be desirable to have an efficient algorithm that could determine, for any L, I, and A, whether I is a proper abstraction. This may be possible in some special cases, but unfortunately, the general case is easily shown to be hard; the details are given in [37].

Thus, if a proper abstraction is required, it is up to the operator to ensure it, for example by setting sufficient joint limits in I. However, even improper abstractions can be useful when combined with the constraint prioritization features of our system (Section 4.3)—if the closure invariants in A are given higher priority than those of the crossing joints between I and A, then the motion of A will remain feasible. It will not exactly match the driving links in I, but because they are virtual, this can be acceptable.

4.2 Local Assembly and Differential Control

We now turn to the simulation of linkage motions, which we achieve by a task-priority solver we call prioritized damped least squares (PDLS), along with a specific design of five priority levels.

We define two motion problems. *Local assembly* needs to be solved after the linkage structure is changed (Figure 8)—the prior linkage state will not generally be compatible with the new structure, and needs to be moved to satisfy the new constraints. The *differential control* (adapted from [16]) problem generalizes both forward and inverse-kinematic control, and supports various modes of operating mixed real/virtual structures including click-and-drag interaction (Figure 9) and trajectory following.

Fig. 8. The local assembly problem.
A virtual revolute joint has been added, closing a chain between the "ankles" of two adjacent ATHLETE limbs. The chain is initially inconsistent (a) because the pose that was specified for the new joint relative to the limbs differs from the actual current relative pose. A red curve indicates the error. The system solves the local assembly problem by moving the legs to remove the error, as shown in (b) and (c). Once the new topology is assembled, it can be operated (d) as originally intended.

Fig. 9. The differential control problem.
Here the operator uses click-and-drag interaction with the mouse to interactively pose the last link in an ATHLETE limb. The hex deck is locked, so the system solves the differential control problem corresponding to the single-limb inverse kinematics.

The PDLS algorithm uses the differentiability of the exponential map representation of joint pose to make local linear approximations of the forward kinematic mapping, and then iteratively solves local assembly, differential control, and kinostatic simulation by gradient descent in a least-squares sense with an SVD-based pseudoinverse. Nullspace projection is used for prioritizing the various constraints, which may conflict.

PDLS derives from the task-priority approach first proposed by Siciliano and Slotine in [32], with more recently improvements from Baerlocher and Boulic in [1]. We do not repeat these results here, see [1, 39] for details. Instead we focus on our contribution: a set of five specific priority levels that together solve the local assembly and differential control problems (in [39] we also insert a sixth level that simulates quasistatic potential energy minimization).

Because the time complexity of PDLS is typically quadratic in the number of joints [39], it is important to break up large linkages into independently solvable components when possible. These are then solved independently in ordered *rounds*, which arise due to the driving/driven relationships of sub-linkages. We have developed an algorithm that analyzes the topology and structure of a linkage and attempts to break it into smaller parts which can be handled separately; details are given in [37, 39].

4.3   The Five Priority Levels

The inputs to PDLS are, for each priority level l, a *level height* h_l, a *residual vector* $e_l \in \mathbb{R}^{h_l}$, and an $h_l \times f$ level Jacobian matrix J_l, which relates a small change in the tree state $x \in \mathbb{R}^f$ (f is the total number of DoF of the linkage) to a change in the level residual. We next derive the residuals and Jacobians for each of our five proposed priority levels: invariant, limit, lock, target, and posture. Table 1 summarizes them.

Table 1. The five SOLVE priority levels in order from highest to lowest.

level l	height h_l	residual e_l	Jacobian J_l
invariant	i	e_i (Eq. 11)	J_i (Eq. 14)
limit	m	e_m (Eq. 27)	J_m (Eq. 28)
lock	k_c	e_k (Eq. 30)	J_k (Eq. 31)
target	t	e_t (Eq. 17)	J_t (Eq. 18)
posture	p_c	e_{pc} (Eq. 22)	J_{pc} (Eq. 23)

The highest priority *invariant* level solves the local assembly problem by minimizing the *invariant residual* e_i. The forward kinematic mapping FK (Eq. 10) and the invariant error projection matrix Π_i multiply to give the forward invariant error mapping FK_i (Eq. 9) taking the current tree state x to the resulting invariant error e_i:

$$FK_i : \mathbb{R}^f \to \mathbb{R}^i \quad e_i = FK_i(x) = \Pi_i FK(x). \tag{11}$$

FK_i' is locally modeled in the neighborhood of x by its Jacobian matrix

$$\underset{[i \times f]}{J_i'(x)} = \frac{\partial}{\partial x} FK_i'(x). \tag{12}$$

Writing $J_i'(x)$ stresses the dependence of the Jacobian matrix both on FK_i' and on x, but for brevity we usually write simply J_i', with the latter dependency implied. J_i' can be computed by multiplying the Jacobian J_y' of the full forward kinematic mapping $FK'(x)$ by the projection matrix Π_i:

$$\text{with } J_y' = \frac{\partial}{\partial x} FK'(x) \tag{13}$$

$$J_i' = \frac{\partial}{\partial x} FK_i'(x) = \Pi_i \frac{\partial}{\partial x} FK'(x) = \Pi_i J_y'. \tag{14}$$

We will cover the limit and lock levels below; for now, skip to the *target* level, which solves the differential control problem. Consider the motion of a linkage after a user has moved a link (Figure 9). If the link is not part of any closed chains, then the forward kinematic mapping will not be affected. Otherwise, to keep e_i (breakage of closure joints) minimized, the system may be required to not (fully) move the link as the user has requested, and/or it may be necessary to move some joints to compensate.

Let the total number of DoF of the closure joints in a linkage L with closure joints C be $d = 6|C| - i$, and define a new $[d \times 6|C|]$ binary projection matrix Π_{fc} which compliments Π_i by selecting only the DoF from $\boldsymbol{y}$. Using this, form a vector $\boldsymbol{z}$ comprising all the DoF of both the tree and the closure joints in a linkage L:

$$\boldsymbol{z}^T = \left(\boldsymbol{x}^T, (\Pi_{fc}\boldsymbol{y})^T\right) \in \mathbb{R}^{f+d=z}. \tag{15}$$

In general, the operator is not restricted to move any single joint, but may manipulate any subset of joint DoF. Further, they may also operate *link DoF*, which are defined simply as the DoF of a general (6-DoF) virtual closure joint automatically attached between the link and the ground, neatly handling inverse kinematic control. Let t be the total number of DoF under control, and let $\boldsymbol{t} \in \mathbb{R}^t$ be the current *target waypoint*. To follow a path, $\boldsymbol{t}$ can be periodically replaced by the operator—e.g. by dragging the mouse—with the system either interpolating intermediate waypoints or just switching directly from one to the next.

Let Π_t be a $[t \times z]$ binary projection matrix selecting the t current target DoF from the z total DoF. The *target residual* is

$$\boldsymbol{e}_t = \Pi_t \boldsymbol{z} - \boldsymbol{t} \tag{16}$$

and the *forward target error mapping* FK$_t$ is

$$\text{FK}_t : \mathbb{R}^f \to \mathbb{R}^t \quad \boldsymbol{e}_t = \text{FK}_t(\boldsymbol{x}) = \Pi_t \begin{bmatrix} \boldsymbol{x} \\ {\scriptstyle [f\times 1]} \\ \Pi_{fc}\text{FK}(\boldsymbol{x}) \\ {\scriptstyle [d\times 1]} \end{bmatrix} - \boldsymbol{t}. \tag{17}$$

The *target Jacobian* is the $[t \times f]$ matrix of partial derivatives of 17:

$$\underset{[t\times f]}{J_t} = \frac{\partial \boldsymbol{e}_t}{\partial \boldsymbol{x}} = \frac{\partial}{\partial \boldsymbol{x}}\text{FK}_t(\boldsymbol{x}) = \underset{[t\times z]}{\Pi_t} \begin{bmatrix} \mathbf{I}_f \\ {\scriptstyle [f\times f]} \\ \Pi_{fc}J_y \\ {\scriptstyle [d\times f]} \end{bmatrix} \tag{18}$$

with $\mathbf{I}_f$ the $[f \times f]$ identity matrix.

The lowest-priority *posture* level derives from the posture variation technique described in [1], drawing the joints toward a lowest-priority default pose. However, [1] only covered tree joint DoF; our formulation, which handles closure as well as tree joints, parallels target tracking. The *posture waypoint* is a vector $\boldsymbol{p} \in \mathbb{R}^p$ and Π_p is a new $[p \times z]$ binary projection matrix s.t.

$$\boldsymbol{e}_p = \Pi_p \boldsymbol{z} - \boldsymbol{p}. \tag{19}$$

However, in this case the Jacobian will not be a $[p \times f]$ matrix, because the posture variation components corresponding to tree joint DoF will be handled directly by the final nullspace projection [1]. $\boldsymbol{e}_p$ must be partitioned into two sub-vectors, one for closure joint DoF and one for tree joint DoF, which will be handled separately:

$$\boldsymbol{e}_p^T = \left(\boldsymbol{e}_{pt}^T, \boldsymbol{e}_{pc}^T \right) \tag{20}$$

s.t. p_t, p_c are the lengths of $\boldsymbol{e}_{pt}$ and $\boldsymbol{e}_{pc}$, respectively, with $p = p_t + p_c$. Then the posture variation vector is

$$\Delta \boldsymbol{x}_p = -\boldsymbol{e}_{pt}. \tag{21}$$

$\boldsymbol{e}_{pc}$ is the *posture residual* vector and J_{pc} the *posture Jacobian*, computed by the *forward posture error mapping* FK_{pc}

$$\mathrm{FK}_{pc} : \mathbb{R}^f \to \mathbb{R}^{p_c} \quad \boldsymbol{e}_{pc} = \mathrm{FK}_{pc}(\boldsymbol{x}) = \Pi_{pc} \Pi_{fc} \mathrm{FK}(\boldsymbol{x}) - \boldsymbol{p}_c \tag{22}$$

$$\underset{[p_c \times f]}{J_{pc}} = \frac{\partial \boldsymbol{e}_{pc}}{\partial \boldsymbol{x}} = \frac{\partial}{\partial \boldsymbol{x}} \mathrm{FK}_{pc}(\boldsymbol{x}) = \Pi_{pc} \Pi_{fc} J_y. \tag{23}$$

Π_{pc} and $\boldsymbol{p}_c$, used in the above equations, are taken from corresponding tree/closure partitions of Π_p and $\boldsymbol{p}$:

$$\underset{[p \times z]}{\Pi_p} = \begin{bmatrix} \underset{[p_t \times f]}{\Pi_{pt}} & \\ & \underset{[p_c \times d]}{\Pi_{pc}} \end{bmatrix} \tag{24}$$

$$\boldsymbol{p}^T = \left(\boldsymbol{p}_t^T, \boldsymbol{p}_c^T \right). \tag{25}$$

We now return to the *limit* level. As for posture variation, [1] also presented a technique for handling joint limits, but only on tree joints; we extend this to also handle limits on closure joints by constructing a penalty function to drive limit-exceeding closure DoF back into compliance. No penalty is assigned to in-limits DoF, but upon exceeding a limit the penalty increases quadratically.

Let m be the total number of closure DoF with set limits, and let Π_m be a $[m \times d]$ binary projection matrix selecting the m limited DoF from the d total closure joint DoF. Let $\boldsymbol{l}_c$ and $\boldsymbol{u}_c$ be vectors of the m lower and upper closure DoF limits, respectively. The limit excess function is

using $\boldsymbol{v}^T = (v_0, \ldots, v_{m-1}), \ \boldsymbol{l}^T = (l_0, \ldots, l_{m-1}), \ \boldsymbol{u}^T = (u_0, \ldots, u_{m-1})$

$$\mathrm{EXCESS}(\boldsymbol{v}, \boldsymbol{l}, \boldsymbol{u}) = (s_0, \ldots, s_{m-1})^T \text{ s.t. } s_i = \begin{cases} v_i - u_i \text{ if } v_i > u_i \\ 0 \qquad \text{if } u_i \geq v_i \geq l_i \\ v_i - l_i \text{ if } v_i < l_i, \end{cases} \tag{26}$$

and the *closure limit residual* vector e_m is the square of the excess for each DoF, computed by the *forward limit error mapping* FK$_m$:

$$\text{FK}_m : \mathbb{R}^f \to \mathbb{R}^m \quad e_m = \text{FK}_m(\boldsymbol{x}) = \text{DIAG}(\boldsymbol{s}_c)\boldsymbol{s}_c \tag{27}$$

$$\text{with } \boldsymbol{s}_c = \text{EXCESS}(\Pi_m \Pi_{fc}\text{FK}(\boldsymbol{x}), \boldsymbol{l}_c, \boldsymbol{u}_c).$$

The *limit Jacobian* is the $[m \times f]$ matrix of partial derivatives of 27:

$$\underset{[m \times f]}{J_m} = \frac{\partial e_m}{\partial \boldsymbol{x}} = \frac{\partial}{\partial \boldsymbol{x}}\text{FK}_m(\boldsymbol{x}) = 2\,\text{DIAG}(\boldsymbol{s}_c)\Pi_m \Pi_{fc}J_y. \tag{28}$$

By inserting the limit level just below the (highest-priority) invariant level, the solver will enforce joint limits to the greatest extent possible without violating the DoI of any closure joints, and target and posture waypoints will only be pursued to the extent possible without exceeding any DoF limits.

The final *lock* level enables freezing the pose of selected DoF, with priority between target tracking and limiting. Locked closures are driven to hold the pose they had when originally locked, in an approach similar to target tracking. Locked tree joints are removed from the problem (considered rigid).

Let Γ be an ordering of the set of locked joints, partitioned into subsequences of tree and closure joints,

$$\Gamma \subset J = (\Gamma_t, \Gamma_c) \text{ with } \Gamma_t = \Gamma \cap T \text{ and } \Gamma_c = \Gamma \cap C. \tag{29}$$

Let k be the total number of DoF of joints in Γ, and let k_t and k_c be same for Γ_t and Γ_c, so $k = k_t + k_c$. Finally, let $\boldsymbol{k}$ be a vector of the values of the k_c locked closure DoF at time of locking. A $[k_c \times d]$ binary projection matrix Π_k selects the k_c locked DoF out of all closure DoF, and the *lock residual* vector e_k is computed by the *forward lock error mapping* FK$_k$

$$\text{FK}_k : \mathbb{R}^f \to \mathbb{R}^k \quad e_k = \text{FK}_k(\boldsymbol{x}) = \Pi_k \Pi_{fc}\text{FK}(\boldsymbol{x}) - \boldsymbol{k}. \tag{30}$$

The *lock Jacobian* is the $[k_c \times f]$ matrix of partial derivatives of 30:

$$\underset{[k_c \times f]}{J_k} = \frac{\partial e_k}{\partial \boldsymbol{x}} = \frac{\partial}{\partial \boldsymbol{x}}\text{FK}_k(\boldsymbol{x}) = \Pi_k \Pi_{fc}J_y. \tag{31}$$

4.4 Interacting with Model State

We already briefly presented a comprehensive set of operations for evolving the *structure* of a linkage model (table 2). Corresponding operations to mutate *state* are relatively simple because the heavy lifting—propagating the effects of the mutations—is done by the PDLS action for differential control.

Forward kinematic state change operations simply change the target or posture values for joint DoF. Again, for inverse kinematics, just apply the changes to a general (6-DoF) closure joint attached between the link to be manipulated and the ground link. Rather than set joint DoF individually, it is helpful to provide more intuitive means such as

Table 2. The structure mutation primitives.

primitive name	description	primitive name	description
ADDLINK	add new link	SETTYPE	change joint type
SETLINKNAME	change name	SETLIMITS	change joint limits
ADDJOINT	add new joint	INVERT	invert a joint
SETJOINTNAME	change name	MAKEGROUND	switch ground link
MAKETREE	change disposition	REPOSITIONLINK	move link
MAKECLOSURE	change disposition	REPOSITIONJOINT	move joint
SETPARENT	re-attach a joint	SPLIT	insert new link and joint
SETCHILD	re-attach a joint	MERGE	delete joint and child link
REMOVEJOINT	delete a joint	ENCAPSULATE	demarcate sub-linkage
REMOVELINK	delete a link	DISSOLVE	merge sub-linkage

- *click-and-drag* manipulation—the operator selects links or joints with the mouse and drags them. Because the mouse only supports two axes of translation, attention needs to be paid when the manipulated object has more than two DoF. For example, different mouse buttons and/or keyboard modifiers can be used to select the DoF to operate. Computationally, drag gestures are discretized and used to repeatedly set new target waypoints for the manipulated DoF.
- *special purpose input devices*—mouse-based manipulation is general-purpose and requires no special hardware, but can only control two axes of motion simultaneously. This limitation can be addressed, for example, by providing application-specific input devices with mobility that more closely matches that of the manipulated linkage model. One example of such a device is our TRACK apparatus for ATHLETE (Figure 12).
- *trajectory following*—here the operator specifies a script for the motion of a set of DoF, for example, by entering a sequence of waypoints and the time intervals that should elapse from one to the next. The system then effects the motion by successively generating target waypoints along the specified trajectory.

5 Operating High-DoF Robots

We now turn to the application of the mixed real/virtual interface in operating high-DoF robots. The core idea is to use virtual articulations to (1) constrain and (2) parameterize the intended motion (Figures 1 and 2). The operator is presented with a graphical model of the robot and a catalog of available virtual joints (Figure 5). To constrain motion for a particular task, the operator instantiates joints from this catalog and interconnects them to the links of the actual robot—and/or to newly instantiated virtual links—thereby constructing arbitrary virtual extensions to the actual robot kinematics. Virtual joints can also parameterize specific task DoF; for example the long prismatic virtual joint in Figure 13 parameterizes the length of a trenching motion.

The system computes local assembly motions as virtual articulations are constructed. From there, the operator can move any joint or link, and the system interactively solves the differential control problem in real-time to find a compatible motion for all joints which best satisfies all constraints. For example, in the trenching task, the operator can effectively command "trench from -0.9m to +0.4m" by operating the corresponding virtual prismatic joint, or they may simply drag the constrained end effector with the mouse. The motions can be validated in simulation and then executed on the hardware, as we have done for ATHLETE (Section 5.2).

This approach to high-DoF operations is generally applicable to kinematic operations in articulated robots of any topology, including both open- and closed-chains as well as both over- and under-constraint. Section 5.3 demonstrates this topology independence in an example with a modular robot. Two layers of structure abstraction are used in this case, which help organize the motion specification problem for the operator, and which also enable the system to hierarchically decompose the full structure into smaller independently solvable parts (Section 4.1).

Two key challenges in high-DoF operations are handling under- and over-constraint; these are discussed next, followed by the ATHLETE and self-reconfiguring robot examples

5.1 Handling Under- and Over-Constraint

Under-constrained systems are especially common in high-DoF robots because they are often used in tasks which involve only a few degrees of constraint, whereas there may be many more degrees of freedom available in the robot. Unlike some prior approaches, which are aimed towards fully constraining the motion [19] (sometimes referred to as "redundancy resolution" [4]), our system allows the operator to add as much or as little constraint as desired.

At the lowest level, the least-squares PDLS solver gives a basic ability to find reasonable motions in any under-constrained system: at each iteration a shortest step is taken in joint space resulting in incrementally minimal motion. Said another way, at a fine scale, the system will produce piecewise linear moves from one configuration to the next in joint space. A more roundabout trajectory might also be feasible, but would likely be surprising.

On top of this foundation, the operator may construct chain-closing virtual articulations to express specific motion constraints, and thus reduce redundancy as much as desired. Figure 2 gives a basic demonstration, though this technique is pervasive in the our approach and is used in all examples.

Over-constraint, where the feasible configuration space is actually empty, is also a possibility. While at first this might seem to be a serious issue, in the presence of virtual articulations, it may be allowable—virtual closure joints can be broken if necessary. Our system also permits over-constraint, and provides the operator with tools to handle it at two levels.

The lowest-level handling of over-constraint is again given by the least-squares PDLS. If any individual priority level is over-constrained, then the least-squares solution will minimize error across all constraints in the level. This is useful and can produce intuitive behavior from the operator's perspective. For example, when multiple closure joints need to be broken, the amount of breakage can be balanced across them all.

The PDLS priority levels (Section 4.3) provide a higher-level tool for handling over-constraint: satisfaction of constraints at a lower priority level will not compromise satisfaction at higher levels, even when the constraints conflict. Consider the object inspection task in Figure 12. In this case we use our mimicking hardware interface *TRACK* (Tele-Robotic ATHLETE Controller for Kinematics) to pose the limb holding the inspection camera. But there is also a virtual spherical joint constraining the camera. TRACK has no haptic feedback, so while the operator should generally try to pose it near to a feasible configuration, invariably this will diverge from the strict spherical constraint surface, over-constraining the limb. The spherical joint constraint is modeled at the invariant level, and TRACK's pose is modeled at the target level, so the system will sacrifice the latter for the former. The overall effect is as if the virtual spherical joint was physically present and rigidly constraining the motion, with an elastic connection between the pose of TRACK and the pose of the actual limb.

While under- and over-constraint is allowable, physical systems that are under-actuated or or inconsistently over-constrained are still possible. Essentially, to the extent that the under- or over-constraint can be restricted to just the virtual articulations, they can be controlled as desired. Many robots are used in fully-actuated tasks and are thus not physically under-constrained; the examples studied in this chapter all fall into this category. While such systems can still be over-constrained, if they can be assembled then at least the over-constraint is self-consistent.

5.2   Experiments with ATHLETE

The All-Terrain Hex-Legged Extra-Terrestrial Explorer is under development for use in future Lunar missions, where it will potentially aid human explorers in various ways [41]. Figure 10 shows two instances of the current hardware, along with its representation in our system. ATHLETE weighs about 1000kg and has six identical limbs, each with six revolute joints and a terminal wheel. All joints have harmonic drivetrains and active-off brakes. The limbs attach to a hexagonal deck which is about 2.5m in (circumscribed) diameter and about 1.8m above the ground in nominal pose (foreground in the figure).

The six limbs of the robot are intended both for locomotion (both rolling and walking) and also for manipulation and inspection tasks. Other researchers are

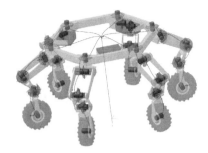

Fig. 10. All-Terrain Hex-Legged Extra-Terrestrial Explorer (ATHLETE).
Left: hardware (image courtesy NASA/JPL/Caltech); right: model.

considering the locomotion problems [17, 34]; here, we focus on manipulation and inspection. The original operations interface for these kinds of motions, based on traditional approaches, consisted of three primitives:

MOVE_JOINTS—Essentially equivalent to a board with 36 knobs, each commanding the position of one joint (forward kinematic control).
MOVE_TOOL—inverse kinematics (IK) for one leg (body locked in space).
MOVE_BODY—IK for the body (wheels locked in space).

While MOVE_TOOL is obviously useful, because it can only control one leg at a time, any task requiring multi-limb motion would need to be coordinated by some additional mechanism. Both MOVE_JOINTS and MOVE_BODY can move all the limbs together, but the former has obvious shortcomings (it simply puts the full burden of motion specification on the operator), and the latter is limited to controlling body posture alone.

Even in the case of single-limb motion, MOVE_TOOL is not always ideal. For example, some tasks do not constrain all 6 DoF of the EE pose; consider a drilling task—rotation of the EE about the drill axis may be unconstrained. And there are often multiple discrete solutions for fully constrained poses, i.e., ATHLETE's "elbows" can generally kink in different directions to reach the same EE pose. Thus, as an initial contribution, we built a direct-manipulation input device that mimics one ATHLETE limb, called the Tele-Robotic ATHLETE Controller for Kinematics (TRACK) [22] (Figure 12, inset). But this does not solve the whole problem—used alone, it still provides no mechanism for constraining motion or for coordinating motion of multiple limbs.

Virtual articulations can aid the operation of specific whole-robot motions which are rapid for human operators to conceptualize but difficult to express in more traditional interfaces. We implemented five different task scenarios; four on the hardware [36] and one in simulation.

- Figure 11 (simulation): A leg-mounted camera is used to inspect a crew module in a cylindrical scanning motion, with a virtual prismatic joint to control elevation and a virtual revolute joint for radial angle.
- Figure 12 (hardware): a limb-mounted camera inspects a roughly spherical object while maintaining a constant distance
- Figure 13 (hardware): a trench is inspected, with the support legs moving the deck to extend reachable trench length
- Figure 14 (hardware): a fixed camera is made to pan and tilt with the motion both parameterized and constrained by virtual revolute joints
- Figure 15 (hardware): two limbs execute a pinching maneuver with the distance and angles controlled by virtual prismatic and revolute joints.

For the object inspection task the operator models the camera distance constraint using a virtual spherical joint between the object (represented as a virtual link) and the camera. The space of reachable viewpoints is extended, relative to moving only the camera limb alone, by using the five other limbs to lean the hexagonal deck. Because the deck often carries a payload, it should remain flat; this is expressed by a virtual Cartesian-3 joint between the deck and the world frame. After configuring these virtual articulations the operator can drag the camera with the mouse to scan the object.

Descriptions of the other experiments are similar. For the bi-manual experiment the robot was partially supported by an overhead crane, as simultaneously raising two limbs is not supported on the current hardware. The crane served as a safety-backup in the other experiments.

 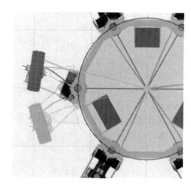

Fig. 11. Simulation of ATHLETE inspecting a cylindrical crew module.

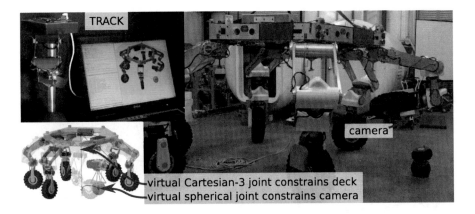

Fig. 12. ATHLETE inspecting an object with a leg-mounted camera.
A leg-mounted camera inspects an object from a range of views, while a virtual spherical joint centered on the object maintains a fixed focal distance. Our system automatically computes postural motions that greatly extend the range of reachable viewpoints vs. actuating a single leg alone. Also, for this task our mimicking hardware interface device *TRACK* was combined with the virtual articulation system.

Fig. 13. ATHLETE performing a trenching motion.
A scoop is mounted and could be used to dig a trench (digging would also require force control). Virtual prismatic joints constrain and parametrize the trench length and depth, virtual revolute joints set the scoop angle, and a Cartesian-3 joint keeps the deck at a constant orientation (e.g. in case that the deck is carrying an additional payload). A similar trenching motion could be operated by inverse kinematics for the active limb alone, but the addition of constrained postural motion approximately doubles the total possible trench length.

Fig. 14. Pan/tilt of an ATHLETE camera via postural motions.
ATHLETE has a number of side-facing cameras rigidly mounted to its deck (no pan/tilt actuators). Postural motions, where all legs cooperatively rotate the deck about the camera's center of focus, can be used to aim them. This can be operated with two crossed-axis virtual revolute joints. With this construction in place, quantified commands—"pan 30 degrees left, tilt 10 degrees down"—have a specific meaning. Our system finds a corresponding whole-robot motion.

Fig. 15. ATHLETE performing bi-manual manipulation motions.
In this bi-manual operations experiment, a virtual link models a movable object that could be grasped by pinching between two adjacent wheels (grasping a physical object would require the addition of force control). The operator can drag this virtual link around in our interface; the system finds compatible kinematic motions for the legs. Motion is automatically limited to the reachable workspace by constraint prioritization—the joint closure constraints are higher priority than the object pose target. Virtual prismatic and revolute joints additionally parametrize the grip.

5.3 Simulation of a Modular Robot

Moving to a second example in high-DoF operations, we now show a simulated modular robot with over 100 joints, demonstrating both the topology independence and scalability of our system. Two layers of structure abstraction are applied in this example, breaking up the operator's motion specification task and also enabling hierarchical decomposition for efficient motion computation.

The system in this section is a *self-reconfiguring* modular robot. Modular SR robots, studied since at least the late 1980s [15], are systems which can globally form arbitrary shapes by re-arranging many smaller interconnected units [23, 30, 31]. This is another class of robot where large numbers of joints are possible—though hardware has so far been limited to a few 10s of modules, work has been done in simulation with larger numbers of up to 1 million [13].

The SR system we use consists of active barbell-shaped modules with two symmetric rotating distal grippers and passive bars [9]. This system is capable of arbitrary topological reconfiguration by attaching and detaching modules; we have previously shown a reconfiguration sequence that builds the tower we study here from a disassembled set of parts [8].

As a *chain-type* modular robot, this system is also capable of *deformation* motions where the inter-module connectivity remains constant, but modules cooperatively use their internal kinematic DoF to effect a global shape-change. Operating this type of motion—i.e. providing an interface where an operator can conveniently specify general deformations in potentially large modular assemblies—has been under-explored; using virtual articulations to constrain and parameterize coordinated motion is a natural fit. By their nature, SR systems can assume arbitrary topologies, and one of the strong points of the virtual articulation approach is topology independence. Structure abstraction (Section 4.1) is also particularly useful because repeated sub-structures are common in large SR constructions. (A related idea in topological reconfiguration is the concept of *meta-modules* [35].)

Our main result is interactive operation of a tower with 120 simulated robot joints and about 150 virtual joints (Figure 17). The tower is built from a repeating block sub-structure (Figure 16). The operator first defines a particular constrained motion at the block level, explained in the next section, and this induces the full-tower motions presented in the following section. Structure abstraction plays a key role, and enables a hierarchical decomposition which keeps an otherwise sluggish motion computation snappy. (Williams and Mahew [42] considered a similar scenario, but that work was for one particular structure, whereas our algorithms are general-purpose.)

As illustrated in Figure 16, the operator wants the tower block to act as if it had two overall DoF: left/right tilting, and up/down expansion and contraction. A highest-level abstraction replaces the block with a chain of one revolute and one prismatic joint. But the actual modules making up the block—the implementation of this abstraction—have more complexity. They

Fig. 16. Operating a tower block with two levels of structure abstraction.
The actual modules comprising a tower block can move in a variety of ways, but
the operator intends only two motions: the block should tilt left/right and expand
up/down, forming the highest-level abstraction (left subfigure). A secondary constraint is that each 4-bar leg should effectively act like a piston (c.f. Fig. 2), so
the legs are further constrained and abstracted (middle and right subfigures; c.f.
Fig. 7).

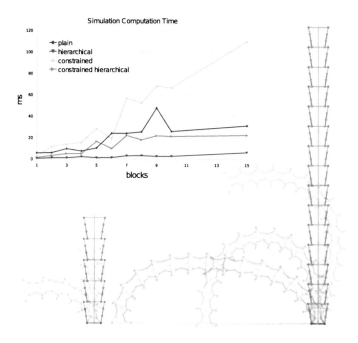

Fig. 17. Operating deformations in 5- and 15-block towers.
Blocks (Fig. 16) are stacked to make towers; a 15-block tower has 271 total joints.
120 represent real joints, the remaining implement virtual constraints and structural
abstractions. Here the operator is dragging the top of the tower and the system
interactively solves the differential control problem to deform the tower. The block-
level abstractions both aid motion design and enable hierarchical decomposition,
keeping computation time below 20ms/iteration. A similar tower operated without
abstractions required over 100ms ("constrained" vs "constrained hierarchical").

form two legs, which can be considered as four-bar linkages. Zooming in to a single leg, the desired motion is like a piston: the end-to-end distance of the constituent chain of modules sets the length of the leg, and the endpoints can also pivot to allow the leg to rotate. This implies a second level of abstraction where each leg is virtually replaced by a piston-like assembly of virtual prismatic and revolute joints. Finally, a Cartesian-2 constraint is added to keep the middle link of each leg parallel with the axis of the leg. In this case, the abstractions can be kept proper (Section 4.1) by sufficiently limiting the range of motion of the joints in the interface linkages.

It is important to note that these particular motion constraints, and this particular set of structure abstractions, are merely the designs of the operator in this instance. Other constraints and abstractions are possible: the idea is that the operator conceptualizes a desired motion, and then expresses it by designing virtual articulations and structure abstractions.

Once the operator is satisfied with the operation of a single block, a number of them are attached together to give towers of varying height. This final assembly is done at the highest level of abstraction, so that the top-level linkage is simply a linear chain of alternating revolute and prismatic joints along the "backbone" of the tower. (This is similar to the backbone-curve method for controlling hyper-redundant robots [6].) The operator can then interactively specify a motion, e.g. by click-and-drag interaction with the mouse, for any link or joint in the top-level linkage. The system will solve the differential control problem in and respond with a deformation as shown in Figure 17.

In this example, structure abstractions help the operator in the same way that traditional algorithmic abstraction aids the architect of a large software system: they break the problem up by enforcing high-level semantic invariants on sub-systems. The designer (in either situation) is thus not required to address the whole detailed problem at once.

These structure abstractions will also result in a hierarchical decomposition (Section 4.1) with three PDLS rounds (Section 4.2): First, the backbone will move as the operator has specified. Second, the top-level abstraction for each block will drive the mid-level abstractions for each leg. Finally, those abstractions will drive the motion of the modules that comprise the legs. The first round will consist of a single PDLS call, and the size of the system will scale linearly as the tower grows in height. The later rounds are different: there, growing the tower will result in a larger number of independent PDLS calls, but the system size in each call will remain constant.

In fact, the tower should move in a very similar way if constructed as a single flat (non-hierarchical) linkage. But in this case there would be only one PDLS round; the size of the system would scale linearly with tower height, but the constant factors would be greater than in the hierarchical case, making a significant difference in computation time. To check this, we conducted an experiment with towers of varying heights, up to 15 blocks, built in four different ways. The resulting measured computation times are comparatively plotted in Figure 17. (Each test was run interactively on a lightly loaded workstation.) The "plain" tower used no hierarchy, and also omitted the constraining Cartesian-2 joints. The "constrained" tower added the constraining joints, and the "hierarchical" tower added the structure abstractions, but not the constraining joints. The "constrained hierarchical" tower included both the abstractions and the constraining joints.

In each case adding the Cartesian-2 constraints increases the computation time, and adding the structure abstractions decreases it. The most interesting comparison is between the "constrained" tower (the worst case) and "constrained hierarchical": the latter remains roughly on-par with the unconstrained versions, at about 20ms per simulation iteration, which is acceptable for interactive response. But all the additional closure joints slow down the system for the non-hierarchical "constrained" tower by forcing PDLS of large systems; performance degrades to over 100ms per iteration for a 15 block tower, making interactive response sluggish.

6 Conclusions

This chapter has presented the theory and implementation of a novel operations system for articulated robots with large numbers (10s to 100s) of DoF, based on virtual articulations and kinematic abstractions. Operating such robots is a challenging problem because they are capable of many different kinds of motion, but often this requires coordination of large numbers of joints. Prior methods exist for specifying motions at both low and high-levels of detail; the new method fills a gap in the middle by allowing the operator to be as detailed as desired in specifying the motion, and is independent of any particular robot or task topology.

The main idea is to allow the operator to dynamically add, remove, and interconnect virtual joints and links with a model of the real robot. Virtual joints can both parameterize task motion and can constrain coordinated motion by closing kinematic chains. Virtual links can represent task-relevant coordinate frames, and also serve as intermediate connection points for constructions involving multiple virtual joints.

For topologically large robots, structure abstraction is a kinematic analogue to the algorithmic abstractions commonly used in software systems. This allows the operator to break the motion specification task into smaller sub-problems, and also enables the system to compute motions more rapidly via hierarchical decomposition.

Our approach does have some limitations: speed and force commanding are not directly addressed; and a large class of trajectories will be constructible even given a limited catalog of joints, some will be out of reach. For example, a true helical joint cannot be synthesized from the current joint catalog. Non-holonomic constraints and rolling contact are also not possible.

Portions of the work described in this chapter were funded under the NSF EFRI program and by a Caltech/NASA/JPL DRDF grant. We thank

Daniela Rus, David Mittman, Jeff Norris, and Brian Wilcox for helping to make this project possible. We also thank the RSVP team at NASA/JPL for providing a VRML model of ATHLETE.

References

1. Baerlocher, P., Boulic, R.: An inverse kinematics architecture enforcing an arbitrary number of strict priority levels. Visual Computer 20, 402–417 (2004)
2. Baerlocher, P.: Inverse Kinematics Techniques for the Interactive Posture Control of Articulated Figures. PhD thesis, EPFL (2001)
3. Bruyninckx, H.: Open RObot COntrol Software (OROCOS),
   http://www.orocos.org
4. Bruyninckx, H.: Kinematic Models for Robot Compliant Motion with Identification of Uncertanties. PhD thesis, Katholieke Universiteit Leuven (1995)
5. Chiacchio, P., Chiaverini, S., Sciavicco, L., Siciliano, B.: Closed-loop inverse kinematics schemes for constrained redundant manipulators with task space augmentation and task priority strategy. IJRR 10(4), 410–425 (1991)
6. Chirikjian, G., Burdick, J.: The kinematics of hyper-redundant robot locomotion. IEEE Trans. on Robotics and Automation 11(6), 781–793 (1995)
7. Davis, E.: Approximation and abstraction in solid object kinematics. Technical Report TR1995-706, NYU Computer Science (1995)
8. Detweiler, C., Vona, M., Kotay, K., Rus, D.: Hierarchical control for self-assembling mobile trusses with passive and active links. In: IEEE International Conference on Robotics and Automation, pp. 1483–1490 (2006)
9. Detweiler, C., Vona, M., Yoon, Y., Yun, S., Rus, D.: Self-assembling mobile linkages. IEEE Robotics and Automation Magazine 14, 45–55 (2007)
10. Diaz-Calderon, A., Nesnas, I.A.D., Nayar, H.D., Kim, W.S.: Towards a unified representation of mechanisms for robotic control software. International Journal of Advanced Robotic Systems 3(1), 061–066 (2006)
11. Dobrjanskyj, L., Freudenstein, F.: Some applications of graph theory to the structural analysis of mechanisms. ASME Journal of Engineering for Industry, 153–158 (1967)
12. Featherstone, R., Orin, D.: Robot dynamics: Equations and algorithms. In: IEEE ICRA, pp. 826–834 (2000)
13. Fitch, R., Butler, Z.: Million module march: Scalable locomotion for large self-reconfiguring robots. International Journal of Robotics Research 27(3/4), 331–343 (2008)
14. Flückiger, L.: A robot interface using virtual reality and automatic kinematics generator. In: Int. Symposium on Robotics, pp. 123–126 (April 1998)
15. Fukuda, T., Nakagawa, S.: Dynamically reconfigurable robotic system. In: IEEE ICRA, pp. 1581–1586 (1988)
16. Gleicher, M.L.: A Differential Approach to Graphical Interaction. PhD thesis, Carnegie Mellon University, School of Computer Science (1994)
17. Hauser, K., Bretl, T., Latombe, J.-C., Wilcox, B.: Motion planning for a six-legged lunar robot. In: WAFR, pp. 301–316 (2006)
18. Ivlev, O., Gräser, A.: An analytical method for the inverse kinematics of redundant robots. In: Proceedings of 3rd ECPD Int. Conf. on Advanced Robots, Intelligent Automation and Active Systems, pp. 416–421 (1997)

19. Ivlev, O., Gräser, A.: Resolving redundancy of series kinematic chains through imaginary links. In: Proceedings of CESA 1998 IMACS Multiconference, Computational Engineering in Systems Applications, pp. 477–482 (1998)

20. Kokkevis, E.: Practical physics for articulated characters. In: Game Developers Conference (2004)

21. Liégeois, A.: Automatic supervisory control of the configuration and behavior of multibody mechanisms. IEEE Transactions on Systems, Man, and Cybernetics, SMC 7(12), 868–871 (1977)

22. Mittman, D., Norris, J., Powell, M., Torres, R., McQuin, C., Vona, M.: Lessons Learned from All-Terrain Hex-Limbed Extra-Terrestrial Explorer Robot Field Test Operations at Moses Lake Sand Dunes, Washington. In: AIAA Space (2008)

23. Moll, M., Rus, D.: Special issue on self-reconfiguring modular robots. International Journal of Robotics Research 27(3/4) (March/April 2008)

24. Nakaoka, S., Nakazawa, A., Yokoi, K., Hirukawa, H., Ikeuchi, K.: Generating whole body motions for a biped humanoid robot from captured human dances. In: IEEE ICRA, pp. 3905–3910 (2003)

25. Phillips, C.B., Zhao, J., Badler, N.I.: Interactive real-time articulated figure manipulation using multiple kinematic constraints. In: Proceedings of SIGGRAPH, pp. 245–250 (1990)

26. Pratt, J., Chew, C., Torres, A., Dilworth, P., Pratt, G.: Virtual model control: An intuitive approach for bipedal locomotion. IJRR 20(2), 129–143 (2001)

27. Pratt, J.E.: Virtual model control of a biped walking robot. Master's thesis, Massachusetts Institute of Technology (1995)

28. Pryor, M.: Task-Based Resource Allocation for Improving the Reusability of Redundant Manipulators. PhD thesis, University of Texas at Austin (2002)

29. Pryor, M.W., Taylor, R.C., Kapoor, C., Tesar, D.: Generalized software components for reconfiguring hyper-redundant manipulators. IEEE/ASME Transactions on Mechatronics 7(4), 475–478 (2002)

30. Rus, D., Butler, Z., Kotay, K., Vona, M.: Self-reconfiguring robots. Communications of the ACM 45(3), 39–45 (2002)

31. Rus, D., Chirikjian, G.S.: Special issue on self-reconfiguable robots. Autonomous Robots 10(1) (January 2001)

32. Siciliano, B., Slotine, J.-J.E.: A general framework for managing multiple tasks in highly redundant robotic systems. In: Fifth International Conference on Advanced Robotics, pp. 1211–1216 (1991)

33. Smith, R.: Open dynamics engine (2008), http://www.ode.org

34. Smith, T., Barreiro, J., Smith, D., SunSpiral, V., Chavez-Clemente, D.: ATHLETE's Feet: Multi-Resolution Planning for a Hexapod Robot. In: ICAPS (2008)

35. Vassilvitskii, S., Kubica, J., Rieffel, E., Suh, J., Yim, M.: On the general reconfiguration problem for expanding cube style modular robots. In: IEEE ICRA, pp. 801–808 (2002)

36. Vona, M., Mittman, D., Norris, J., Rus, D.: Using virtual articulations to operate high-DoF manipulation and inspection motions. In: FSR (2009)

37. Vona, M.: Hierarchical decomposition and kinematic abstraction with virtual articulations. In: Advances in Robot Kinematics, pp. 33–43 (2010)

38. Vona, M.: MSim: Mixed Real/Virtual Simulator and Interface (2011), http://www.ccs.neu.edu/research/gpc/msim

39. Vona, M.A.: Virtual Articulation and Kinematic Abstraction in Robotics. PhD thesis, EECS, Massachusetts Institute of Technology (August 2009)
40. Welman, C.: Inverse kinematics and geometric constraints for articulated figure manipulation. Master's thesis, Simon Fraser University (1993)
41. Wilcox, B., Litwin, T., Biesiadecki, J., Matthews, J., Heverly, M., Morrison, J., Townsend, J., Ahmad, N., Sirota, A., Cooper, B.: ATHLETE: A cargo handling and manipulation robot for the moon. Field Robotics 24, 421–434 (2007)
42. Williams, R., Mayhew, J.: Control of truss-based manipulators using virtual serial models. In: ASME DETC (1996)
43. Wood, G.D., Kennedy, D.C.: Simulating mechanical systems in simulink with SimMechanics. Technical report, The Mathworks (2003)
44. Zanganeh, K.E., Angeles, J.: A formalism for the analysis and design of modular kinematic structures. IJRR 17(7), 720–730 (1998)

GPU-Based Real-Time Virtual Reality Modeling and Simulation of Seashore

Minzhi Luo[1], Guanghong Gong[2], and Abdelkader El Kamel[1]

[1] LAGIS, Ecole Centrale de Lille, Cite Scientifique,
BP 48 F59651, Villeneuve D'Ascq, France
{minzhi.luo,abdelkader.elkamel}@ec-lille.fr
[2] School of Automation Science and Electrical Engineering,
Beijing University of Aeronautics and Astronautics, Beijing 100083, China
ggh@buaa.edu.cn

As a important natural exterior scene, virtual seashore is widely used as a part of virtual environment for marine simulation, various simulators, games, etc. Nowadays, brilliant achievements about the deep ocean simulations have been made, however, concerning the real-time simulation of seashore, lots of work still need to be further studied. This chapter is devoted to efficient algorithms for real-time rendering of seashore which take advantage of both the CPU calculation and programmable Graphics Processing Unit (GPU). With regard to the modeling of seashore, a concept model based on Unified Modeling Language (UML) as well as precise mathematical models of seashore are presented. While regarding the simulation of seashore, optics effects imitation and the simulation of foam and spray are realized.

1 Introduction

In the field of Synthesis Nature Environment(SNE) simulation, marine simulation is one of the most important parts for military simulation and virtual reality application. The scene of seashore is a usual component of virtual environment, not only for the marine simulation but also exists in the other research areas, like the simulators for navigation, aviation or driving, the generic virtual training or analysis, the games and so on. All of the above researches need us to provide natural exterior scenes which should be realistic and real-time.

Although a number of work present ready-to-use simulations about the visualization of ocean, some points are still defective and need to be further studied: such as ocean models integrating both shallow and deep ocean, real-time simulations of coastline and breaking waves, special but typical natural phenomena near to the coast, etc. Some commercial products like MultiGen Vega Prime [1] or some free 3D engines like Ogre [2] also need to be enriched with the above points.

T. Gulrez, A.E. Hassanien (Eds.): Advances in Robotics & Virtual Reality, ISRL 26, pp. 307–332.
springerlink.com © Springer-Verlag Berlin Heidelberg 2012

In this chapter, the methods for real-time modeling and simulation of seashore with GPU programming are discussed. We apply an integrated ocean wave model and focus on the real-time simulation of coastline and breaking waves. This chapter is organized as follows:

- In section 2, the background is introduced. It describes the concept of Synthesis Nature Environment simulation, the natural phenomenon of the sea, the development status of virtual reality and the related work in ocean simulation.
- In section 3, the concept modeling and the mathematic modeling of seashore are studied. The UML has been used to help us design this seashore simulation system from an global perspective. While the precise mathematic modeling of seashore comprises ocean wave modeling as well as coastline and breaking wave modeling.
- In section 4, the simulation of seashore is proposed following the modeling step. We present the methods to realize the optics effects imitation and the simulation of foam and spray.

2 Background

2.1 Synthesis Nature Environment

SNE is the representation of the natural environment including terrain, ocean, atmosphere and space in the distributed simulation environment, and it is an important component of the complex simulation system [3]. A simple SNE concept model has been depicted in Fig. 1, which includes four parts.

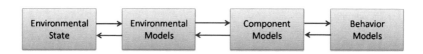

Fig. 1. A simple SNE concept model

In this model, environment state is used to describe the measurable properties of natural environment at one moment, such as atmospheric temperature, atmospheric pressure, etc. Environment models mean the algorithms which could create continued environment states during a period, such as the temperature changes in a day. While component models are used to simulate the static properties of the components in natural environment and behavior models work for describing their behavioral properties.

The application of SNE is extensive and significant. It supports the distributed simulation systems simulating under various natural environments, and through the evaluation of natural environment impact on these systems, corresponding adjustments could be done. SNE could be applied in the field of military simulation which covers navy, army and air force, as well as some civil simulations like the driving simulator (car, bus, subway, train, etc.), the automated highway system simulation, and so on.

The constitution of SNE requires several techniques: environment data models, standardization of environmental data, environmental database and resource library, modeling and simulation of SNE, and finally visualization of SNE.

2.2 Heterogeneous Sea

When describing the phenomena of the heterogeneous sea, it is necessary to consider many factors, like the climate, the wave, the currents, the ledge, etc. These elements could influence the sea behavior by changing the wave's wavelength, speed, direction of propagation, localization, camber and so on.

Wind Waves

The wind is mainly responsible for the generation of waves and it is irregular when blowing across the water. The puffs or blasts could last several seconds and/or locate over small regions, while the global wind blows above the overall of the water. As the wind blows, pressure and friction forces disarrange the equilibrium of the water surface and the energy would be transferred from the air to the water to form the waves. The result of the wind action in the ocean, called ocean surface waves, or more precisely, wind-generated ocean surface waves. Waves in the ocean can travel thousands of miles before reaching land and range in size from small ripples to huge rogue waves. The wind scale and the corresponding phenomenon observed on the sea have been listed in Table 1 [4].

As the ocean surface waves transfer from the deep ocean to the coast, they slow down and their wavelength becomes shorter with the amplitude increasing meanwhile.

Breaking Waves

When closing to the coast, especially there is a slope or steepness ratio, if a wave is too great, breaking is inevitable. In more detail, wave breaks when its amplitude reaches a critical level (much larger than its wavelength) or the wind grows strong enough to blow the crest off the base of the wave.

Different levels of breaking waves could be defined according to their effects on the ocean surface. The first one is called Spilling breakers, which are the safest wave and could be found in most areas with relatively flat shorelines. The next is Plunging breakers, and this phenomena occurs when the ocean floor is steep or has sudden depth changes and could be caused by strong off-shore winds and long wave periods. The crest of the Plunging waves is much steeper than Spilling waves. As the third level of breaking waves, Surging

Table 1. Aspect of the sea associated with an average speed of wind

Wind scale	phenomenon observed	wind speed
0	as a mirror	$\leq ms^{-1}$
1	wrinkles, like fish scales, no foam	1.5 to $2.5ms^{-1}$
2	ripples, do not scroll	3.0 to 4.0 ms^{-1}
3	very small waves, crests begin to break, sometimes the foams appears	4.5 to $6.0ms^{-1}$
4	small waves, lots of foams	6.5 to $8.0ms^{-1}$
5	moderate waves that much elongated, many foams and little spray	8.5 to $10.5ms^{-1}$
6	bigger wave forms; the white foams on the crests are more extensive and persistent	11.0 to $13.0ms^{-1}$
7	the white foams from breaking waves blow in the direction of the wind	13.5 to $15.5ms^{-1}$
8	top edge of ridges separates into spindrift, the foams streak in the wind	16.0 to $18.5ms^{-1}$
9	large waves begin to roll over, the thick foams streak	19.0 to $21.0ms^{-1}$
10	breaking waves become intense and brutal, poor visibility due to permanent spray, the water surface is all white	22.0 to $25.0ms^{-1}$
11	all the sea is covered with white patches of foam, the waves become exceptionally high, very low visibility	25.5 to $28.5ms^{-1}$
12	air is filled with foam and spray, almost zero visibility	$\geq 29ms^{-1}$

waves form on steeper beaches, the energy of the waves could be reflected by the bottom back into the ocean and causing standing waves. Eventually, the biggest breaking waves, called tsunami or tidal waves, are usually attributed to either earthquakes, landslides or volcanic eruptions. Fig. 2 shows the natural breaking waves. Comparing with the normal breaking waves in the left, the one in the right picture is much bigger.

Foam and Spray

Another common phenomenon occurred in the seashore would be foam and spray. At most of the time, waves break into foaming whitecaps or sprays when meeting the land or breaking. The larger the wave, the more the foams or sprays. Fig. 2 also shows us this phenomenon.

Fig. 2. Natural breaking waves

2.3 Virtual Reality

Nowadays, several disciplines have contributed to the development of Virtual Reality (VR): robotics, computer graphics, man-machine interface, real-time computing, ergonomics, physical and physiological and more recently psychology and cognitive ergonomics. Although VR is most often related to computer science, the contribution of other disciplines in recent years can not be overlooked. The VR has taken this multidisciplinary approach to construct itself as a research field [5].

The term Virtual Environment (VE) was used as a synonym of VR in the early 90's by researchers at the Massachusetts Institute of Technology (MIT) [6]. It is a dynamic description of the objects within a simulation. Designing an VE requires a set of data and models, for example, the virtual scene, the behavior of objects, behavioral models for defining relationships between objects, physical models to control the laws of interaction and so on.

The representation of an VE is primarily visual. Technological advances in the field of computer graphics and image synthesis allow to improve the visual quality of VE. Image synthesis consists of two main phases: a modeling phase to reconstruct the environment and a rendering phase which aims to visualize the modeled environment. Each stage uses different techniques. Progress in the field of computer graphics for the VR are around these techniques, but also around the hardware to optimize each of the two phases (physical memory, CPU, GPU, etc ....). Briefly, the aim is to improve the quality of images displayed in real time and allow the immersion of the user in the VE.

GPU, short for Graphics Processing Unit, is a specialized microprocessor for rendering graphics. Thanks to their highly parallel structure, GPUs are very efficient at manipulating computer graphics and more effective than general-purpose CPUs for a range of complex algorithms. Comparing with that the CPU core has only one thread, GPU processor has about 32 threads [7], which could be used selectively or simultaneously to carry out the same task, which makes the computing more efficient. Since the term was defined and popularized by Nvidia in 1999, GPU technology has grown up in the past ten years: from low-level machine language to the high-level shading language (programming in vertex shader and fragment shader)and now to the CUDA; the pipeline function from simple to complex; the computing speed from high to even higher...

2.4   Related Work

Water surface simulation or ocean wave generation is a very popular direction in virtual reality research and there are many approaches to this problem. Some of them are based on solving the Navier-Stokes equations in 2D or 3D. Some of them employ simplified water models. One of the first descriptions of water waves in computer graphics was by Fournier and Reeves [8], who modeled a shoreline with waves coming up on it using a water surface model called Gerstner waves. In the same area, Darwin Peachey [9] presented a variation on this approach using basis shapes other than sinusoids. Recently, the key laboratory of marine simulation and control in China Dalian Maritime University [10] presents a method for simulating shallow-water for the marine simulation. The modeling of sea surface height is achieved by solving 2D Boussinesq type equations. However, their shallow-water simulation includes neither the coastline nor the breaking waves. In fact, various methods for ocean wave generation have been studied: P. Y. Ts'o and B. A. Barsky [11] modeled shallow water waves using different basis shapes; Gonzato [12] created a phenomenological model of ocean wave based on physical considerations; G. A. Mastin [13] has used a noise synthesis approach to simulate the appearance of the ocean surface seen from a distance; J. Tessendorf [14] has presented a realistic deep ocean model based on Fast Fourier Transformation (FFT), etc.

Some researches also have been done in the area of breaking waves simulation. Keyyong Hong [15] proposed the wave energy analyzes of breaking waves, and Viorel Mihalef [16] creats the three-dimensional breaking waves shapes based on the two-dimensional wave library. In addition, a two-dimensional multi-scale turbulence model is proposed to study breaking waves by Qun Zhao [17], to produce a relatively accurate model with moderate computer requirements. Nevertheless, the visualization of breaking waves would become more complicated when it is requested to be real-time, because finding out the balance between speed and effectiveness is difficult.

Concerning the real-time rendering, with the development of GPU, some traditional approaches of optic effects simulations (like ray-tracing) can be adapted to real-time requirements thanks to the parallel calculations in GPU. Hence, optic effects of ocean water can be rendered with the help of vertex shader and fragment shader which could decrease the simulation time enormously. Y. Kryachko [18] used vertex texture displacement for realistic water rendering. This method calculated the wave mesh in vertex shader, whereas it required a great performance of graphics card.

3 Modeling of Seashore

3.1 UML-Based Design

The UML is a specification defining a graphical language for visualizing, specifying, constructing and documenting the artifacts of distributed object systems [19]. It is widely accepted by industry and it has become the standard for object-oriented modeling and design.

In this part, a use-case diagram, a class diagram as well as an activity diagram have been created to describe the seashore simulation via both static and dynamic aspects.

Use-Case Diagram

A use case describes a sequence of actions that provide something of measurable value to an actor, and a use-case diagram displays the relationship among actors and use cases. It helps us to clarify exactly what the system is supposed to do.

In our design, this virtual seashore application is supposed to be an intelligent part of VE for marine simulation or driving simulator, etc. Users could affect the scene through several actions. See Fig. 3.

Besides observing the virtual seashore with the ocean waves' sound, users could also do the screenshot to store image data, change viewpoint to get different scenes, change weather/time or wind scale/direction to see the influence on the seashore waves, and finally, switch the rendering mode between Line mode and Fill mode.

Class Diagram

Several classes have been designed to construct the seashore application, see Fig. 4. The relations among them could be described by three types of association:

- Aggregation: indicates the fact that X "is part of"Y. Represented by a white lozenge at the end of an arrow, it consists in a weak membership.
- Dependence: shows the relationship that X "depends on"Y. Represented by a dotted line with arrow.
- Use: means that X "uses"Y. Represented by a dotted line with arrow and the word "use"above.

The whole seashore scene consists of the modeling of Sky&Terrain, Ocean waves, Coastline, Breaking waves and the rendering of all the above models. The Rendering Window is the graphics interface that shows users the virtual seashore visually. We use the class Font to display related real-time information on the rendering window. In the simulation, class Texture provides the texture materials and the class Foam&Spray is embedded in class Coastline and class Breaking waves to perfect the simulation effects.

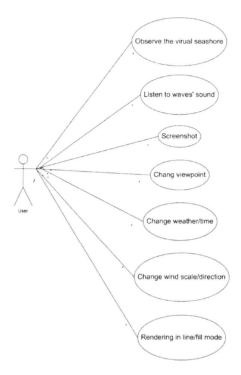

Fig. 3. Use-case diagram

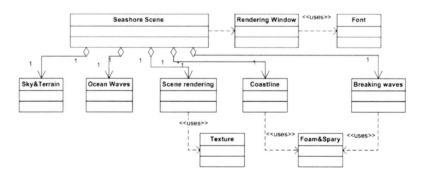

Fig. 4. Class diagram

Activity Diagram

Belonging to the behavior diagrams, the activity diagrams have a wide number of uses, from defining basic program flow, to capturing the decision points and actions within any generalized process. Here, the activity diagram (Fig. 5) with two swim lanes has been drawn to clarify which calculation is prepared to be done in GPU and which one would be done in CPU.

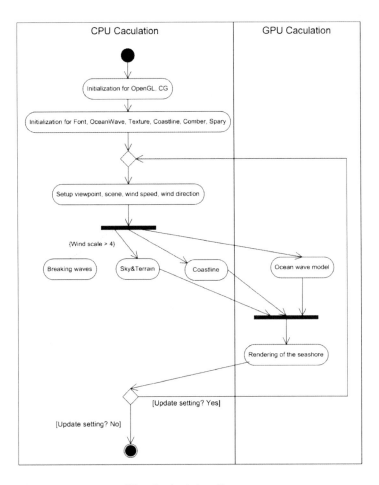

Fig. 5. Activity diagram

At the beginning of the simulation, the main work is the initialization of OpenGL and CG (for GPU programming), the establishment of the liaison between CPU and GPU, and the initialization of all the classes in the application. After the above preparations, the simulation loop starts.

First, we could setup manually the viewpoint, the weather/time, the wind speed/direction, etc. in CPU, and then our application starts modeling the sky, the terrain, the coastline, and the breaking waves with a restriction that if the wind scale is more than 4. While in GPU, the ocean wave mesh is constructed in Vertex Shader and the rendering of the whole seashore scene is done in Fragment Shader.

At any time of the simulation, the parameters could be updated with the help of key-board or mouse. When all the experiments have been finished, the application would be ended by users.

3.2 Ocean Wave Modeling

In this part, a ocean wave model based on Gerstner waves is presented to describe both shallow and deep ocean. Ocean depth and some simple atmospheric conditions like wind velocity and wind direction are considered. The wave mesh was generated in vertex shader(of GPU), which is proven to be more efficient.

Gerstner Waves

As described in [14], Gerstner waves is a physical model that describes the water surface in terms of the motion of individual points on the surface. If a point on the undisturbed surface is labeled $\mathbf{x_0} = (x_0, z_0)$ and the undisturbed height is $y_0 = 0$, then when a single wave with amplitude A passed by, the point on the surface is displaced at time to:

$$\begin{cases} \mathbf{x} = \mathbf{x_0} - (\mathbf{K}/k)A\sin(\mathbf{K} \cdot \mathbf{x_0} - wt) \\ y = A\cos(\mathbf{K} \cdot \mathbf{x_0} - wt) \end{cases} \tag{1}$$

In this expression, $\mathbf{K}$ is called the wavevector, which is a horizontal vector that points in the direction of travel of the wave, and has magnitude k related to the length of the wave λ by $k = 2\pi/\lambda$.

In order to make this equation more clear, a new variable $\mathbf{I}$, the unit vector of $\mathbf{K}$, was presented. Expression (1) could be changed to (2) due to the formula $\mathbf{K} = k \cdot \mathbf{I}$.

$$\begin{cases} \mathbf{x} = \mathbf{x_0} - \mathbf{I}A\sin(k \cdot \mathbf{I} \cdot \mathbf{x_0} - wt) \\ y = A\cos(k \cdot \mathbf{I} \cdot \mathbf{x_0} - wt) \end{cases} \tag{2}$$

Actually, the wave profile could be seen as the curve generated by a point P at a distance A from the center of a circle of radius $1/k$ rolling over a line at distance $1/k$ under the X axis, see Fig. 6.

It is necessary to note that, as A getting close to $1/k$, the wave becomes steeper and steeper, which is very suitable to simulate the ocean waves, while as $A > 1/k$, a loop would form at the tops of the wave, which is not a particularly desirable or realistic effect. Fig. 7 shows two example wave profiles, each with a different relationship between A and $1/k$.

As presented so far, Gerstner waves are rather limited because they are a single sine wave horizontally and vertically. However, summing a set of waves can generalize to a more complex profile. One picks a set of unit wavevectors $\mathbf{I_i}$, amplitudes A_i, frequencies w_i, for $i = 1...N$, to get the expression:

$$\begin{cases} \mathbf{x} = \mathbf{x_0} - \sum_{i=1}^{N} \mathbf{I}_i A_i \sin(\mathbf{I}_i \cdot \mathbf{x_0} \cdot k_i - w_i t) \\ y = \sum_{i=1}^{N} A_i \cos(\mathbf{I}_i \cdot \mathbf{x_0} \cdot k_i - w_i t) \end{cases} \tag{3}$$

In this way, realistic and complex shapes could be obtained.

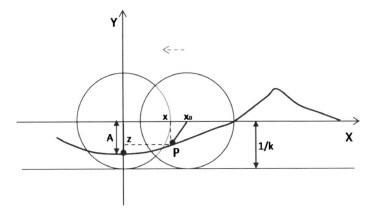

Fig. 6. The generalization of cycloid

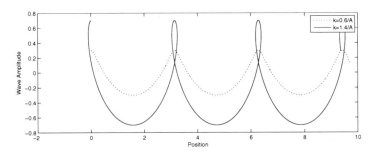

Fig. 7. Profiles of two single-mode Gerstner waves

Depth of Ocean

As the aim is to present an integrated ocean wave model, the depth of ocean h should be taken into account. The first effect of the depth is to alter the wavelength. It is assumed, and largely true, that the period is not affected. We call k_∞ as the wave number at infinite depth, and a well-know relationship between these frequencies and the magnitude of their corresponding wavevectors k_∞ could be used, the relationship is:

$$w^2 = gk_\infty \tag{4}$$

The parameter g is the acceleration of gravity, nominally $9.8m/sec^2$. The bottom will have a retarding affect on the waves, when it is relatively shallow compared to the length of the waves. If we call k as the wave number at finite depth, the dispersion relation is:

$$w^2 = gk\tanh(kh) \tag{5}$$

Then a good approximation for the wave number k at depth h is:

$$k\tanh(kh) = k_\infty \tag{6}$$

A character of the function tanh should be known, when $x \to 0$ then $\tanh(x) \to x$. So at small depth, the relation becomes:

$$k^2 h = k_\infty \quad (k = \sqrt{k_\infty / h}) \tag{7}$$

When $x \to \infty$, then $\tanh(x) \to 1$, therefore at large depth, the relation becomes:

$$k = k_\infty = w^2 / g \tag{8}$$

That is to be hoped. Note that $kh = 2\pi h/\lambda$, a ratio h/λ of 0.5 gives an argument of π for the tanh, which makes it practically equal to 1. Thus, we could suppose that, as the ratio h/λ is greater than 0.5, it is deep ocean, and (8) comes into existence; as the ratio h/λ is smaller than 0.5, it becomes shallow ocean. A good approximation for (6), valid for the entire range within 5% is:

$$k = \frac{k_\infty}{\sqrt{\tanh(k_\infty h)}} \tag{9}$$

Considering (4), (9) can be changed to:

$$k = \frac{w^2}{g\sqrt{\tanh(w^2 h / g)}} \tag{10}$$

Now, the relationship between k, w and h has been established, and for a given w and h, the wave number k would be determined. It is evident that k becomes the crucial factor to distinguish the deep ocean and the shallow ocean. When describing the deep ocean, we could use (8) to replace the k in the ocean model (3), and when describing the shallow ocean, using (10) to replace the k in (3).

Final Ocean Wave Model

Obviously, different points on the ocean surface have different value of h, so choosing $h_{(x,z)}$ as the symbol of ocean depth would be better. Expression (11) was imported to transform the vector to scalar quantity in (3).

$$\begin{cases} \mathbf{I}_i = (I_{i1}, I_{i2}) \\ \mathbf{x} = (x, z) \\ \mathbf{x}_0 = (x_0, z_0) \end{cases} \tag{11}$$

The final ocean wave model can be described as (12):

$$\begin{cases} x = x_0 - \sum_{i=1}^{N} I_{i1} A_i \sin((I_{i1} x_0 + I_{i2} z_0) k_i - w_i t) \\ z = z_0 - \sum_{i=1}^{N} I_{i2} A_i \sin((I_{i1} x_0 + I_{i2} z_0) k_i - w_i t) \\ y = \sum_{i=1}^{N} A_i \cos((I_{i1} x_0 + I_{i2} z_0) k_i - w_i t) \end{cases} \tag{12}$$

In this equation, when $h_{(x,z)}/\lambda_i < 0.5$, which means the shallow ocean, expression $k_i = \dfrac{w_i^2}{g\sqrt{\tanh(w_i^2 h_{(x,z)}/g)}}$ is used; and when $h_{(x,z)}/\lambda_i > 0.5$, which means the deep ocean, expression $k_i = k_{\infty i} = w_i^2/g$ is used.

As a supplement, a approximate expression between w and wind speed V has been shown [8], see (13).

$$w = \frac{g}{V}\sqrt{\frac{2}{3}} \tag{13}$$

In this way, our final ocean wave model could connect with the atmospheric conditions perfectly: A_i, w_i are used for representing the wind power and $\mathbf{I_i}$ could represent the wind direction. However, it is necessary to avoid the formation of loop on the tops of wave, when setting the parameters' value.

This ocean wave model has characteristics of simpler, universal, and also considerate, so it can perfectly match the requirement of real-time.

Glossary

The mathematical symbols used in this section have been listed in table 2.

Table 2. Mathematical symbols

Symbols	Description
k	Wave number
k_∞	Wave number constant of deep ocean
$\mathbf{k}$	Wave vector (k_x, k_y)
$\mathbf{I}$	Unit wave vector (I_1, I_2)
$\mathbf{x}$	2D spatial position of simulation point (x, z)
$\mathbf{x_0}$	Initialization of simulation point (x_0, z_0)
A	Wave amplitude
T	Period of wave
w	Angular frequency
t	Time
h	Height of ocean water
λ	Wavelength
g	Gravitational constant
V	Wind speed

3.3 Coastline and Breaking Waves Modeling

When modeling the seashore, the interaction between waves and shore can be counted as the most important task. The waves rush and flap the beach which forms a natural and variable curve called coastline. It could not be determined due to the dynamic nature of tides and some waves undergo a phenomenon called "breaking", when they run into shallow water. In this part, we present a method to structure the coastline geometrically and modeling the real-time breaking waves by 3D Bézier curved surface using the technologies of metamorphosing and key-frame animation in considering the influences caused by wind velocity or direction at the same time. This method is of high efficiency and can meet the requirement of real-time. The concepts of techniques are first introduced and then the specific modeling of coastline and breaking waves will be shown.

Metamorphosing and Key-Frame Animation

Metamorphosing which is also called morphing is a special effect in motion pictures and animations that changes or morphs one configuration (or image) into another through a seamless transition. If treating two configurations as two key frames, and making one of them morphing to another, a simple key-frame animation was created. The difference between metamorphosing and key-frame animation is that, metamorphosing pays more attention to the changing process but key-frame animation pays more attention to the "frame". In a key-frame system, the main job of the animator is creating important frames of a sequence, and then the software fills in the gap automatically following the methods designed by animator.

In this part, these two techniques were used together in three-dimensional space.

Coastline Design

The coastline discussed here is not expected to be macro like in a map but to be a specific one in the coast whose parameters can be controlled, like shape, position etc. In that way, as our further researching, certain statistical data that come from oceanography or some other sources could be applied to coastline. And these potentialities make our design more significant.

As the statistical data have not been obtained, the coastline was designed randomly in this paper. Fig. 8(left) shows the curve designed as the coastline and only these points whose y-value is positive were used. The original curve (points locate on the borderline of ocean mesh) was treated as the first key-frame and the curve mentioned above as the second key-frame, using linear interpolating, the animation that ocean waves frapping the beach with the coastline formed differently very time could be created. Some stochastic quantities have also been added into the coastline to get a better effect.

Until now, the mutative coastline animation has been shown, but it is not realistic enough. Because in real scene, there are layers of waves flapping the beach. Instead of using texture to simulate this phenomenon, a layer of mesh was added to deal with it (Fig. 8(right)). Hence, the animation becomes that one ocean wave flaps on the shore and when it is fading, the other ocean wave starts to flap and these two waves keep alternating again and again. Fig. 9 shows the wave mesh of seashore with the designed coastline.

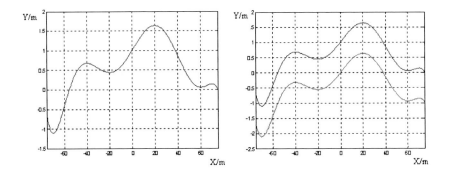

Fig. 8. The coastline designed in MATLAB

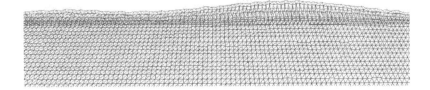

Fig. 9. The effect of coastline

In our application, this animation could be controlled to adapt to different wind speeds by two parameters: the distance of waves toward the shore $d_{coastline}$ and the speed of waves toward the shore $v_{coastline}$. The greater the wind, the greater the values of $d_{coastline}$ and $v_{coastline}$.

Breaking Waves Design

Since the traditional formal methods to create the breaking waves are usually too complex, here a geometrical method was presented. The choice of a mathematical representation for the breaking waves is critical to be efficient and controllable. As we know in the field of Computer Graphics, parametric cubic Bézier curves are commonly used to model smooth curves that can be scaled indefinitely, so it can be a good choice.

Normally, four points P_0, P_1, P_2 and P_3 in the plane in 3D space define a cubic Bézier curve. The curve starts at P_0 going toward P_1 and arrives at P_3 coming from the direction of P_2. The parametric form of the curve is:

$$B(t) = (1-t)^3 P_0 + 3(1-t)^2 t P_1 + 3(1-t)t^2 P_2 + t^3 P_3$$
$$t \in [0, 1] \tag{14}$$

Generalizations of Bézier curve to higher dimensions are called Bézier surface, which is defined by a set of control points. A given Bézier surface of order (n, m) is defined by a set of $(n + 1)$ $(m + 1)$ control points $P_{i,j}$. A two-dimensional Bézier surface can be defined as a parametric surface where the position of a point Q as a function of the parametric coordinates u, v is given by equation (15):

$$Q(u, v) = \sum_{i=0}^{n} \sum_{j=0}^{m} B_i^n(u) B_j^m(v) P_{i,j} \tag{15}$$

in this equation, $B_i^n(u) = \begin{pmatrix} n \\ i \end{pmatrix} u^i (1 - u)^{n-i}$ is a Bernstein polynomial, and $\begin{pmatrix} n \\ i \end{pmatrix} = \frac{n!}{i!(n-i)!}$ is the binomial coefficient.

When observing breaking waves on the shore deliberately, some orderliness can be found out. From the appearance of a breaking wave to its disappearance on the beach, the process can be divided into five important stages, and each stage has a special shape which can be described by two connected Bézier patches (each patch has 16 control points), see Fig. 10. We treat these five shapes as five key-frames, adding linear interpolating and making these shapes morphing from one to another in sequence, whereupon the key-frame animation of breaking waves could be produced.

Frame A shows us two Bézier patches connected with C^1-continuity, which is taken for the premier stage of the generation process of breaking waves. This stage can persist a few moments, and then the head of the wave becomes much sharper (Frame B, C^0-continuity). As the wind force acts on the wave, the terrain resistance whose orientation is opposite to the wind force effects on the bottom of the wave, and simultaneously, the top of the wave would move rapider than the bottom (Frame C, C^0-continuity). On account of gravity, the wave falls down (Frame D, C^0-continuity) and flaps on the beach (Frame E). Finally, it returns to the ocean and the second breaking wave start to take place.

As the shape of Bézier patch is controllable, various heights of breaking waves (h_{bw}) could be defined to simulate different types of breaking waves. Furthermore, the duration of the animation (t_{bw}) could be changed according to different weather condition.

3.4 Summary of Wind Influence

As previously described, the wind could influence the ocean wave to a great extent. when modeling the seashore, various parameters must be set. The selection of all the related parameters has been listed to correspond with different wind scales as well as wind direction, see Table 3.

When the wind scale changes from 1 to 4, the parameters belong to the ocean wave model (A_i and $\bar{\omega}$) and the ones of flap waves ($d_{coastline}$ and

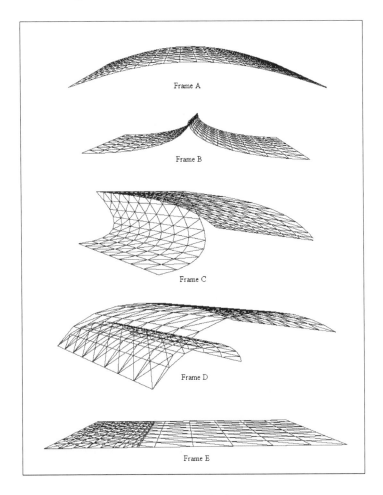

Fig. 10. Key-frames of a breaking wave

$v_{coastline}$) are selected. As the wind continues to grow, breaking waves start to appear, the amplitude of waves is even higher and the foams and spray are come to being. In this situation, our basic ocean wave model becomes insufficient, hence, we should add the virtual real-time breaking waves with corresponding parameters (h_{bw} and t_{bw}) to simulate the scene. Eventually, the wind direction influence could be simulated with different evaluation of $\mathbf{I_i}$.

4   Simulation of Seashore

In this section, the rendering of the whole seashore scene will be performed. The optics effects of the ocean surface are firstly realized by modulating a reflection component and a refraction component using Fresnel equations.

Table 3. Related parameters for wind influence

Related parameters	Wind scale (1-4)	Wind scale (5-12)	Wind direction
A_i	√	√	
$\bar{w}$	√	√	
$d_{coastline}$	√	√	
$v_{coastline}$	√	√	
h_{bw}		√	
t_{bw}		√	
$\mathbf{I_i}$			√

Then, the foam and spray are simulated by applying the graphics techniques alpha-blending and particle system. Finally, the seashore scene under different atmospheric conditions have been displayed.

4.1 Optics Effects Imitation

Reflection and Refraction

The reflection color $C_{reflected}$ caused by the environment can be divided to sky reflected light $C_{skylight}$ and sunlight $C_{sunlight}$. Sky reflection color is based on the sky color, and sunlight can be obtained by the following expression:

$$C_{sunlight} = C_{suncolor}((V_{eye} + V_{sun}) \cdot V_{normal})^{sunshineness} \quad (16)$$

The default of $C_{suncolor}$ is near to yellow, V_{eye} means the vector from the observer's eye to the ocean surface point, V_{sun} means the vector of the sunlight and V_{normal} means the normal vector of the ocean surface point. The "sunshineness" denotes the intensity of the sun. Thus, the color of one point on the ocean surface could be calculated, see (17).

$$C_{reflected} = C_{skylight} + C_{suncolor}((V_{eye} + V_{sun}) \cdot V_{normal})^{sunshiniess} \quad (17)$$

The refraction color $C_{refracted}$ is determined by the scattering object color ($C_{objectcolor}$) inside the water and the water color ($C_{watercolor}$).

$$C_{refracted} = C_{objectcolor} + C_{watercolor} \quad (18)$$

After gathering all the points and doing the color interpolation, the reflection and the refraction colors of the whole ocean surface could be obtained.

A technique called environment normal mapping was used to calculate $C_{reflected}$ and $C_{refracted}$. This technique actually is a mix of environment mapping and normal mapping. Environment mapping technique simulates an object reflecting and refracting its surroundings. Normal mapping is a technique that has the ability of faking the lighting of bumps and dents. In this way, some thin ripples can be added to the ocean surface based on the ocean dynamic waves to achieve realistic effects.

When applying this technology, with the help of shading language programming, the calculation of ocean color could be carried out in fragment shader, but in respect that the calculation cannot be done if the above corresponding vectors are not in the same space, the space switching becomes the hardest mission.

According to the transform pipeline in the graphics card, if a virtual 3D object could finally be shown in the screen, it has to be transformed through several spaces using 4×4 matrix, see Fig. 11.

Fig. 11. Transform pipeline

In our software application, the original state of ocean mesh is object space, while the light vectors, the reflection vectors, and the refraction vectors are in the world space. The normal vectors are come from a bump map, whose original state is tangent space (the tangent space is made up of binormal B, normal N and tangent T, see Fig. 12). To reduce the computation, we have chosen to transform all the above vectors from its original space to the world space.

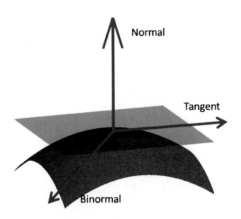

Fig. 12. Tangent space

The transformation of vertex positions can be done by multiplying the world matrix, while the transformation of the normal vectors is complex. The first step is from tangent space to object space, and the translation matrix,

BTN matrix must be calculated. The second step is from object space to world space, which can be realized by multiplying the IV world matrix (the inverse of world matrix).

The world matrix can be calculated by multiplying transfer matrix and rotate matrix; BTN matrix can be acquired by (19):

$$M_{BTN} = \begin{pmatrix} B_x & T_x & N_x \\ B_y & T_y & N_y \\ B_z & T_z & N_z \end{pmatrix} \tag{19}$$

Again, B, T and N are respectively binormal, tangent, and normal of each point on the ocean mesh. This matrix can be calculated by (20) in which we demand our final ocean wave model on the x_0 and z_0 partial derivatives.

$$N = B \times T$$

$$M_{BTN} = \begin{pmatrix} \frac{\partial x}{\partial x_0} & \frac{\partial x}{\partial z_0} & N_x \\ \frac{\partial y}{\partial x_0} & \frac{\partial y}{\partial z_0} & N_y \\ \frac{\partial z}{\partial x_0} & \frac{\partial z}{\partial z_0} & N_z \end{pmatrix} \tag{20}$$

The variables of (20) can be obtained from (21):

$$\begin{cases} \frac{\partial x}{\partial x_0} = 1 - \sum\limits_{i=1}^{N} [I_{i1} A_i \cos(k_i(I_{i1}x_0 + I_{i2}z_0) - w_i t) \cdot k_i I_{i1}] \\ \frac{\partial y}{\partial x_0} = - \sum\limits_{i=1}^{N} [A_i \sin(k_i(I_{i1}x_0 + I_{i2}z_0) - w_i t) \cdot k_i I_{i1}] \\ \frac{\partial z}{\partial x_0} = - \sum\limits_{i=1}^{N} [I_{i2} A_i \cos(k_i(I_{i1}x_0 + I_{i2}z_0) - w_i t) \cdot k_i I_{i1}] \\ \frac{\partial x}{\partial z_0} = - \sum\limits_{i=1}^{N} [I_{i1} A_i \cos(k_i(I_{i1}x_0 + I_{i2}z_0) - w_i t) \cdot k_i I_{i2}] \\ \frac{\partial y}{\partial z_0} = - \sum\limits_{i=1}^{N} [A_i \sin(k_i(I_{i1}x_0 + I_{i2}z_0) - w_i t) \cdot k_i I_{i2}] \\ \frac{\partial z}{\partial z_0} = 1 - \sum\limits_{i=1}^{N} [I_{i2} A_i \cos(k_i(I_{i1}x_0 + I_{i2}z_0) - w_i t) \cdot k_i I_{i2}] \end{cases} \tag{21}$$

Fresnel Effect

In general, when light reaches the interface of two materials, some light reflects off the surface, and some light refracts through the surface. This phenomenon is called Fresnel effect which adds realism to the images.

The Fresnel equations, which quantify the Fresnel effect, are complicated [20]. As our aim is to create images that look plausible, not necessarily to describe accurately the intricacies of the underlying physics, thus, instead of using the equations themselves, we are going to use the empirical approximation in (22), which gives good results with significantly less complication:

$$fastFresnel = r + (1 - r) \times pow(1.0 - dot(\mathbf{I}, \mathbf{N}), 5.0) \tag{22}$$

r is the reflection coefficient from air to water, whose value usually be 0.2037. Hence, the final color $C_{finalcolor}$ of the ocean surface could be got by using the reflection coefficient to mix the reflected and refracted contributions according to (23).

$$C_{finalcolor} = fastFresnel \times C_{reflected} \\ + (1 - fastFresnel) \times C_{refracted} \tag{23}$$

Apparently, the range of the reflection coefficient is clamped to the range [0, 1]. Fig. 13 shows us the optics effects of deep ocean and shallow ocean respectively.

(a)   (b)

Fig. 13. Optics effects of deep ocean(a) and shallow ocean(b)

4.2 Foam Simulation

In a 32-bit graphics system, each pixel contains four channels: three 8-bit channels for red, green and blue(RGB) and one 8-bit alpha channel. The main function of alpha channel is to show translucency; it specifies how the pixel's colors should be merged with another pixel when the two are overlaid. White and black represent opaque and fully transparent, while various gray levels represent levels of translucency. In computer graphics, rendering overlapping objects that include an alpha value is called alpha-blending.

In our virtual seashore scene, foams randomly disperse on the shallow water surface, and also exist along the coastline. As the alpha-blending is the process of combining a translucent foreground color with a background color, here a foam texture is treated as the foreground color and the ocean surface color is the background color. The texture that we have chosen is a type of TGA, which has an additional alpha channel to store the alpha value of every pixel, see Fig. 14.

The pattern of the alpha channel is pre-designed optionally and in the simulation, its value changes according to the action of waves. When the waves rush to the beach, foams form, the white areas alter from black to white; when the waves recede, the foams disappear into the sand, the white areas back into the black from white.

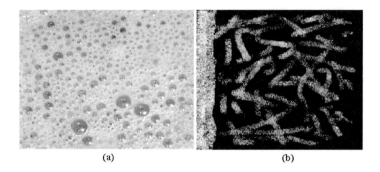

(a)   (b)

Fig. 14. Foam texture(a) and its alpha channel(b)

The foam effect is rendered in the fragment shader. While it should be noted that we render the foam after the optics effects simulation of ocean surface due to the influence of Z-buffer in 3D graphics card [21]. Fig. 15(a)(b) shows two layers of wave with their foams on the beach, and Fig. 15(d) shows the foams is fading with the receding of waves.

4.3   Spray Simulation

Generally, spray could be considered that it is composed of many water particles. Thus, a graphics technology called particle system is very appropriate for spray simulating.

A particle system is a collection of many particles that model certain fuzzy objects, such as fire, smoke, water, etc. An object represented by a particle system is not deterministic, its shape and form is not completely specified. For each frame, particles move, new particles are created and old particles are destroyed [22].

In our application, about 5000 particles are created to simulate the spray when a wave falls into the water or beach. The initial states (like the velocity, the acceleration, the position, etc.) of each particle have been designed as random quantities within a certain range. After being emitted, the motion of particles would be influenced by the gravity or wind force following the physical laws. The spray will finally disappear when scattering in the sea or sand.

Each particle is affixed a texture to make them less regular, the alpha-blending is reused here to mix the spray color with the background color. Fig. 15(e) shows the spray effect.

4.4   Simulation Results

The seashore scenes under different weathers (sunny, cloudy, stormy) and time conditions (early morning, morning, noon)have been rendered, see Fig. 15. The sound of the seashore also has been added into the simulation to increase the immersion.

Fig. 15. Different seashore scenes : early morning (a); cloudy noon (b); sunny (c); Frame C of breaking waves (d); Frame D of breaking waves (e); morning (f)

With 250×250 grid, and 1280×968 pixels, we got a steady frame rate: about 60 frames per second with the graphic card of Nvidia GeForce 9800 GTX/9800 GTX+ and 30 frames per second with the graphic card of ATI Radeon X600 Series. All the experiment results have achieved the real-time requirements.

5 Conclusion and Perspectives

Currently, several commercial computer-vision software and open source 3D engines are developed to simulate various natural scenes, which also include ocean simulations. However, the ocean wave model integrating both shallow and deep ocean as well as the real-time model of coastline and breaking waves are not taken into account in these software or engines. For resolving the problem mentioned above, in this chapter, we present our own methods for real-time modeling and simulation of seashore based on GPU. These methods could bring us rapid and realistic seashore scene.

As most of the calculation is implemented as shader programs in graphics hardware, our program permits to realize a real-time simulation of seashore scene with a sufficient frame rate. However it still has some limits and further investigations could be made to improve the visualization quality. For instance, more control points could be used to generate detailed Bézier surface and more stochastic quantities could be added to the Bézier surface in order to make the key-frames less regular and more realistic. Alternatively we can design key-frames with more than two Bézier patches. Nevertheless the above improvement would make the simulation frame rate lower as calculations become more complicated.

With regard to the influence of atmospheric conditions, environmental models could be constituted to obtain more reasonable effects rather than defining directly the environmental states. Certainly, these environmental models should be based on real environmental data.

Finally, moreover optical effects could also be strengthened by applying better algorithms or running simulations with more powerful graphics card.

References

1. MultiGen-Paradigm, Vega Prime Programmers Guide Version 2.2 (2007), http://www.presagis.com/
2. The OGRE Team, OGRE Manual Version 1.6. California, USA (2008), http://www.ogre3d.org/
3. Weihua, L.I.U.: Research on Key Technologies of Synthetic Natural Environment (SNE). PHD thesis. Beijing University of Aeronautics and Astronautics, China (2004)
4. Parenthoen, M.: Phenomenological Animation of the Sea. PHD thesis, Computer Science Laboratory for Complex Systems, France (2004)
5. Chellali, M.-E.-A.: A study on human-human interactions for common frame of reference development within Collaborative virtual environments. PHD thesis, Université de Nantes UFR Sciences et Techniques, France (2009)
6. Ellis, C.A., Gibbs, S.J., Rein, G.L.: Groupware: some issues and experiences. Communications of the ACM 34(1), 38–58 (1991)

7. Nvidia team, Nvidia CUDA Programming Guide, Version 2.3.1,
   `http://www.nvidia.com/object/cuda_home_new.html`
8. Fournier, A., Reeves, W.T.: A simple model of ocean waves. In: Computer Graphics (SIGGRAPH Proceedings), vol. 20, pp. 75–84 (1986)
9. Peachey, D.R.: Modeling waves and surface. In: Computer Graphics (SIGGRAPH Proceedings), vol. 20, pp. 65–74 (1986)
10. Li, Y., Jin, Y., Yin, Y., Shen, H.: Simulation of Shallow-Water Waves in Coastal Region for Marine Simulator. The International Journal of Virtual Reality 8(2), 65–70 (2009)
11. Ts'o, P.Y., Barsky, B.A.: Modeling and rendering waves: wave-tracing using beta-splines and reflective and refractive texture mapping. ACM Transactions on Graphics 6(3), 191–214 (1987)
12. Gonzato: A phenomenological model of coastal scenes based on physical considerations. In: 8th Eurographics Workshop on Computer Animation and Simulation, pp. 137–148 (1997)
13. Mastin, G.A., Watterberg, P.A., Mareda, J.F.: Fourier synthesis of ocean scenes. IEEE Computer Graphics and Applications (1987)
14. Tessendorf, J.: Simulating Ocean Water. In: SIGGRAPH 2001 Course Notes (2001)
15. Hong, K., Liu, S.: Nonlinear Analysis of Wave Energy Dissipation and Energy Transfer of Directional Breaking Waves in Deep Water. OCEANS - Asia Pacific (2006)
16. Mihalef, V., Metaxas, D., Sussman, M.: Animation and Control of Breaking Waves. In: Eurographics/ACM SIGGRAPH Symposium on Computer Animation, pp. 315–324 (2004)
17. Zhao, Q., Armfield, S., Tanimoto, K.: Numerical simulation of breaking waves by a multi-scale turbulence model. Coastal Engineering 51, 53–80 (2003)
18. Kryachko, Y.: Using vertex texture displacement for realistic water rendering. GPU Gems 2, ch. 18 (2005)
19. Object Management Group, Introduction to OMG's Unified Modeling Language (2011), `http://www.uml.org/`
20. Huygens, Young, Fresnel: The wave theory of light, pp. 79–145. American Book Company (1900)
21. Watt, A.: 3D Computer Graphics. Addison-Wesley Longman Publishing Co. Inc., Boston (1999)
22. Reeves, W.T.: Particle Systems - A Technique for Modeling a Class of Fuzzy Objects. In: ACM SIGGRAPH Proceedings (1983)
23. Mitchell, J.L.: Real-Time Synthesis and Rendering of Ocean Water. ATI ResearchTechnical Report:121-126, Paris (2005)
24. Johanson, C.: Real-time water rendering. MA thesis, Lund University, Sweden (2004)
25. Smith, B.W.: Realistic Simulation of Curling and Breaking WavesMA thesis, University of Maryland Baltimore County (2004)
26. Gonzato, J., Le Saec, B.: A phenomenological model of coastal scenesbased on physical considerations. Computer Animation and Simulation, 137–148 (1997)

27. Kass, M., Miller, G.: Rapid Stable Fluid Dynamic for Computer Graphics. Computer Graphics 24(4), 49–56 (1990)
28. Chiu, Y.-F., Chang, C.-F.: GPU-based Ocean Rendering. In: IEEE International Conference on Multimedia & Expo (ICME), Toronto, Canada (2006)
29. Jeschke, S., Birkholz, H., Schmann, H.: A procedural model for interactive animation of breaking ocean waves. In: International Conferences in Central Europe on Computer Graphics, Visualization and Computer Vision (WSCG), Plzen - Bory, Czech Republic (2003)
30. Yang, X., Pi, X., Zeng, L., Li, S.: GPU-Based Real-time Simulation and Rendering of Unbounded Ocean Surface. In: Ninth International Conference on Computer Aided Design and Computer Graphics, Hong Kong (2005)

Data Processing for Virtual Reality

Csaba Szabó, Štefan Korečko, and Branislav Sobota

Department of Computers and Informatics,
Faculty of Electrical Engineering and Informatics,
Technical University of Koisice, Slovakia
{csaba.szabo,stefan.korecko,branislav.sobota}@tuke.sk

Abstract. Digitization of real objects into 3D models is a rapidly expanding field, with ever increasing range of applications. The most interesting area of its application is in the creation of realistic 3D scenes, i.e. virtual reality. On the other hand, virtual reality systems are typical examples of highly modular systems of data processing. These modules are considered as subsystems dealing with selected types of inputs from the modeled world, or producing this world including all providing effects respectively. This chapter focuses on data processing tasks within a virtual reality system. Processes considered are as follows: 3D modeling and 3D model acquisition; spoken term recognition as part of the acoustic subsystem; and visualization, rendering and stereoscopy as processes of output generation from the visual subsystem.

1 Introduction

Modeling and 3D visualization is indispensable for the design, engineering and development of production systems, the implementation of changes to production processes and introduction of new technologies and products. With recourse to virtual reality (VR) one can create realistic 3D environment available for the practical purpose of further development and testing of the product or even training of the product usage and/or production. The aim is to include real 3D objects rather scanned than modeled into the VR scene.

A VR system represents an interactive computer system, creating the illusion of non-existing space, more precisely one can speak about perfect simulation within the environment of close human-computer relation.

By using a term of a 3D model, we refer to a numerical description of an object that can be used to render images of the object from arbitrary viewpoints and under arbitrary lighting conditions [1] e.g. for 3D user interfaces of information systems [21, 22].

In the last few years 3D modeling became more and more popular. Almost every big city in Europe has started to work on 3D virtual map projects of their cities. Its purpose is to bring closer real imagination of the city to common people and it should improve orientation of visitors in specific areas and also offer extension of basic map collections. There is a plenty of specialized 3D software and number

T. Gulrez, A.E. Hassanien (Eds.): Advances in Robotics & Virtual Reality, ISRL 26, pp. 333–361.
springerlink.com

of modeling techniques, which a 3D model can be created with. It is necessary to choose the right modeling technique and specialized software to get proper model.

Nowadays, 3D (laser) scanning technologies are more popular because scanning devices have become more reliable, cheaper, faster and smaller. One of the most common applications of this technology is the acquisition of 3D models (such as a cloud of polygons [7]) of real 3D objects. One important example is the Digital Michelangelo project, which includes full scans of all Michelangelo's sculpture. Interesting application of 3D laser scanning is a digital library or virtual museum in which the value of 3D artifacts can be digitized and displayed in a 3D web environment. Other applications of these technologies involves the reverse engineering, rapid prototyping, archeology, forensic imaging and so on.

The raw output from the 3D scanner is usually a point-cloud. Next, methods for transformation of these points to usable model (usually, the target model is a surface model based on triangles) are needed. It is possible to execute this process in parallel environment too – either for triangulation or for visualization or both [18, 19]. These trends, coupled with increasing Internet bandwidth, make the use of complex 3D models accessible to a much larger audience.

Main goal of this chapter is to show how important data processing is for virtual reality systems. VR systems consist of several subsystems that could be considered as separate agents working in parallel with a huge amount of data. These data are gained from preparation phases or real-time. Both types of data need further processing to produce the required effect of virtual world of the VR system.

The structure of the chapter is as follows. First, the model for 3D scene creation for VR systems is presented. The section includes description of VR subsystems, roles, properties and processes with emphasis on data processing. The next two sections (Sec. 3 and 4) deal with VR object creation and processing data gained as VR system inputs. Section 5 discusses data processing for VR object visualization. Both input and output data processing descriptions are illustrated by examples. The last section of the chapter concludes current work and shows future directions.

2   3D Scene Creation Model for Virtual Reality Systems

2.1   VR Scene Elements

Virtual reality applications create a projection of a generated environment around the actors of the scene. This VR scene consists of the following base elements:

- static real objects, e.g. a table
- movable real objects, e.g. a pen, a paper sheet
- static virtual objects, e.g. walls
- move-able virtual objects, e.g. a toy dog
- audio effects
- interaction tools

Example scene objects are shown on Fig. 1.

(a) Scanned toy dog   (b) Real object – a model printed in 3D

(c) Scene model animation

Fig. 1. Example scene objects

2.2 Levels of Virtual Reality

Before virtual reality systems implementation specification, it is important to present everything falling into the area of VR. Except the listed levels of characteristics of classification, it is also possible to consider telerobotics respectively other types of telepresentations (i.e. a remote action sharing, virtual education or virtual constructor-office).

From a technical point of view, partially also by programming resources needed for accomplishing appropriate VR results, we can classify VR systems into:

- Entry VR,
- Basic VR,
- Advanced (medium) VR, and
- Immerse VR.

To create a perfect illusion in virtual reality (VR) it is needed to integrate numerous combinations of special technologies, which can be divided into the following categories: visual displays, tracking systems, input devices, sound systems, haptic devices, graphic and computing hardware and software tools.

The role of the display device can be given to head-mounted and/or surround-screen projection. Surround-screen projection was used for first time already in the 70-es of previous century as a part of a flying simulator for pilots' training. Application of manifold projection in VR was for the first time used in 1995, in a project named CAVE (Cave Automatic Virtual Environment). It gives the illusion of virtual world projected onto all four walls and a ceiling of a cube shaped room. This system is based on projection devices, separately for each wall. Coordination of all displayed figures is regulated by high- performance computer system. One actor is directly connected to the virtual world (e.g. by using data gloves or cyberpuck) and he can inference it.

2.3   Subsystems of VR Systems

Subsystems of VR-systems categorization is given especially according to senses, which are effected by individual VR system parts:

- Visual subsystem,
- Acoustic subsystem,
- Kinematics and statokinetic subsystems,
- Subsystems of touch and contact, and
- Other senses (e.g. sense of smell, taste, sensibility to pheromones, sensibility when being ill, pain, sleep or thoughts).

From the point of view of VR-systems implementation, it is necessary to consider the above mentioned subsystems.

2.4   Specification of VR Roles

Specification of VR roles from the point of view of their implementation [17] consists of following actions:

- input or processed information defining
- VR objects defining and placement of individual input information into these VR objects
- creating or completing these VR objects by relational equation and database (relational equations and database describe features of individual VR objects, and the objects in the interrelation to other objects)
- earmarking the way of work with VR objects, i.e. earmarking the operations for individual VR objects - encapsulation (i.e. object-oriented programming)
- earmarking the possibility of parallel branches of computation of individual VR objects

The issue of interaction level between the observer and the environment is of great importance when specifying this problem.

2.5   The Object-Oriented Property of VR Systems

The whole VR system can be defined as strictly object system. This means that everything that exists in the VR system is an object. Even the VR system itself can be defined as an object that contains other objects, i.e. there is a notable object hierarchy.

In terms of such a definition, it is possible to exploit the features of heredity, encapsulation and polymorphism.

As seen on Fig. 2, VR objects contain frames, which are divided according to basic subsystems of the VR system. Each of these frames contains internal information invisible for other frames. For frames of a specific VR object, global information exist as well, which are necessary for each of these frames and thus, they are mutually shared. However, the problem concerning creating data structures is very extensive. Its optimization is of great importance. It is obvious, when one considers the following:

- the whole system consists of large number of objects
- each of these objects must be defined in terms of its shape and visual features
- each of these objects must be defined in terms of its mechanical features
- each of these objects must be defined in terms of its sound features
- each of these objects must be defined in terms of its time characteristics

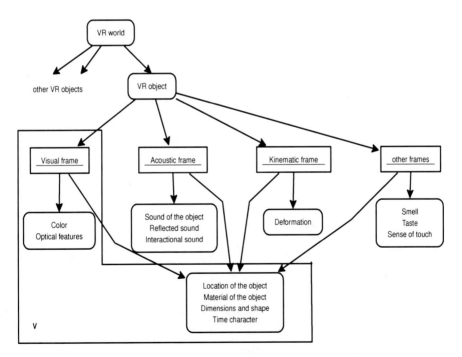

Fig. 2. VR object frames from the point of view of definition

Considering only the elements for 3D visualization (i.e. excluding other VR subsystems), virtual objects need to have an internal representation. This internal representation could be gained by modeling or scanning process. On Fig. 3, the process of 3D scene creation is presented with focus on the role of 3D scanning. Besides the main processes, model stages are shown too.

3 Processing Input Data

In this section, mainly the visual subsystem is discussed being the most representative part of the VR system. The other reason for the selection is the area of research interest of the authors. Partially, the acoustic subsystem is also mentioned.

3.1 Scanning

The 3D model acquisition process [1, 2, 5, 6, 14, 31] shown on Fig. 3 consists of two main stages:

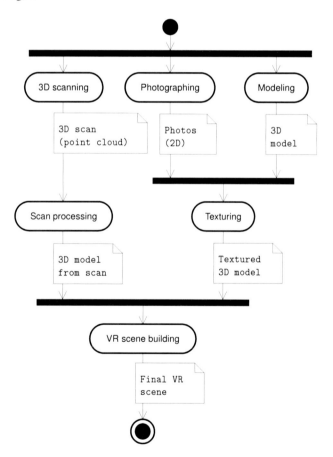

Fig. 3. The role of 3D scanning in 3D scene creation

- 3D scanning and
- data processing.

Digitizing objects is possible thanks to scanning devices, and devices that allow us to transfer the real three-dimensional objects into digital form are 3D scanners. The main operating principle of these devices is based on sensing surface of the object in discrete points. Such a process results into a large number of points in space, i.e. point cloud, that could be processed on a digital computer. Individual scanners differ in the technology of the acquisition of the points of the object surface.

The category of most frequently used scanners are optical, laser, mechanical and magnetic-resonance (scanning also internal shapes) scanners. The right choice of the scanner type depends on the requirements that are imposed on the accuracy of the correlation between the real and the digitized model. The main criterion for choosing the correct type of scanner accuracy. Another important criterion for choosing the correct type of scanner is scanning time (time at which the scanner is able to capture the object). The fastest scanners group (scanners to capture short-term) include laser scanners. One of the factors to choose the right scanning equipment, is the size of scanning area, as well as the accuracy of scanning equipment.

According to the scanner points recognition, scanners can be divided into the following groups:

1. contact
2. optic
3. laser
4. destructive
5. x-ray
6. ultrasound

Another classification puts the scanners into two categories – whether touching the surface of the object being scanned(1, 4) or not(2, 3, 5, 6).

Contact scanners require a physical contact with the object being scanned. Although they are usually very precise, they are also much slower (order of 103 Hz) than non-contact scanners (order of 105 Hz). A typical example of a 3D contact scanner is a coordinate measuring machine (CMM).

Non-contact scanners use radiation to acquire required information about objects. They are of two basic types: passive and active. The main advantage of passive scanners is that they are cheap as they do not require so specialized hardware to operate. To scan objects, they only use existing radiation in its surroundings (usually visible light). In contrast, active scanners are equipped with their own radiation emitter (usually emitting laser light). While the latter are considerably more expensive, they are also much more accurate and able to scan over much bigger distances (up to few km).

Besides these, there are scanners and imaging systems for the internal geometry. By using 3D destructive scanners we make total destruction of the scanned object, which on the other hand, allows us to have accurate scanning of the complex internal geometry.

The next stage of the model acquisition process is data processing. It consists of several parts. First, the point cloud has to be meshed (Fig. 4), i.e. the points have to be connected into a collection of triangles (called faces) [20].

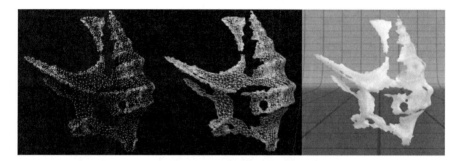

Fig. 4. The meshing process

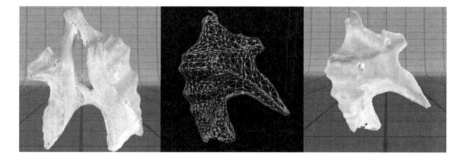

Fig. 5. The simplifying process

The next step is to align the scans from various angles to create the whole object surface. The aligned scans then have to be merged into one continuous mesh, so that no overlapping parts occur. The merging process also involves filling the eventual "holes" (unscanned parts) in the model. Additionally, there is an optional step to simplify the mesh (Fig. 5, 478 000 triangles, scan-time 60 min.), which consists of reducing the number of triangles in order to save memory needed to visualize the final 3D model.

3.2   Grid-Based 3D Scanning Example

3.2.1   Problem Simplification

The main goal of this 3D scanning application is scanning objects in a three-dimensional coordinate system. This application requires a PC running on windows operating system with framework, web camera and a data projector. Scanning area should be in a dark place with a white background, see Fig. 6.

3.2.2   The Scanning Process

The scanning process is as follows:

1. After starting the application user should set up the web camera and set it view point to the scanning area. Then setup the data projector and set it projection also to the scanning area. Then user can set up the output grid lines (their density, color, position). Output Grid window should be shown at data projector screen. Output grid setup window is used for setting up the out grid lines. User can set their position, density, colors and sensitivity of scanning algorithm on colors.

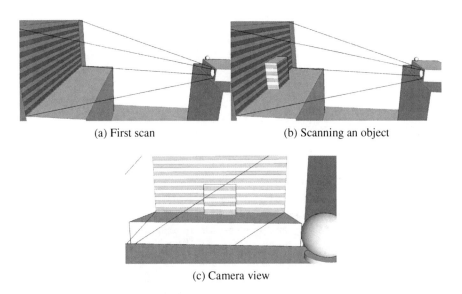

(a) First scan   (b) Scanning an object

(c) Camera view

Fig. 6. Grid-based 3D scanning

2. After all setup user can start the scanning. For the first scan the scanning area should be empty, and the first scan can be made by pressing the "Catch first" button. The first scan is very important because the scanning algorithm will compare those scanned lines with scanned lines made on object. At first scan the scanning area is without scanning object. Only data projector is projecting the output lines and camera at top of the data projector is taking pictures from the scanning area.
3. After first scan is finished we can put the scanning object in the scanning area and start the scan. As one can see from the view of the camera, lines in the scanning area projected on the object are little bit below. That is how scanning algorithm can tell that there is an object and which shape it is.

4. The calculation of 3D points is made by comparing changes between lines made by first scan and lines made by scan with the scanning object. When the comparing algorithm is finished a point cloud is created. That is a set of vertices in a three-dimensional coordinate system. These vertices are defined by X, Y and Z coordinates. Then the Delaunay triangulation algorithm is used to compute the triangles of the scanned object. This is very important step for better view of scanned points.

Definition 1. *Delaunay triangulation for a set P of points in the plane is a triangulation DT(P) such that no point in P is inside the circumcircle of any triangle in DT(P). Delaunay triangulations maximize the minimum angle of all the angles of the triangles in the triangulation, they tend to avoid skinny triangles.*

5. When the triangulation is finished a VRML file is created.

Limitations of the above described process are in its basic idea: each additional light source adds unwanted shadows onto the background that introduces fault points and shapes into the scan.

3.3 Acoustic Data Processing

Acoustic data processing is in fact identification and analysis of voice input data, e.g. audio commands.

To react to the audio commands of the user there are several choices. Speech recognition [3, 9, 13] could be performed, in which process we transcribe all uttered sentences to their more-or-less accurate transcription. Then this textual representation can be parsed quite easily, and it gives a way to react to the user's commands in a context-dependent way by using some kind of dialog model.

For this approach, some freely available speech recognition systems like HTK [32] or Sphinx [30] could be incorporated. The drawback of this approach is that even current speech recognition techniques cannot provide close to 100% accurate transcriptions, especially for large-vocabulary tasks. Moreover, their language modeling is somewhat limited: it is based on a statistic of large amount of texts, so it is usually domain-independent, which might differ from our actual requirements. To construct a language model on some specific domain, one have to collect several hundred sentences from this topic, which could be quite expensive. And finally, this approach is very vulnerable to out-of-vocabulary (OOV) words: words not occurring in the training set of texts cannot be detected, which more or less affects the whole sentence they are uttered in.

Another option is to use only a number of keywords and concentrate only on their occurrences. This approach has the advantage that – similarly to the early days of speech recognition – this can be done in a computationally cheap way.

It is also closely related to the recently popular task of Spoken Term Detection (STD) [4, 10]. Moreover, in contrast to traditional speech recognition, the list of command words used can be updated easily even later. But there are a number of drawbacks as well, which all originate from the fact that in speech recognition, language model is not just a limitation, but – similarly to the case of humans – it also helps providing a correct transcription. As in this second approach there is not present any application of any kind of language modeling (except the list of possible command words), we have to rely only on the acoustic information, which leads to a number of false keyword detections due to the presence of similarly-sounding words or one word containing another one. On the other hand, due to pronouncement variations and not 100% accurate acoustic modeling, we are likely to miss a number of uttered command words. Of course there is a trade-off between the two types of errors, so by adjusting the acceptance level one can reduce the number of one while increasing the frequency of the other, to suit our expectations.

A kind of primitive type of dialog modeling like menus could be introduced as well to aid spoken term recognition [8]. On other hand, using type-2 fuzzy sets and stochastic additive analog to digital conversion [12, 26] could also improve elimination of problems related to pronouncement variations.

4 Modeling of 3D Objects

4.1 3D Polygonal Modeling

There is a number of modeling techniques in 3D computer graphics such as constructive solid geometry, implicit surfaces, NURBS modeling, subdivision surfaces and polygonal modeling[29]. Polygonal modeling which uses polygons for representing surfaces of object is faster than other modeling techniques. While a modern graphics card can show a highly detailed scene at a frame rate of 60 frames per second or higher, raytracers, the main way of displaying non-polygonal models, are incapable of achieving an interactive frame rate (10 fps or higher) with a similar amount of details. That is why this modeling technique is proper for modeling parts of cities and for common usage.

3D computer graphics are different from 2D computer graphics in that a three-dimensional representation of geometric data is stored in the computer for the purposes of performing calculations and rendering 2D images. Such images may be for later display or for real-time viewing. Despite these differences, 3D computer graphics rely on many of the same algorithms as 2D computer vector graphics in the wire frame model and 2D computer raster graphics in the final rendered display. In computer graphics software, the distinction between 2D and 3D is occasionally blurred; 2D applications may use 3D techniques to achieve effects such as lighting, and primarily 3D may use 2D rendering techniques. The process of creating 3D computer graphics can be sequentially divided into three basic phases: 3D modeling which describes the shape of an object, Layout and Animation which describes the motion and placement of objects within a scene, and 3D rendering which produces an image of an object.

3D modeling is the creation of 3D computer graphics based on wire frame modeling via specialized software. The modeling process of preparing geometric data for 3D computer graphics is similar to plastic arts such as sculpting, whereas the art of 2D graphics is analogous to 2D visual arts such as illustration.

In 3D computer graphics, polygonal modeling is an approach for modeling objects by representing or approximating their surfaces using polygons. Polygonal modeling is well suited to scan-line rendering and is therefore the method of choice for real-time computer graphics. Alternate methods of representing 3D objects include NURBS surfaces, subdivision surfaces, and equation-based representations used in ray tracers.

4.1.1 Polygon, Mesh, Texture and Texture Mapping

The basic object used in mesh modeling is a vertex, a point in three dimensional space. Two vertices connected by a straight line become an edge. Three vertices, connected to the each other by three edges, define a triangle, which is the simplest polygon in Euclidean space. More complex polygons can be created out of multiple triangles, or as a single object with more than 3 vertices. Four sided polygons (generally referred to as quads) and triangles are the most common shapes used in polygonal modeling. A group of polygons, connected to each other by shared vertices, is generally referred to as an element. Each of the polygons making up an element is called a face.

In Euclidean geometry, any three points determine a plane. For this reason, triangles always inhabit a single plane. This is not necessarily true of more complex polygons, however. The flat nature of triangles makes it simple to determine their surface normal, a three-dimensional vector perpendicular to the triangle's edges. Every triangle has two surface normals, which face away from each other. In many systems only one of these normals is considered valid the other side of the polygon is referred to as a back-face, and can be made visible or invisible depending on the programmers desires.

It is possible to construct a mesh by manually specifying vertices and faces; it is much more common to build meshes using a variety of tools. A wide variety of 3D graphics software packages are available for use in constructing polygon meshes. One of the more popular methods of constructing meshes is box modeling, which uses two simple tools:

- The subdivide tool splits faces and edges into smaller pieces by adding new vertices. For example, a square would be subdivided by adding one vertex in the center and one on each edge, creating four smaller squares, component.
- The extrude tool is applied to a face or a group of faces. It creates a new face of the same size and shape which is connected to each of the existing edges by a face. Thus, performing the extrude operation on a square face would create a cube connected to the surface at the location of the face, component.

Once a polygonal mesh has been constructed, the model must be texture mapped to add colors and texture to the surface and it must be given an inverse kinematics skeleton for animation. Meshes can also be assigned weights and centers of gravity for use in physical simulation.

In geometry a polygon is a plane figure that is bounded by a closed path or circuit, composed of a finite number of sequential line segments. The straight line segments that make up the boundary of the polygon are called its edges or sides and the points where the edges meet are the polygon's vertices or corners. The interior of the polygon is its body.

Texture refers to the properties held and sensations caused by the external surface of objects received through the sense of touch. Texture mapping is a method of adding detail, surface texture, or color to a computer-generated graphic or 3D model. A texture map is applied (mapped) to the surface of a shape. This process is akin to applying gift wrapping paper to a plain white box.

4.2   Large Graphical Data Set Example - City Model

A city modeling process is based on 4 steps:

- Model preparation
- Polygonal modeling
- Texture preparation
- Texture composition and filtering

4.2.1   Cadaster Map Conversion to 3D Model

Modeling of streets is highly dependent on cadastral map of the city, because of approaching the real city. It includes information about ground plan and right orientation of the buildings. It brings problems, because model should be as simple as it is possible because of low requirements of hardware performance. Model is designed in 3D specialized software which enables to build and modify 3D models quickly and easily. Ground plan of the building is drawn in modeling software and the whole building is modeled using extrude tool. (see Fig. 7)

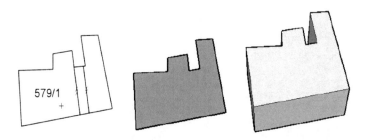

Fig. 7. Cadaster map, ground plan and 3D model of a building

The simplest polygon in Euclidean space is triangle which means that polygon is defined by three edges. The more edges polygon contains the more difficult polygon is which means more requirement for hardware performance. The most useful polygon for buildings is quadrilaterals polygon denned by four edges. Model (Fig. 7) is mesh - group of 12 polygons connected together by shared vertices. 9 polygons are quadrilaterals, and 2 polygons are heptagons polygon defined by seven edges. Adding a roof to this model would be not very easy and might add other difficult polygons to the model which is ineligible. This model could be modeled simpler as a cube with parallel edges using the front edge and one of the side edge from ground plan (Fig. 8).

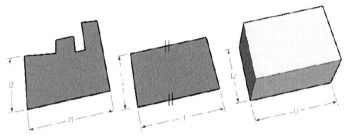

Fig. 8. Ground plan simplification

4.2.2   Building Modeling

Invisible parts of the building from the street are modified and result is a cube with realistic ground plan and just 6 quadrilaterals, which is simpler than previous model. Adding roof is very simple by two triangles and two quadrilaterals. Final model with roof is modeled from 9 polygons which is simpler than previous model with 12 polygons. (see Fig. 9)

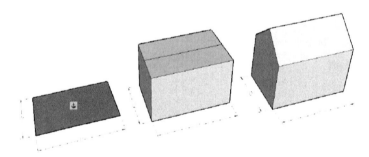

Fig. 9. The simplified 3D model of the building

4.2.3   Texture Preparation, Composition and Mapping

Detail modeling of every part of the building would lead to big amount of polygons and to performance demands which is ineligible. On the visible face of the mode is mapped a texture which is made from real picture taken on the street.

Texture is in .jpg format which offers good quality of the picture and small size of the file. After mapping a texture the whole complexity of the model is unchanged. The process of texture preparation, composition and final mapping give a result is realistic looking model with very low requirements for hardware performance. (see Fig. 10, Fig. 11, Fig. 12).

Fig. 10. Process of texture preparation

Fig. 11. A texture composition process example

Fig. 12. Texture-to-building mapping example

5 Processing Data for Scene Visualization

For visualizing in visual frame of our VR-system is used graphical visualizing engine. Graphical engines are often based on serial using of information. On the input side is world model and on output side are correct colored pixels on the screen.

Between these two stages there are some other stages, which partially change the input information. The basic idea of implementation of graphical system is to minimize the time used by each stage for real time system. In multiprocessor systems it is able to apply some stages in parallel, which is real speedup. On the following diagram (see Fig. 13) there are the basic stages of our visualizing kernel.

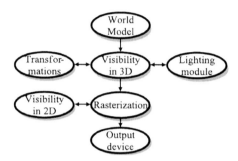

Fig. 13. Visualizing kernel structure

The main information part is world model [17]. It is now based on convex polygons (in the future there will be a parallel implementation of NURBS computing as a base unit for this visualizing kernel), which can lie arbitrary in the space. The world model internally is represented by 3D BS/BSP tree. There are included 3D dynamically animated objects, which are stored separately of the static world model. There are some special parts of the world model: texture manager (which manages used textures in memory) and light manager (which manages lights).

The next part of the pipeline is computing visibility in 3D. Visibility of static world is done by BS/BSP tree traversal, which computes the polygon list on which polygons are sorted from front to back and the vertices are two dimensional (screen coordinates). This module uses the transformation module, which computes all the necessary transformation (including transforming to the screen). And lighting module is used by this 3D visibility module, too. There the correct lighting for all the visible polygons is computed using three types of light: directional, ambient and point light. All the lights can change every frame, it means they can change the position, orientation and light intensity. Visibility for dynamic objects is computed too, but the polygon list is stored in different set.

Computed data from 3D visibility module are used in rasterization module. Here every polygon is divided into horizontal spans, which are displayed on the screen and the z-buffer is set accordingly. After displaying the static world, it becomes displaying of dynamic objects. They are divided into horizontal spans, but every pixel in the span is tested with z-buffer. If it is visible, the screen pixel color and z- value are set accordingly. Rasterization module uses 2D visibility module, which manages the z-buffer and span buffer.

5.1   Visibility Problem Solving

The significant part of visualizing calculation is the invisible part removing. This is the argument for search optimal algorithm for visibility problem solving. There are two ways how to solve visibility problem in real time:

- using fast parallel computers or special hardware (this is extensive way).
- using fast and optimal algorithms.

If one can implement the VR-system on a low-end workstations or on a PC, then the second way is very interesting. The new BS algorithm [17], which is used in our visualizing kernel, is suitable for these systems. This algorithm has some constraints. These constraints are as follow: the algorithm is applicable only on polygonal surfaces and observer's view is limited to side and up or down directions.

5.1.1   The BS Algorithm

This algorithm solves visibility problem behalf object consisting of polygonal surfaces. The principle consists into appropriate vertex encoding for every polygon. The encoding principle is displayed on Fig. 14.

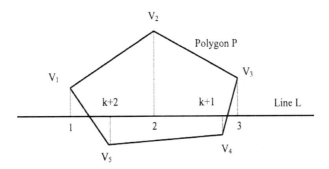

Fig. 14. Vertices encoding

Through polygon P is routed line L. This is the projection line. At least one of polygon vertices must be on other side of projection line than others. Next, codes to polygon vertices are assigned. The vertices above the projection line will have codes form 1 to k. And vertices below the projection line will have codes form k+1 to 2k. The order of vertices encoding must be in the same direction for every polygon in scene. After this, projections of vertices into projection line are created. Then, a list of codes of polygon vertices is made. This operation is repeated for every polygon in 3D scene. At the end, a list of all lists of codes is created. This is pre-processing phase.

The solving of visibility problem by this algorithm is the following. There is list of lists of codes. Observer's direction of view is determined, then the projection of polygons into observer's projection plane is made. A new list of potentially visible polygons with the following method is created. The first element (first code) of list of polygon codes created for polygon in projection plane with the first element of list of corresponding polygon codes created in pre-processing phase is compared. If the elements are the same (equal), then the polygon will be added into the list of potentially visible polygons. In opposite case, the polygon will be not added into this list. After that, the visibility of polygons in this list is solved using conventional algorithm (for example with Painter's algorithm or Z-buffer algorithm). Example of this method is on Fig. 15.

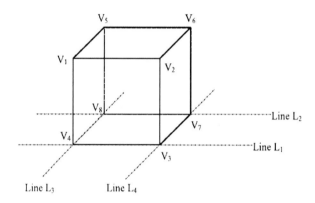

Fig. 15. BS algorithm usage example

There is an object consisting of four polygons on Fig. 15. The front polygon consists of these vertices: V_1, V_2, V_3, V_4. The next one is left side polygon consisting of following vertices: V_5, V_1, V_4, V_8. Another one is the back polygon (vertices V_6, V_5, V_8, V_7). And the last polygon is the right side polygon with vertices V_2, V_6, V_7, V_3. There are four projection lines L_1–L_4. For every polygon there is one projection line. The list of lists of vertices codes created in pre-processing phase is the following:

- front polygon – $[1, k+2, 2, k+1]$;(vertex V_1 has code 1, V_2 has code $k+2$, etc.)
- left polygon – $[1, k+2, 2, k+1]$
- back polygon – $[1, k+2, 2, k+1]$
- right polygon – $[1, k+2, 2, k+1]$

Resultant list of lists: $[1, k+2, 2, k+1], [1, k+2, 2, k+1], [1, k+2, 2, k+1], [1, k+2, 2, k+1]$.

The following new codes for the projection polygon are gained executing the next phase of the algorithm using the results of the pre-processing phase:

- front polygon – $[1, k+2, 2, k+1]$
- left polygon – $[2, k+1, 1, k+2]$

- back polygon – $[2, k+1, 1, k+2]$
- right polygon – $[1, k+2, 2, k+1]$

Resultant list of lists – $[1, k+2, 2, k+1], [2, k+1, 1, k+2], [2, k+1, 1, k+2], [1, k+2, 2, k+1]$.

The left and back polygons have different first elements. It means that these polygons are not visible. And the front and right polygons are potentially visible. Only these polygons can be drawn by conventional visibility algorithm (Painter's algorithm or Z-buffer, as is listed in previous chapter).

5.2 Texture Mapping Computation

The basic idea is to recall that a picture is being drawn (better known as 2D bitmap or texture) onto some predefined 3D convex polygon. Now, the process involved in drawing a texture onto some polygon is simply known as texture mapping. That means, texels are retrieved from the texture to polygon onscreen pixels. The algorithm behind it iterates over each destination pixel by computing the appropriate texel, see the following listing.

Listing 1. Texel computation iteration

```
byte *TextureBitmap, *ScreenPointer;
for (int j = PolygonYMinScreenValue; j <
    PolygonYMaxScreenValue; j++)
{ for (int i = ScanlineXStart; i < ScanlineXEnd; i
    ++)
    { int u = TextureMapUValue(i, j);
      int v = TextureMapVValue(i, j);
      ScreenPointer[j][i] = TextureBitmap[v][u];
    }
}
```

As far as mapping is concerned, definitions of functions `TextureMapUValue (i,j)` and `TextureMapVValue(i,j)` enhance the type of texture mapping currently performed.

5.2.1 Texture Mapping Types

Texture mapping involves a whole lot of math computation, thus scarcely allowing fast real-time rendering. But as for beginners, concern first raises for perfect or so called perspective correct texture mapping and of course approximations thereof.

There is a 3D to 2D projection of texture source to the frame to be displayed. One of the means to result to this is to define the texture in 3D space coordinates defining the polygon's environment.

From Fig. 16, let O, U, and V be vectors defining texture space whereby O is the origin, and U and V are vectors for the *uv* space. As you can see:

$$O = vertex3$$
$$U = vertex0 - vertex3 \qquad (1)$$
$$V = vertex2 - vertex3$$

Notice that these vectors should be in eye view space, which is defined as being the 3D space from where you would achieve 3D to 2D transformation as:

$$i = focus * x/z + HorCenter$$
$$j = -focus * y/z + VerCenter \qquad (2)$$

This means that even the given *uv* space should have primarily been transformed to eye coordinate system, hence O, U, and V must be scaled accordingly by multiplying their x components by focus and their y components by $-focus$.

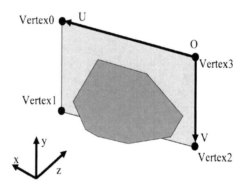

Fig. 16. Polygon and texture in 3D

Then for each polygon we compute three following vectors, often referred as "magic"

$$VA = CrossProduct(O, V);$$
$$VB = CrossProduct(U, O); \qquad (3)$$
$$VC = CrossProduct(V, U);$$

We are attempting to find a texel `TextureBitmap(v, u)` for a given polygon screen location `(i, j)`, we then complete the mapping by computing the following:

$$S = SetVector(i, j, 1);$$
$$va = DotProduct(S, VA);$$
$$vb = DotProduct(S, VB); \qquad (4)$$
$$vc = DotProduct(S, VC);$$

From which:
$$u = TextureWidth * va/vc;$$
$$v = TextureHeight * vb/vc;$$
(5)

Then:
$$ScreenPointer[j][i] = TextureBitmap[v][u];$$
(6)

From hereon, only hints, tips and trickery are left to the programmer to tune this code faster. But we are afraid, as we above mentioned that space restricts our moves although we will recall one of them without code.

Second used texture mapping type is bilinear interpolation. This approach of texture mapping mainly emphasizes its capabilities by reducing the divides and multiplications of the inner loop by breaking polygonal horizontal scan lines into spans of 8 or 16 pixels renderable at higher speed than normal, but at the cost of perfection. Accuracy is only computed at the beginning and end of these sub spans that is why for most part, this technique can produce results indistinguishable from perfect mapping

The only disadvantage is the visible distortion at regular intervals as the polygon is being viewed at sharped and narrowed angles. There is a way out of it, though, but this notwithstanding we will not mention it over here.

5.3 Rendering in Cluster Environment

Cluster-based rendering in general can be described as the use of a set of computers connected via a network for rendering purposes, ranging from distributed non-photo-realistic volume rendering over ray tracing and radiosity-based rendering to interactive rendering using application programming interfaces (APIs) like OpenGL or DirectX. We use the same terminology as is used by X-Windows. A client runs the application while the server renders on the local display.

Most of the recent research on cluster-based rendering focuses on different algorithms to distribute the rendering of polygonal geometry across the cluster. Molnar et al. [11] classified these algorithms into three general classes based on where the sorting of the primitives occurs in the transition from object to screen space. The three classes are sort-first, sort-middle and sort-last.

In sort-first algorithms, the display is partitioned into discrete, disjoint tiles. Each rendering node of the cluster is then assigned one or more of these tiles and is responsible for the complete rendering of only those primitives that lie within one of its tiles. To accomplish this, primitives are usually pre-transformed to determine their screen space extents and then sent only to those tiles they overlap with. The required network bandwidth can be high when sending primitives to the appropriate render server, but utilizing knowledge of the frame-to- frame coherency of the primitives can reduce the amount of network traffic significantly. Sort-first algorithms suffer from load balancing due to primitive clustering. Samanta et al. [16] investigated methods to improve load balancing by dynamically changing the tiling. Sort-first algorithms also do not scale well when the number of nodes in the cluster increases. Every primitive that lies on the border of two tiles must be rendered by both tiles. As the number of tiles increases, the number of these primitives increases. Samanta et al. [15] solved this problem by using a hybrid sort-first, sort-last approach.

Sort-middle algorithms begin by distributing each graphics primitive to exactly one processor for geometry processing. After the primitive has been transformed into screen space, it is forwarded to another processor for rendering. Similar to the sort-first approach the screen space is divided into tiles, but each processor is only responsible for rasterization of primitives within that tile. This approach requires a separation of rasterization engine and rendering pipeline, so that primitives can be redistributed. Currently this approach can only be implemented using specialized hardware, such as SGI's InfiniteReality engine. In sort-last approaches, each primitive is sent to exactly one node for rendering. After all primitives have been rendered, the nodes must composite the images to form the final image. This usually requires a large amount of bandwidth because each node must send the entire image to a compositor.

Tiled displays lead naturally toward a sort-first approach. The screen is already partitioned into tiles, with each tile being driven by a single cluster node. Other approaches require distributing the primitives to the rendering nodes and distributing the final image to the nodes responsible for tile-rendering. For these reasons the majority of software systems designed for rendering on a tiled display implements a sort-first algorithm.

5.4   Stereoscopy Implementations

There are several methods of 3D visualization [21]. The most common ones are stereoscopy, autostereoscopy, volumetric visualization and holography. This chapter only deals with stereoscopy, since it offers a fairly impressive 3D effect at low cost and relatively easy implementation.

Stereoscopy is based on the way human brain perceives the surrounding objects. When a person looks at an object, it is seen by each eye from a slightly different angle. Human brain processes the information and enables stereoscopic vision. In the stereoscopic visualization, human eyes are replaced by two cameras "looking" at an object from different angles. A stereoscopic image (or a video) is a composition of left and right images captured by the two cameras. The image is then presented to the viewer in such manner that each eye can see the image captured only by one of the cameras. In this way human brain interprets a 2D image as a 3D scene. There are two common ways of implementing this method.

The first one is known as anaglyph. It has become quite a popular stereoscopic method, since it only requires usage of very cheap red-cyan glasses. The anaglyph is a composition of a red channel from the left image and green and blue channels (together creating a "cyan channel") from the right image. When looking at the anaglyph through red-cyan glasses, the left eye can only see the red part of the image and the right eye the rest.

The second way of the implementation of the stereoscopy is based on light polarization. It requires two projectors with the ability to project polarized light and special polarized glasses. Through the glasses, the left eye can see only the scene from one projector and the right eye from the other one.

5.5   Stereoscopic Image Quality

There are several factors affecting image quality, which ones have to be considered when implementing stereoscopic visualization in order to achieve high quality images.

The most important one is stereo base (separation), i.e. the distance between the two cameras capturing images. The wider the stereo base, the farther the visualized object appears from the projection plane (Fig. 17). Therefore, the stereo base has to be adjusted to the viewing distance. Too large value could result in uncomfortable viewing, while too small could lead to an almost unnoticeable 3D effect. For example, when an image is to be viewed on a regular PC monitor (viewing distance around 50 cm), the stereo base should be narrower than when an image is to be projected on a large screen (viewing distance more than 2,5 m).

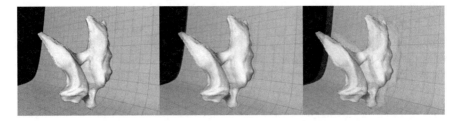

Fig. 17. Varying stereo base: zero, narrow and wide

Another factor is the need for reference objects. The problem is that the 3D effect is not so apparent when there is only one object in the image, because human brain has no other objects to compare distances with. For example, on Fig. 18, there is an object on a black background. It is much more difficult to see the 3D effect in this picture than in the one with some additional objects (Fig. 17)

Fig. 18. Anaglyph with no reference objects

Finally, there is also a quality affecting factor concerning image (video) compression and resolution, since compression artifacts can significantly degrade the visual quality of the 3D stereoscopic effect. Therefore, it is recommended to use high resolutions and quality compression formats for stereoscopic visualizations. For images, Portable Network Graphics (PNG) appears to be a very suitable format, since it produces images, which do not take up too much hard-drive space. For stereoscopic videos, a suitable solution would be to use either uncompressed video or some of the modern, high quality codecs, such as x264, with high bitrate settings.

5.6   Data Preparation and Preprocessing – GIS Example

There is a plenty of specialized 3D software and number of modeling techniques with which 3D model of selected city agglomeration can be created [22]. It is necessary to choose right modeling technique and specialized software to get proper model. Second problem is a visualization of these models with additional information data displaying. Because the issue of 3D city agglomeration visualizations is wide-ranging, work described in this chapter was divided into these three parts: modeling, presentation and visualization. Main goal of this chapter is description of the presentation part.

Work on presentation part was divided into the following parts:

- Develop a 3D model of a selected part of an urban agglomeration ,
- Implement software using the scripting language that allows the smooth redraw of model given by the transition path.

The choice of the city agglomeration area, which was visualized, was the Technical University of Kosice bounded by the adjacent streets.

Panoramic images creation consists of the following steps:

1. loading of the input data and coordinates in script,
2. parameters setting for script,
3. creation of animation using script,
4. parameters setting for photorealistic lighting,
5. rendering of all frames of animation,
6. creation of panoramic images .

Before rendering, it was important to decide, how many computers will be used for rendering: single PC or multiple PC connected by network (network rendering).

There also appears a need to compare this approach with the method that uses real photography to see the level of photorealistic visualization in praxis.

5.6.1   Script Description

The most important part is the script that allows the smooth redraw of model given by the transition path. Since 3D Studio Max [33] contains its own scripting language, there was no need to seek other solutions [34]. The created script consists of the following parts:

- loading of input matrix,
- setting start and end coordinates,
- setting basic parameters,
- creation of animation,
- sorting out frames of animation.

5.6.2 Lighting

Creation of realistic lighting is one of the key conditions for creating photorealistic images. It is therefore necessary to choose the most realistic lighting, but also take rendering time into consideration. It is necessary to create a realistic sky for better impression of the real lighting as well.

Creation of lighting was divided into several points:

- selection of a suitable light source,
- sky creation,
- remove of color bleeding,
- lighting parameters setting.

5.6.3 Rendering

Because rendering with single PC would take long time, network rendering was used. Render network (render farm) was build from 20 PC. Configuration of each PC:

- CPU AMD Athlon(tm) 64 Processor 3700+
- Memory 1GB DDR
- HDD Maxtor 160 GB
- GPU 256 MB GeForce 7300 GT

Since 3D Studio Max [33] contains its own network manager (Backburner Manager) for network rendering, it was not necessary to use any other software for network rendering. Rendering time for all images was approximately 10.5 days.

5.6.4 Panoramic Image Creation

When all images were rendered and sorted, it was necessary to create panoramic images from them. For panoramic images creation the program Panorama Maker Pro 4 (trial version) [35] was used. Each one panoramic image was composed from 16 rendered pictures. When a panoramic image was created, its resolution was adjusted to 4096x1024 to speed-up the loading and processing of them on the graphic card.

6   Conclusion

This chapter gave an overview on VR systems creation process presenting VR roles, subsystems, data acquisition and processing, 3D modeling, the object oriented property of VR systems, and visualization tasks solving. Several examples were presented to illustrate these processes.

VR system division into subsystems is related to the senses that are used to create the effect of illusion. Visual subsystem is used for presenting the virtual world using visual effects, rendering and stereoscopy. This visualization requires data preparation and processing.

3D data acquisition is implemented using various 3D scanning techniques. The scanned data are usually point cloud that needs further processing such triangulation and simplification.

As an example, the design and implementation of an application tool for 3D scanning process was presented. The implementation is based on projection of raster on the background that follows by the capture by the camera. The output is in VRML format. It can be further used in the process of 3D scene creation. VRML (Virtual Reality Modeling Language) is a standard file format for representing 3D interactive vector graphics, designed particularly with the World Wide Web in mind. VRML is a text file format where, e.g. vertices and edges for a 3D polygon can be specified along with the surface color, UV mapped textures, shininess, transparency, and so on. URLs can be associated with graphical components so that a web browser might fetch a web page or a new VRML file from the Internet when the user clicks on the specific graphical component.

Another way to obtain 3D models is 3D modeling. This process uses architectural data and textures. Both kinds of data need to undergo normalization and optimization. Some pre-processing steps could be run in parallel environment such clusters that increase process efficiency.

At the DCI FEI TU Kosice a VR system for immerse visualization of complex datasets is being built. Typical dataset represents the main street in Kosice. The system will provide not only nice textured geometry, but also some information about buildings. This VR system can be used like information immerse interactive system. For better experience is good to stimulate also sense of hearing. For this purpose it is designing sound subsystem on the open source library OpenAL. First task for preparation the immerse presentation is to design a nice scene. When the geometry part is completely textured, lights and fog are added to the scene. After this step, presentation is almost complete. If one wants only walking through the virtual street it is enough, but if the user excepts more, one can add to the scene metadata with historical information, e.g. when the user is near to the Elizabeth Cathedral, he can read on the screen the story about history of this site.

This system is in the development and it has these main features: hierarchical graph scene representation, support main 3D models formats such as (3DS, OBJ, VRML), support script controlling (Ruby, Python), visualization engine with multi-screen and cluster support.

After analyzing model requirements, polygonal modeling appears to be an optimal modeling technique for this kind of project. Its superiority in the sphere of performance allows bringing 3D virtual map project to common people and offers possibility to expand in public usage. An authentically looking model provides almost realistic view to the city and offers opportunity for common people to touch virtual reality in practical usage. The solution is based on parallel processing in grid environment.

The most small-sized objects could be used in the scenes of the visual subsystem in the form of scanned 3D models rather than modeled ones. These could be also printed as outputs from a 3D editor as part of the VR system.

The acoustic subsystem implementation needs more analysis on existing solutions – building one from scratch is time consuming and not effective. Audio command navigation is therefore of less importance, much higher interest is in visualization optimization using verification techniques and formal methods such Petri nets.

Performance tuning could be provided by using a significantly larger cluster or a data flow solution [27, 28].

For a network based implementation, the usage of panoramic pictures as scene backgrounds offers the open area of web based systems (e.g. geographical information systems). There is a limitation given by technology and tools, but using pre-processed data opens the perspective.

The idea of creating a 3D laboratory that offers a web based interface as well will be analyzed and discussed too. The most involved and investigated areas include mobile devices and mobile networks.

Acknowledgement. This work was supported by VEGA grant project No. 1/0646/09: "Tasks solution for large graphical data processing in the environment of parallel, distributed and network computer systems."

References

1. Bernardini, F., Rushmeier, H.: The 3D Model Acquisition Pipeline. Computer Graphics Forum 21(2), 149–172 (2002)
2. Fang, S., George, B., Palakal, M.: Automatic Surface Scanning of 3D Artifacts. In: International Conference on Cyberworlds (2008),
   http://ieeexplore.ieee.org/stamp/
   stamp.jsp?tp=&arnumber=4741319&isnumber=4741260
3. Gosztolya, G., Dombi, J., Kocsor, A.: Applying the Generalized Dombi Operator Family to the Speech Recognition Task. Journal of Computing and Information Technology 17(3), 285–293 (2009)
4. Gosztolya, G., Tóth, L.: Spoken Term Detection Based on the Most Probable Phoneme Sequence. In: Proceedings of the 2011 International Symposium on Applied Machine Intelligence and Informatics (SAMI) (IEEE), Smolenice, Slovakia, pp. 101–106 (2011)
5. Hahn, D.V., Duncan, D.D., Baldwin, K.C., Cohen, J.D., Purnomo, B.: Digital Hammurabi: design and development of a 3D scanner for cuneiform tablets. In: Proc. SPIE, vol. 6056, pp. 130–141 (2006)
6. Kari, P., Abi-rached, H., Duchamp, T., Shapiro, L.G., Stuetzle, W.: Acquisition and Visualization of Colored 3D Objects. In: Proceedings of the 14th International Conference on Pattern Recognition, August 16-20, vol. 1. IEEE Computer Society, Washington, DC (1998)
7. Katz, S., Tal, A., Basri, R.: Direct Visibility of Point Sets. In: ACM SIGGRAPH 2007 Papers, SIGGRAPH 2007, August 05-09. ACM, New York (2007)

8. Kocsor, A., Gosztolya, G.: The Use of Speed-Up Techniques for a Speech Recognizer System. The International Journal of Speech Technology 9(3-4), 95–107 (2006)

9. Kuljić, B., Simon, J., Szakáll, T.: Mobile robot controlled by voice. In: Proc. of the 7th International Symposium on Intelligent Systems and Informatics (SISY), Subotica, Serbia, pp. 189–192 (2007)

10. Mamou, J., Ramabhadran, B., Siohan, O.: Vocabulary Independent Spoken Term Detection. In: Proc. of SIGIR, Amsterdam, The Netherlands (2007)

11. Molnar, S., Cox, M., Ellsworth, D., Fuchs, H.: A Sorting Classification of Parallel Rendering. IEEE Computer Graphics & Applications 14(4), 23–32 (1994)

12. Nagy, K., Takács, M.: Type-2 fuzzy sets and SSAD as a Possible Application. Acta Polytechnica Hungarica 5(1), 111–120 (2008)

13. Rabiner, L., Juang, B.-H.: Fundamentals of Speech Recognition. Prentice-Hall, Englewood Cliffs (1993)

14. Rusinkiewicz, S., Hall-holt, O., Levoy, M.: Real-Time 3D Model Acquisition. ACM Transactions on Graphics (Proc. SIGGRAPH). 21(3), 438–446 (2002)

15. Samanta, R., Funkhouser, T., Li, K., Singh, J.: Hybrid Sort-First and Sort-Last Parallel Rendering with a Cluster of PCs. In: Eurographics/SIGGRAPH Workshop on Graphics Hardware, pp. 99–108 (2000)

16. Samanta, R., Zheng, J., Funkhouser, T., Li, K., Singh, J.: Load Balancing for Multi- Projector Rendering Systems. In: SIGGRAPH/Eurographics Workshop on Graphics Hardware (1999)

17. Sobota, B., Janošo, R., Babušíková, D.: Visibility problem solving in Virtual Reality Systems. In: Proc. of the International Workshop on Digital Metodologies and Applications for Multimedia and Signal Processing DMMS, Budapest, Hungary, pp. 139–145 (1997)

18. Sobota, B., Straka, M.: A conception of parallel graphic architecture for large graphical data volume visualization in grid architecture environment. In: Proc. of Grid Computing for Complex Problems, Bratislava, Institute of Informatics SAS, pp. 36–43 (2006)

19. Sobota, B., Straka, M., Perháč, J.: An Operation for Subtraction of Convex Polyhedra for Large Graphical Data Volume Editing. In: Proc. of the 9th International Scientific Conference on Engineering of Modern Electric System, EMES 2007, Oradea, Romania, pp. 107–110 (2007)

20. Sobota, B., Straka, M., Vavrek, M.: The preparation of contribution of 3D model input using triangulation. In: MOSMIC 2007 - Modeling and Simulation in Management, Informatics and Control, Zilina, Slovakia, pp. 21–26 (2007)

21. Sobota, B., Szabó, C., Perháč, J., Ádám, N.: 3D Visualization for City Information System. In: Proceedings of International Conference on Applied Electrical Engineering and Informatics AEI 2008, Athens, Greece (2008)

22. Sobota, B., Szabó, C., Perháč, J., Myšková, H.: Three-dimensional interfaces of geographical information systems. In: ICETA 2008: 6th International Conference on Emerging eLearning Technologies and Applications: Information and Communications Technologies in Learning: Conference Proceedings, The High Tatras, Slovakia (2008)

23. Szabó, C., Fesič, T.: Database Refactoring for GIS. In: Electrical Engineering and Informatics, Proceeding of the Faculty of Electrical Engineering and Informatics of the Technical University of Košice, pp. 347–350 (2010)

24. Szabó, C., Korečko, Š., Sobota, B.: Processing 3D Scanner Data for Virtual Reality. In: Proceedings of the 2010 10th International Conference on Intelligent Systems Design and Applications, Cairo, Egypt, pp. 1281–1286. IEEE, Los Alamitos (2010)

25. Szabó, C., Sobota, B.: Some Aspects of Database Refactoring for GIS. In: Proceedings of the Fourth International Conference on Intelligent Computing and Information Systems ICICIS 2009, Cairo, Egypt, pp. 352–356 (2009)

26. Takács, M.: Fuzziness of rule outputs by the DB operators-based control problems. In: Rudas, I.J., Fodor, J., Kacprzyk, J. (eds.) Towards Intelligent Engineering and Information Technology. Studies in Computational Intelligence, vol. 243, pp. 653–663. Springer, Heidelberg (2009)

27. Vokorokos, L., Danková, E., Ádám, N.: Parallel scene splitting and assigning for fast ray tracing. Acta Electrotechnica et Informatica 10(2), 33–37 (2010)

28. Vokorokos, L., Danková, E., Ádám, N.: Task scheduling in distributed system for photorealistic rendering. In: SAMI 2010: 8th International Symposium on Applied Machine Intelligence and Informatics, Herlany, Slovakia, pp. 43–47 (2010)

29. Wald, I., Slusallek, P., Benthin, C.: Interactive Distributed Ray Tracing of Highly Complex Models. In: Eurographics Workshop on Rendering, pp. 277–288 (2001)

30. Walker, W., Lamere, P., Kwok, P., Raj, B., Singh, R., Gouvea, E.: Sphinx-4: A Flexible Open Source Framework for Speech Recognition. Sun Microsystems Technical Report TR-2004 139 (2004)

31. Weik, S.: A Passive Full Body Scanner Using Shape from Silhouettes. In: Proc. of the International Conference on Pattern Recognition, ICPR, vol. 1. IEEE Computer Society, Washington, DC (2000)

32. Young S.: The HMM Toolkit (HTK) (software and manual) (1995),
http://htk.eng.cam.ac.uk/

33. 3ds Max 9 User Reference,
http://usa.autodesk.com/adsk/servlet/
item?siteID=123112&id=10175188&linkID=9241177

34. MAXScript Reference 9.0,
http://usa.autodesk.com/adsk/servlet/
item?siteID=123112&id=10175420&linkID=9241175

35. Panorama Maker 4 Pro.,
http://www.arcsoft.com/products/panoramamakerpro/

Virtual Reality Technology for Blind and Visual Impaired People: Reviews and Recent Advances

Neveen I. Ghali[1], Omar Soluiman[2], Nashwa El-Bendary[3],
Tamer M. Nassef[4], Sara A. Ahmed[1], Yomna M. Elbarawy[1],
and Aboul Ella Hassanien[1]

[1] Faculty of Science, Al-Azhar University, Cairo, Egypt
   nev_ghali@yahoo.com
[2] Faculty of Computers and Information, Cairo University, Cairo, Egypt
   aboitcairo@gmail.com
[3] Arab Academy for Science, Technology, and Maritime Transport
   P.O. Box 12311, Giza, Egypt
   nashwa_m@aast.edu
[4] Faculty of Engineering, Misr University for Science and
   Technology (MUST) Giza, Egypt
   tamer.nassef@k-space.org

Abstract. Virtual reality technology enables people to become immersed in a computer-simulated and three-dimensional environment. In this chapter, we investigate the effects of the virtual reality technology on disabled people such as blind and visually impaired people (VIP) in order to enhance their computer skills and prepare them to make use of recent technology in their daily life. As well as, they need to advance their information technology skills beyond the basic computer training and skills. This chapter describes what best tools and practices in information technology to support disabled people such as deaf-blind and visual impaired people in their activities such as mobility systems, computer games, accessibility of e-learning, web-based information system, and wearable finger-braille interface for navigation of deaf-blind. Moreover, we will show how physical disabled people can benefits from the innovative virtual reality techniques and discuss some representative examples to illustrate how virtual reality technology can be utilized to address the information technology problem of blind and visual impaired people. Challenges to be addressed and an extensive bibliography are included.

Keywords: Virtual reality, assisitive technology, disabled people, mobility system, computer games, accessibility of e-learning.

1 Introduction

Virtual reality technology is the use of graphics systems in conjunction with various display and interface devices to provide the effect of immersion in the interactive three dimensional computer-generated environment which is called a virtual environment. Virtual reality and environment applications

T. Gulrez, A.E. Hassanien (Eds.): Advances in Robotics & Virtual Reality, ISRL 26, pp. 363–385.
springerlink.com   © Springer-Verlag Berlin Heidelberg 2012

covers a wide range of specific application areas, from business and commerce to telecommunications, entertainment and gaming to medicine. It is a multidisciplinary technology that is based on engineering and social sciences, and whose possibilities and progress largely depend on technical developments. In accordance with hardware and software, virtual reality enhances the involvement of the user in a more or less immersive and interactive virtual human experiment [49].

Virtual reality refers to the use of interactive simulations to provide users with opportunities to engage in environments that may appear and feel similar to real-world objects and events and that may engender a subjective feeling of presence in the virtual world. This feeling of presence refers to the idea of being carried into another, virtual world. Virtual reality is of special interest for persons with disabilities showing dysfunction or complete loss of specific output functions such as motion or speech, in two ways. First, disabilities people can be brought into virtual scenarios where they can perform specific tasks-either simply to measure their performance such as duration or accuracy of movements or to train certain actions for rehabilitation purposes. Second, virtual reality input devices like a data glove, originally developed for interaction within virtual environments, can be used by disabled people as technology tools to act in the real world [5, 6].

Recently, more virtual reality applications have been developed which could minimize the effects of a disability, improve quality of life, enhance social participation, and improve life skills, mobility and cognitive abilities, while providing a motivating and interesting experience for children with disabilities. One of the main deprivations caused by blindness is the problem of access to information. Virtual reality is an increasingly important method for people to understand complex information using tables, graphs, 3D plots and to navigate around structured information.

Spatial problems can also be experienced by children with physical disabilities that limit their autonomous movement. Due to this, they are not likely to be able to fully explore environments, which is also a problem for those who need some form of assistance. They may also have limited access, and may rely on their assistant to choose routes. Studies on cognitive spatial skills have shown that children with mobility limitations have difficulty forming effective cognitive spatial maps [3].

This chapter investigates the effects of the virtual reality technology on disabled people, in particular blind and visually impaired people (VIP) in order to enhance their computer skills and prepare them to make use of recent technology in their daily life. As well as, they need to advance their information technology skills beyond the basic computer training and skills. Moreover, we describe what best tools and practices in information technology to support disabled people, such as deaf-blind and visual impaired people in their activities such as mobility systems, computer games, accessibility of e-learning, web-based information system, and wearable finger-braille interface for navigation of deaf-blind.

This chapter has the following organization. Section (2) reviews some of the work that utilize the intelligent mobility systems for blind and visually impaired people. Section (3) introduces the accessibility of e-learning including Visual impairment categories, its considerations and requirements for the blind people. Section (4) discusses the wearable finger-braille interface for navigation of deaf-blind people. Section (5) discusses the computer games for disabled children including audio games/sonification and tactile games. Section (6) discusses the web-based information system. Finally, challenges are discussed in Section (7) and an extensive bibliography is provided.

2   Mobility Systems for Blind and Visually Impaired People

Disability people needs to become as independent as possible in their daily life in order to guarantee a fully social inclusion. Mobility is a persons ability to travel between locations safely and independently [4]. Blind mobility requires skill, effort and training. Mobility problems for a blind person can be caused by changes in terrain and depth; unwanted contacts;and street crossings (which involve judging the speed and distance of vehicles and may involve identifying traffic light colour). The most dangerous events for a blind or partially sighted person are drop offs (sudden depth changes, such as on the edge of a subway platform) and moving vehicles [4]. Making unwanted contact with pedestrians is also undesirable as it can be socially awkward and may pose a threat to a persons safety.

Mobile and wireless technologies, and in particular the ones used to locate persons or objects, can be used to realize navigation systems in an intelligent environment[7]. A wearable system presents the development of an obstacle detection system for visually impaired people, for example. The user is alerted of closed obstacles in range while traveling in their environment. The mobility system can detects obstacle that surrounds the deaf people by using multi-sonar system and sending appropriate vibro-tactile feedback, which, serves as an aid by permitting a person to feel the vibrations of sounds. It is known as a mechanical instrument that helps individuals who are deaf to detect and interpret sounds through their sense of touch.

Such system works by sensing the surrounding environment via sonar sensors and sending vibro-tactile feedback to the deaf people of the position of the closest obstacles in range. The idea is to extend the senses of the deaf people through a cyborgian voice interface. This means that the deaf people should use it, after a training period, without any conscious effort, as an extension of its own body functions. It determine from which direction the obstacles are coming from. Localization on the horizontal plane is done by appropriate combination of vibration according to the feedback of the sensors [14].

This section will introduces the fundamental aspects of wireless technology for mobility and reviews some of the work that utilize the intelligent mobility systems for blind and visually impaired people.

2.1 Wireless Technology for Mobility

The radio virgilio/sesamonet is information and communication technologies system intends to improve blind and low vision users' mobility experience by coupling tactile perceptions with hearing aids. To this extent, wireless technologies including radio frequency identification (RFID) and Bluetooth, hand-held devices (PDA), smart phones, and specific system and application software for mobile device including text-to-speech and database are combined together[8].

Radio virgilio/sesamonet provides a suitable audio output helps the autonomous mobility of blind and visual impaired people. It has the following features:

- keeps the people inside a safe path;
- provides information about turns and obstacles on the path
- checks the right direction
- provides general and specific environmental information on demand
- provides on-line help and assistance via global system for mobile communications

Atri et al. in [7] and Ugo et al. in [8] developed and designed RFID system to help blind and visually impaired people in mobility. The system shows a disabled interaction with the system simultaneously using three different devices: a RFID Cane Reader, a PDA, and a Bluetooth headset. Together with the tag grid, they constitute the physical architecture of the system. The RFID cane reader reads the tag ID and sends this number, via Bluetooth, to the PDA; the PDA software associates the received ID to mobility information and, after converting it in a suitable message, sends it to the headset. Radio Virgilio/SesamoNet software runs on any Windows CE based portable device having a Bluetooth antenna to communicate with the RFID cane. The system interface contains the following modules:

- **Bluetooth connection manager** which keeps the BT connection channel open between the RFID reader and the PDA for tag ID string transmission.
- **Navigation data interface** it retrieves navigational data from a local database, providing the navigation logic with extended data related to a tag ID.
- **Navigation logic interface** which is the core software which handles navigation and tag data in order to provide the user with mandatory safety-related or on environmental navigation information. It also checks if the direction is right and not reverted. This module tells the user he/she is probing the central tag or one on the right or the left hand side of the path.

We have to note that, the user is not supposed to perform any task when navigation session starts or the BT manager is resumed after any loss of connection with the cane and it can send a text string containing more complex navigation or environmental information to the text to speech component.

Focussing on this problem, many successful works have been addressed and discussed. For example, Gerard and Kenneth in [1] describe an application of mobile robot technology to the construction of a mobility for the visual impaired people. The robot mobility physically supports the person walking behind it and provides obstacle avoidance to ensure safer walking. Moreover, the author summarize the main user requirements for robot mobility and describe the mechanical design, the user interface, the software and hardware architectures as well. The obtained results was evaluated by mobility experts and the user.

The ability to navigate space independently, safely and efficiently is a combined product of motor, sensory and cognitive skills. This ability has direct influence in the individuals quality of life. Lahav and Mioduser in [2] developed a multi-sensory virtual environment simulating real-life space enabling blind people to learn about different spaces which they are required to navigate. This virtual environment comprises two module of operation:(a) Teacher module and (b) Learning module. The virtual environment editor is the core component of the teacher module, which includes 3D environment builder; force feedback output editor; and (c) audio feedback editor. While, the learning module includes user interface and teacher interface. The user interface consists of a 3D virtual environment, which simulates real rooms and objects and then the user navigates this environment. During this navigation varied interactions occur between the user and the environment components. As a result of this interactions the user get haptic feedback through environment navigates. This feedback includes sensations such as friction, force fields and vibrations.

3 Accessibility of E-learning for Blind and Visually Impaired

Vision, of the body's five sensory inputs, is the essential sense used in learning. Vision also modifies or dominates the interpretation from the other senses where there is variance between the inputs from more than one sense [22]. Low vision and totally blind students must rely on input from physical senses other than sight; however, most e-learning environments generally assume the learner has sight [20]. Considering e-learning, simulations, active experimentation, discovery-learning techniques, questioning with feedback, video, animations, photographs, and practical hands-on skills, can be utilized for virtual teaching [23]. This may be the case for sighted students; however, vision impaired students do not have the sight needed to access many of these multi-media sources of delivery. Lack of accessibility in the design of e-learning courses continues to act as an obstacle in the learning way of students with vision impairment. E-learning materials are predominantly vision-centric, incorporating images, animation, and interactive media, and as a result students with severe vision impairment do not have equal opportunity to gain tertiary qualifications or skills relevant to the marketplace and their disability [20, 21].

This section presents an overview with addition to introducing considerations and requirements in order to obtain accessible e-learning environment for blind and visual impaired people.

3.1 Visual Impairment Categories

The term vision impairment refers to a vision disability resulting in little or no useful vision [19]. There are different eye conditions, which can affect sight in various ways. These conditions include short- and long-sightedness, color blindness, cataracts, which is responsible for almost half of all cases of blindness worldwide, and the world's leading preventable cause of blindness that is known as glaucoma [19].

A useful distinction can be made between the congenitally blind (those who are blind from birth) and the adventitiously blind (those who developed blindness later in life, perhaps as a result of accident, trauma, disease, or medication). There is a significant difference between these two visual impairment groups. For example, learners who have been blind from birth may find it more difficult making sense of tactile maps than adventitiously blind learners. That's due to the fact that learners who have been blind from birth have not previously acquired spatial awareness through visual interaction with their environment. Also, the degree of visual impairment can vary considerably. Generally, the range of impact can run the entire scale from total blindness through low vision to minor impairment. A broad distinction is usually made between blind people and partially sighted people. These two groups may exhibit rather different study patterns and difficulties, and may require different kinds of support [19].

3.2 E-learning Accessibility Considerations

There are two vital aspects, namely technological and methodological, to be taken into consideration for obtaining a fully accessible e-learning environment [20]. For technological issue, the most frequently used network technologies in e-learning are email and the Web. Because email is the most used communication service on the Internet, it is very useful to include it as part of any e-learning environment. Although email presents no significant problems for users with disabilities, some research proposes techniques for improving email accessibility [20]. Likewise, the web is the most used tool for accessing information on the Internet and it is the best solution for distributing educational material for e-learning. Producing a document overview is one of the main issues to be considered in an application for surfing the Web, which has a vocal interface.

Regarding methodological issue, the content interaction, in an educational context designing, is extremely important in order to reach a learning goal. Moreover, in online learning the methodology is crucial. For example, a tool may meet technical accessibility requirements, but it may be unusable for a blind student because it is designed with a visual interface in mind. Likewise, the design of a lesson could be perfect if it is delivered using a multimedia system, but may be poor if it uses adaptive technologies like a speech synthesizer. Therefore, it is very important to redesign traditional pedagogical approaches by integrating information, virtual reality and communication technologies into courses [20].

3.3 Web Accessibility Requirements for the Blind People

The web design considerations which meeting several requirements in the web development process and improves web accessibility for all blind web users. These consideration are (1) Assistive technologies for blind people, such as screen readers, have plain interfaces that sequentially express in words Web content in the order it is structured in the source code [25, 26]. Therefore a text only version of the entire website can improve the access speed for blind users, (2) Web designers should provide a text alternative for every visual element (textual description that conveys the same function or purpose as the visual element) and avoid elements that cannot be presented in this form [27, 62, 25] so that screen readers can adapt text into audio formats for the blind users to access, (3) Skip navigation links [24] enable users of screen readers to skip repeated or peripheral content and go straight to the main content. This saves time and improves usability for the blind [29], (4) Web designers should mark up different sections of Web pages with predefined semantics (such as main, heading, navigation and adverts) [26] in order to make it possible for blind people to navigate to different sections of the website including the ability to skip certain sections, (5) Designers should identify table headers in the first row and first column. Complex data tables are tables with two or more logical levels of row or column headers. They can be made accessible by associating heading information with the data cell to relate it to its heading [28, 62], (6) Frame elements should have meaningful titles and name attributes explaining the role of the frame in the frame set. Alternatively, designers can provide alternative content without frames [26, 63], (7) Web designers should explicitly and programmatically associate form labels with their controls (e.g. place text information for text entry fields and put the prompt As the blind Web users rely on keyboards as their primary input device, designers have to be sure that all parts of a Web application are usable with keyboard only access. for a checkbox or radio button to the right of the object). For forms used for search functionality, designers should ensure that results from the search are accessible with the keyboard and screen reader for the functionality. and (8) As the blind Web users rely on keyboards as their primary input device, designers have to be sure that all parts of a Web application are usable with keyboard only access.

The development of the web content accessibility guidelines by the World Wide Web Consortium (W3C) and the Section 508 standards for the federal government were both guided by universal design principles. The web content accessibility guidelines tell how to design Web pages that are accessible to people with disabilities. More recently, in response to the Section 508 amendments, the access board created standards to be used by the federal government to assure the procurement, development, and use of accessible electronic and information technology, including Web pages. Some institutions that sponsor e-learning programs that are not strictly required to comply with Section 508 adopt the Section 508 standards or W3C's guidelines. Others develop their own list of accessibility requirements for their programs [31].

4   Wearable Finger-Braille Interface for Navigation of Deaf-Blind People

Globally, deaf-blindness is a condition that combines varying degrees of both hearing and visual impairment; people who are blind or deaf-blind (i.e., with severe vision impairments, or both severe vision and hearing impairments) cannot easily read, where only 10% of the blind children receive instruction in Braille. Researchers are focalized on navigation rather than environment disclosure; reading is one of the problems that related to information transmission. A portable reader position is discovered through either the identification of the closest reference tag surrounding it; wearable devices are distinctive from portable devices by allowing hands-free interaction, or at least minimizing the use of hands when using the device. This is achieved by devices that are actually worn on the body such as head-mounted devices, wristbands, vests, belts, shoes, etc.

Reading and travel are targets for wearable assistive devices that developed as task-specific solutions for these activities; they try to open new communication channels through hearing and touch. Devices are as diverse as the technology used and the location on the body. Fig. 1.1, as given in [32] overviews the body areas involved in wearable assistive devices as described by Velazquez [32] fingers, hands, wrist, abdomen, chest, feet, tongue, ears, etc. have been studied to transmit visual information to the blind.

Touch becomes the primary input for the blind people as receipt of non-audible physical information, where the 3D objects can rapidly and accurately identify by touch. There are three main groups of sensors organized by biological function defined by Velazquez in [32] the *Thermoreceptors*, responsible for thermal sensing, the *nociceptors*, responsible for pain sensing and the *mechanoreceptors*, sensitive to mechanical stimulus and skin deformation. Most of the assistive devices for the blind that exploit touch as the

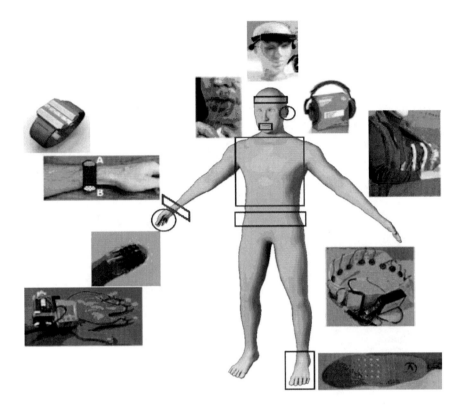

Fig. 1. Overview of wearable assistive devices for the VIP [32]

substitution sense are tactile displays for the fingertips and palms. Typical tactile displays involve arrays of vibrators or upward/downward moveable pins as skin indentation mechanisms.

Communication support technology and position identification technology are essential to support Deaf-Blind people. As communication interfaces, the Japanese proposed wearable Finger-Braille interfaces which are hands-free and can communicate with others in real-time. As positioning technology, they designed the ubiquitous environment for barrier-free applications which consists of network cameras, wireless LAN and floor-embedded RFID tags [33]. Finger-Braille is a method of tapping Deaf-Blind person's fingers to transmit verbal information, which are assigned to the digits of Braille. This method enables Deaf-Blind people to obtain information as if they are listening in real-time. The Japanese Finger-Braille interface is a wearable assistive device to communicate information to the deaf-blind that based on Japanese script, Kanji (the Chinese ideographic script), Hiragana and Katakana (syllabic script) are the three kinds of Japanese writing symbols used (The Association for Overseas Technical Scholarship, 1975) [34]. In this system, the fingers are regarded as Braille dots: 6 fingers, 3 at each hand, are enough to code any 6-dot Braille character. Some examples of translation given in [32] are shown in Fig. 1.

Glove-style interfaces seem to be the most suitable design for the Finger-Braille interface because they are easy to wear, where the first glove-based systems were designed in the 1970s, and since then, a number of different designs have been proposed. However, they cover the palm or the fingertip that has the highest tactile sensitivity.

Authors in [36] compared between two different wearable tactile interfaces, a back array and a waist belt with experimental trials, where the data were analyzed using a three-way repeated- measures ANOVA with tactile interface, screen orientation and visual display. There results suggest that the belt is a better choice for wearable tactile direction indication than the back array, however, their experiments did not seek to tease out which particular features of these two established approaches led to the observed differences.

5 Games for Visually Impaired People

Virtual reality and gaming technology have the potential to address clinical challenges for a range of disabilities. Virtual reality-based games provide the ability to assess and augment cognitive and motor rehabilitation under a range of stimulus conditions that are not easily controllable and quantifiable in the real world. People who cannot use the ordinary graphical interface, because they are totally blind or because they have a severe visual impairment, do not have access or have very restricted access to this important part of the youth culture. Research and development in the field of information technology and the disabled has focused on education rather than leisure [9, 12].

Indeed the mainstream commercial market for computer games and other multimedia products have shown a rather impressive development in the last years. For instance in 2002, costs for the development of games could vary between 300,000 Euros for a game on wearable device, to 30 millions for the biggest productions [12].

Visual impairment prevents player from using standard output devices. So the main problem of visually impaired player is to acquire the game information. Two modalities can be used to replace visual modality: sound modality or tactile modality. Audio computer games are games whose user interface is presented by using audio effects, while in tactile computer games user interface is based on a set of specialized touch-sensitive panels. Communities of the visually impaired tend to favor audio computer games, because there is a number of their members particularly those who were not born blind, but have acquired their blindness later in life, who are not familiar with tactile interfaces. The accessibility of multimedia games to the visually impaired does not rest solely on their ability to be presented in audio or tactile formats, but the language of the game is important as well [11].

5.1   Audio Games/Sonification

Recent progress in computer audio technologies has enhanced the importance of sound in interactive multimedia. The new possibilities to use sound in interactive media are very welcome to computer users that have difficulties in using graphical displays. Today, it is possible to create sound-based interactive entertainment, such as computer games [13, 10] designing of computer games that work for visually impaired children is a Challenge.

Sonification and visualization by sound, has been studied widely for decades and it has also been applied in software for blind people. The aim is to display, for example, graphics, line graphs or even pictures using non-speech sound [64]. By altering the various attributes of sound, for example pitch, volume and waveform, the sound is changed according to its visual counterpart. In the Sound-View application a colored surface could be explored with a pointing device. The idea was that the characteristics of colors, hue, saturation and brightness were mapped into sounds [15, 16, 17].

In 10 years, over 400 audio games have been developed, which is very small as compared to video games. Researchers could have an important role to play in that expansion process, by contributing to develop innovative features like a more pleasant audio rendering, by projecting audio games in a futuristic point of view by using uncommon technology like GPS for instance and by participate in the elaboration of games which can interest both visually and visually impaired community [12].

Current audio games can be studied with three dependency factors (1) dependency on verbal information, (2) dependency on interaction mechanisms based on timing and (3) dependency on the interaction mechanisms of exploration [12].

Action games directly refer to games coming from and inspired by arcade games and require a good dexterity of the player. One example of such games is "Tampokme" [58] is such an accessible audio game. In this game timing is essential. The player has to identify various kinds of audio signals and react accordingly and in fixed time. Adventure games offer players to take part in a quest. This type of game combines three essential features: an interesting scenario, the exploration of new worlds and activities of riddle solving. Example of such games is "Super Egg Hunt" [59] focuses on the move of the avatar on a grid, where the players must locate objects only from audio feedback. A quite clever use of stereo, volume and pitch of the sounds allows an easy and pleasant handling.

There are a few strategy audio games but the manipulation of maps is rather difficult without the visual. The game "simcarriere" [60] ignores the map aspect and focuses on the simulation/management side. The player has to manage a lawyer's office by buying consumable, hiring personal, and choosing the kind of cases to defend. Puzzle audio games are similar to video puzzle games in principle. "K-Chess advance" [61] is an adaptation of the Chess game, but focusing on audio. One can find a good number of audio games. The Swedish Library of Talking Books and Braille (TPB) has published web-based games dedicated to young children with visual impairment. On the other end, Terraformers is the result of three years of practical research in developing a real-time 3D graphic game accessible for blind and low vision gamers as well as full sighted gamers. A quite comprehensive list of audio games can be found in [9].

5.2 Tactile Games

Tactile games are games, where the inputs and/or the outputs are done by tactile boards or by Braille displays, in combination with usually audio feedback. Then during the TiM project [18], funded by the European Commission, several tactile games were developed [9, 12]. One was an accessible version of Reader rabbit's Toddler, where the children can feel tactile buttons on the tactile board, and then drive the game from this board.

Another game developed within the TiM project was FindIt, which was a very simple audio/tactile discovery and matching game. From this game game generator was developed which is currently being evaluated with practitioners working with visually impaired children. The generator allows educators and teachers who work with visually impaired children to design their own scenarios and to associate them with tactile sheets (that they design themselves manually) [9, 12].

6 Web-Based Information System for Blind and Visually Impaired

In this section we introduce Web-based information system accessibility, Web-based city maps support visually impaired people and Email accessibility support visually impaired people.

6.1 Web-Based Information System Accessibility

For visual impaired people, the application of the web content accessibility often might not even make a significant difference in terms of efficiency, errors or satisfaction in website usage. This section presents the development of guidelines to construct an enhanced text user interface as an alternative to the graphical user interface for visual impaired people. And one of the important current major problems for visual impaired people is the provision of adequate access to the WWW. Specialist browsers are beginning to emerge to provide a degree of access, along with guidelines for page design and proxy servers to assist in reorganizing a page to simplify spoken presentation. A further problem in the provision of access to the WWW is the emerging use of 3D images. The method of providing these 3D images for visual impaired people is still in an early stage of development. It is the purpose of this subsection is to identify potential solutions to the problem of access for visual impaired people, and to initiate a dialogue within the WWW developer community about this issue, before the methods of 3D presentation of images become unalterable.

Drishti [53] which is an integrated navigation system for visual impaired people uses the Global Positioning System (GPS) and Geographical Information System (GIS) technologies. It is designed to be used within the university premises and contains a GIS dataset of the university. This contains geographically referenced information for both static and dynamic environments and is referred to as a spatial database. The spatial database is accessible through a wireless network to a wearable device that is carried by the visual impaired people. A differential GPS receiver in the wearable device determines the localization of the user. Drishti is an assistive device which is operable in dynamically changing environments and can optimize routes for navigation when there is an unforeseen obstacle in the path. Like SESAMONET, Drishti gives assistance to the user by means of speech. Drishti may be considered as the first reliable assistive technology system which can help the navigation of visual impaired people in dynamically changing environments. However, there are two limitations with this system. First, the prototype weighs eight pounds. Second, the degradation of the RF signals inside buildings degrades the accuracy of the GPS localization.

Tee et al.[52] proposed a wireless assistive system using a combination of GPS, dead reckoning module and wireless sensor network for improved localization indoors and outdoors. The system is also designed to be light in weight. An important part of the system is the web-based system where system administrator can monitor and give assistance when required. The overview of their proposed system is depicted in Fig. 2. The proposed system consists of three components: SmartGuide hardware devices, mesh wireless sensor network and intelligent assistive navigation management system with the web-based system administrator monitoring system. The web based system enables r the system administrator to access and manage the system remotely with ease. The wireless sensor network is designed to be very low-power and fault tolerant by using a mesh network topology. Where the wireless sensor network has minimal delay in data relaying even with walls and obstacles, thus it is concluded the system would work reliably in indoor environments.

Dougles et al. [37] examined behaviors of 10 visually impaired adults carrying out a copytyping task, the adults with visual impairments had inefficient working habits; such as poor touch typing, rare use of short cut keys, lack of adjustment of equipment, furniture and copy material. Authors in [37] recommended more proactive and creative strategies for improving skills and work techniques and adjusting positions of equipment and furniture.

Dobransky et al. [38] presented technical accessibility problems as one of the extra barriers that people with a visual impaired people need to tackle. On other study by Hackett et al. [39], studied about usability of access for web site accessibility; six visually impaired computer users (two men and

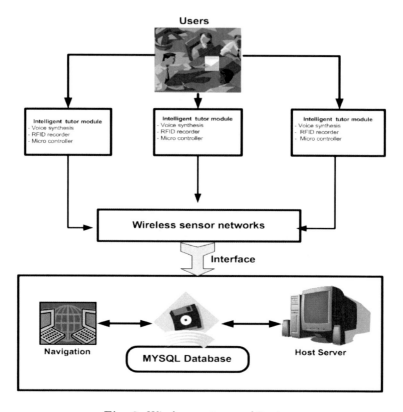

Fig. 2. Wireless system architecture

four women, aged 40 and older) were examined. In their study, six think-aloud assessments were conducted to compare access with the standard web display, with the goal of improving the design of access by identifying usability issues. The results showed that the visual impaired people were more satisfied with transformed web site. Also, Baye et al. [40] found that especially navigation and screen reading posed problems for blind internet users. Another study related to access and use of web by visually impaired students recommends web-site designers to be sensitive to the needs of visually-impaired users when preparing their information sites.

In early user interfaces a person interacting with a computer used a plug board or hard wired commands and saw the output on a pen recorder or electro-mechanical counter. This has rapidly developed into the popular personal computer interface seen today, i.e. a multimedia capability within a Windows, Icon, Mouse and Pointer environment. There is an ever increasing sophistication in the visual processing available at the user interface. The increasing visual sophistication of the human computer interface has undoubtedly improved ease of use. However, it has also reduced access for people with visual disabilities. Sophisticated methods now need to be employed to enable modern Windows, Icon, Mouse and Pointer interfaces to be used by visual impaired people. Most WWW documents are still written to be intelligible without the pictures. However, this is principally to cater for people with low performance computers or slow network links, rather than to provide alternative access for visual impaired people. We should therefore expect the situation for visual impaired people to worsen as the processing power and networking speed of personal computers grows.

The graphical user interface is the most widespread user frontend for applications today and the dominant user interface for websites on the internet. But graphical elements like windows and buttons are designed for sighted users; visual impaired people can neither perceive nor use them. To compensate for this disadvantage, several countries have passed laws to enforce accessibility of websites for handicapped users. Therefore, there is a discrimination against people with disabilities in all aspects of daily life, including education, work and access to public buildings. Some governments like Swiss act have required the government to provide access to all internet services for people with disabilities. Most approaches of developing a new user interface for visual impaired people pursue promising ideas, but lack quantitative empirical evaluation of the benefits for the visual impaired people. Stefan et al. [41] provided a direct quantitative comparison between two interface types.

In order to successfully use navigation, users first have to form a mental model of the underlying navigation space. In order to form a mental model, users have to make sense of the grouping and labeling of navigation items. This sense-making is based on cognitive processes coupled to sensory input. The sensory input leading to mental model formation is completely different for sighted and for visual impaired people: For sighted users, visual grouping and non-audible attributes (such as text size, color and formatting) yield a great deal of insight into the intended grouping of navigation items and hence communicate the intended structure of the navigation space. visual impaired people have to form their mental model of the navigation space based solely on the linear representation of navigation items and audible cues added to visually represented content.

Some qualitative insights and examples from exploratory studies that explains visual impaired people problems with a GUI:

- visual impaired people cannot guess relationships between primary and secondary navigation items if they are expressed visually,
- visual impaired people often only learn by chance which navigation options are recurrent on every page,
- it takes visual impaired people a lot of time to explore navigation options: Whereas trial and error is a valid navigation strategy for sighted users, we never observed blind users applying this strategy.

Thus, theories of navigation and qualitative insights from our exploratory studies demonstrate the necessity of a user interface that is enhanced regarding navigation through and interpretation of the linear representation of content, using user interface elements especially designed for the visual impaired people. The GUI has been introduced to make use of human perceptual abilities in order to reduce demands on working memory: Instead of learning hundreds of commands by heart, users could see all available commands for a selected object and reserve their cognitive resources mainly for decision making. The only reason sighted users can instantly use a computer program or website without training is their ability to see data objects represented on the screen together with the corresponding commands.

Stefan et al.[41] used a text-only interface in order to free visual impaired people from having to listen to additive clutter from visual impaired people interface elements, and impose a structure consisting of additive cues. The most serious limitation of arose from time constraints: Welcoming users to the usability lab, letting them get acquainted with the test setting and recording task execution of two tasks already took well over an hour.

6.2   Web-Based City Maps Support Visually Impaired People

Mental mapping of spaces and of the possible paths for navigating these spaces is essential for the development of efficient orientation and mobility skills. Most of the information required for this mental mapping is gathered through the visual channel. Visually impaired people lack this information, and consequently they are required to use compensatory sensorial channels and alternative exploration methods. Although digital maps become more and more popular, they still belong to those elements of the web which are not accessible to all user groups. So far, visually impaired and especially blind people do not get the chance to fully discover web-based city maps. One way to make web-based maps accessible for people with visual impairments is to describe the map in words. The goal is to develop a semantic description of the urban space that can be generated automatically so that worldwide deployment is possible [54]. Web-based maps support virtual and live discovery of cities, provide spatial information and improve orientation. Web-based city maps can be accessed either from home with a PC or on tour thanks to mobile devices, which may also be connected to GPS. However, this is not the case for blind and visually impaired people. To properly access and view digital maps is often challenging for this user group. Therefore, a methodology based on geographic information technologies is developed to automatically generate a textual spatial description of the map and a user specified interface respecting the requirements of users with visual impairment.

The work in [42] is based on the assumption that the supply of appropriate spatial information through compensatory sensorial channels, as an alternative to the visually impaired people channel, may help to enhance visually impaired people's ability to explore unknown environments [43] and to navigate in real environments. The area of touch-based human-computer interaction has grown rapidly over the last few years. A range of new applications has become possible now that touch can be used as an interaction technique. Most research in this area has been concentrated on the use of force feedback devices for the simulation of contact forces when interacting with simulated objects.

Research on orientation and mobility in known and unknown spaces visually impaired people indicates that support for the acquisition of spatial mapping and orientation skills should be supplied at two main levels: perceptual and conceptual. At the perceptual level, visual information shortage is compensated by other senses, e.g., tactile or auditory information. Tactile and touch information appear to be a main resource for supporting appropriate spatial performance for visually impaired people using a new concept *active touch*. Researches continued this line of research, active touch, focusing on the relationship between vision and touch information for visually impaired people in the context of philosophy and esthetics. Active touch can be described as a concomitant excitation of receptors in the joint and tendons along with new and changing patterns in the skin. Moreover, when the hand is feeling an object, the movement or angle of each joint from the first phalanx of each finger up to the shoulder and the backbone makes its contribution. Touch information is commonly supplied by the hands' palm and fingers for fine recognition of object form, texture, and location, by using low-resolution scanning of the immediate surroundings, and by the feet regarding surface information. The auditory channel supplies complementary information about events, the presence of other object such as machines or animals in the environment, or estimates of distances within a space. The olfactory channel supplies additional information about particular contexts or about people.

At the conceptual level, the focus is on supporting the development of appropriate strategies for an efficient mapping of the space and the generation of navigation paths. For instance, Jacobson [44] described the indoor environment familiarization process by visually impaired people as one that starts with the use of a perimeter-recognition-tactic-walking along the room's walls and exploring objects attached to the walls, followed by a grid-scanning-tactic-aiming to explore the room's interior. Research indicates that people use two main spatial strategies: route and map strategies. Route strategy is based on linear recognition of spatial features, while map strategy is holistic and encompasses multiple perspectives of the target space. Fletcher [45] showed that visual impaired people use mainly route strategy when recognizing and navigating new spaces.For a long period of time the information-technology devices that provided the visually impaired people with information before her/his arrival to an environment were mainly verbal descriptions, tactile maps and physical models. Today, advanced computer technology offers new possibilities for supporting rehabilitation and learning environments for people with disabilities such as visually impaired people. Over the past 30 years, visually impaired people have used computers supported by assistive technology such as audio outputs. The Optacon was one of the first devices to employ a matrix of pins for tactile-vision substitution and was the first device of this kind to be developed as a commercial product [46]. The input to the device is a 6x24 array of photosensitive cells, which detects patterns of light and dark as material is moved underneath. The display part of the device is a 6x24 array of pins on which the user places their fingertip. The output of the camera is represented by vibrating pins on the tactile display. Audio assistive technology includes text-to-speech software and print-to-speech reading machines (e.g., Kurzweil's Reading Machine invented in 1976). The exploration and learning of a new environment by visually impaired people is a long process, and requires the use of special information-technology aids. There are two types of aids: passive and active. Passive aids provide the user with information before his/her arrival to the environment, for instance [47]. Active aids provide the user with information in situ, for instance [48]. Research results on passive and active aids indicate a number of limitations that include erroneous distance estimation, underestimation of component sizes, low information density, or symbolic representation misunderstanding. Basically, Virtual reality has been a popular paradigm in simulation- based training, in the gaming and entertainment industries. It has also been used for rehabilitation and learning environments for people with sensory, physical, mental, and learning disabilities [49, 50]. Recent technological advances, particularly in touch interface technology, enable visually impaired people to expand their knowledge by using an artificially made reality built on touch and audio feedback. Research on the implementation of touch technologies within VEs has reported on its potential for supporting the development of cognitive models of navigation and spatial knowledge with visually impaired people [51]. The exploration process by visually impaired people is collected mainly using the touch and audio channels.

6.3 Email Accessibility Support Visually Impaired People

Email is one of many desktop applications that have become increasingly portable and web-based. However, web-based applications can introduce many usability problems, and technology like Flash and AJAX can create problems for blind users [56]. The blind users are more likely to avoid something when they know that it will cause them accessibility problems, such as the problems often presented by dynamic web content. The 70-75% unemployment rate of individuals who are blind in the US makes it essential that they identify any email usability problems that could negatively impact blind users in the workplace [57].

Brian et al. [55] discussed results of usability evaluations of desktop and web-based email applications used by those who are blind. Email is an important tool for workplace communication, but computer software and websites can present accessibility and usability barriers to blind users who use screen readers to access computers and websites.

To identify usability problems that blind users have with email, 15 blind users tested seven commonly used email applications. Each user tested two applications, so each application was tested by three to five users. From the results, we identify several ways to improve email applications so that blind people can use them more easily. The findings of this study should also assist employers as they make decisions about the types of email applications that they will use within their organizations. This exploratory research can serve as a focus for more extensive studies in the future.

7   Conclusion and Challenges

This chapter discussed and investigate the effects of the virtual reality technology on disabled people, especially blind and visually impaired people in order to enhance their computer skills and prepare them to make use of recent technology in their daily life. As well as, they need to advance their information technology skills beyond the basic computer training and skills. In addition, we explained what best tools and practices in information technology to support disabled people such as deaf-blind and visual impaired people in their activities such as mobility systems, computer games, accessibility of e-learning, web-based information system, and wearable finger-braille interface for navigation of deaf-blind. Moreover, we showed how physical disabled people can benefits from the innovative virtual reality techniques and discuss some representative examples to illustrate how virtual reality technology can be utilized to address the information technology problem of blind and visual impaired people.

The main challenge on VR technique is to enhance accuracy, reduce computational cost, and to improve the proposed technique toward tracking applications. There are many challenge points must covered by any VR systems as, supports accurate registration in any arbitrary unprepared environment, indoors or outdoors. Allowing VR systems to go anywhere also requires portable and wearable systems that are comfortable and unobtrusive. A VR system should track everything: all other body parts and all objects and people in the environment. Systems that acquire real-time depth information of the surrounding environment, through vision-based and scanning light approaches, represent progress in this direction. New visualization algorithms are needed to handle density, occlusion, and general situational awareness issues. Many concepts and prototypes of VR applications have been built but what is lacking is experimental validation and demonstration of quantified performance improvements in a VR application. Such evidence is required to justify the expense and effort of adopting this new technology. Basic visual conflicts and optical illusions caused by combining real and virtual require more study. Experimental results must guide and validate the interfaces and visualization approaches developed for VR systems. Although many VR applications only need simple graphics such as wireframe outlines and text labels, the ultimate goal is to render the virtual objects to be indistinguishable from the real.

This must be done in real time, without the manual intervention of artists or programmers. Some steps have been taken in this direction, although typically not in real time. Since removing real objects from the environment is a critical capability, developments of such Mediated Reality approaches are needed. Technical issues are not the only barrier to the acceptance of VR applications. Users must find the technology socially acceptable as well. The tracking required for information display can also be used for monitoring and recording.

References

1. Lacey, G., Dawson-Howe, K.M.: The application of robotics to a mobility aid for the elderly blind. Robotics and Autonomous Systems 23(4), 245–252 (1998)
2. Lahav, O., Mioduser, D.: Multi-sensory virtual environment for supporting blind persons acquisition of spatial cognitive mapping, orientation, and mobility skills. In: The Third International Conference on Disability, Virtual Reality and Associated Technologies, Alghero, Sardinia, Italy (2000)
3. Lányi, S., Geiszt, Z., Károlyi, P., Magyar, Á.T.V.: Virtual Reality in Special Needs Early Education. The International Journal of Virtual Reality 5(4), 55–68 (2006)
4. Dowling, J., Maeder, A., Boles, W.: Intelligent image processing constraints for blind Mobility facilitated through artificial vision. In: Lovell, B.C., Campbell, D.A., Fookes, C.B., Maeder, A.J. (eds.) 8th Australian and New Zealand Intelligent Information Systems Conference, December 10-12, Macquarie University, Sydney (2003)
5. Kuhlen, T., Dohle, C.: Virtual reality for physically disabled people. Compuf. Bid. Med. 25(2), 205–211 (1995)
6. Klinger, E., Weiss, P.L., Joseph, P.A.: Virtual reality for learning and rehabilitation. In: Rethinking Physical and Rehabilitation Medicine Collection de Lacadémie Européenne de Médecine de RéAdaptation, Part III, pp. 203–221 (2010)
7. D'Atri, E., Medaglia, C.M., Serbanati, A., Ceipidor, U.B., Panizzi, E., D'Atri, A.: A system to aid blind people in the mobility: a usability test and its results. In: The 2nd International Conference on Systems, ICONS 2007, Sainte-Luce, Martinique, France, April 22-28 (2007)
8. Ugo, B.C., D'Atri, E., Medaglia, C.M., Serbanati, A., Azzalin, G., Rizzo, F., Sironi, M., Contenti, M.: A RFID System to help visually impaired people in mobility. In: EU RFID Forum 2007, Brussels, Belgium, March 13-14 (2007)
9. Archambault, D.: People with disabilities: Entertainment software accessibility. In: Miesenberger, K., Klaus, J., Zagler, W.L., Karshmer, A.I. (eds.) ICCHP 2006. LNCS, vol. 4061, pp. 369–371. Springer, Heidelberg (2006)
10. Gaudy, T., Natkin, S., Archambault, D.: Pyvox 2: An audio game accessible to visually impaired people playable without visual nor verbal instructions. T. Edutainment 2, 176–186 (2009)
11. Delić, V., Vujnović Sedlar, N.: Stereo presentation and binaural localization in a memory game for the visually impaired. In: Esposito, A., Campbell, N., Vogel, C., Hussain, A., Nijholt, A. (eds.) Second COST 2102. LNCS, vol. 5967, pp. 354–363. Springer, Heidelberg (2010)

12. Gaudy, T., Natkin, S., Archambault, D.: Pyvox 2: An audio game accessible to visually impaired people playable without visual nor verbal instructions. T. Edutainment 2, 176–186 (2009)
13. Gardenfors, D.: Designing sound-based computer games. Digital Creativity 14(2), 111–114 (2003)
14. Cardin, S., Thalmann, D., Vexo, F.: Wearable system for mobility improvement of visually impaired people. The Visual Computer: International Journal of Computer Graphics 23(2) (January 2007)
15. Van den Doel, K.: SoundView: sensing color images by kinesthetic audio. In: Proceedings of the International Conference on Auditory Display, ICAD 2003, pp. 303–306 (2003)
16. Raisamo, R., Ki, S.P., Hasu, M., Pasto, V.: Design and evaluation of a tactile memory game for visually impaired children. Interacting with Computers 19, 196–205 (2003)
17. Sodnik, J., Jakus, G., Tomazic, S.: Multiple spatial sounds in hierarchical menu navigation for visually impaired computer users. International Journal on Human-Computer Studies 69, 100–112 (2011)
18. Archambault, D.: The TiM Project: Overview of Results. In: Miesenberger, K., Klaus, J., Zagler, W.L., Burger, D. (eds.) ICCHP 2004. LNCS, vol. 3118, pp. 248–256. Springer, Heidelberg (2004)
19. Armstrong, H.L.: Advanced IT education for the Vision impaired via e-learning. Journal of Information Technology Education 8, 223–256 (2009)
20. Harper, S., Goble, C., Stevens, R.: Web mobility guidelines for visually impaired surfers. Journal of Research and Practice in Information Technology 33(1), 30–41 (2001)
21. Kelley, P., Sanspree, M., Davidson, R.: Vision impairment in children and youth. The Lighthouse Handbook on Vision Impairment and Vision Rehabilitation Set, 1111–1128 (2001)
22. Shore, D.I., Klein, R.M.: On the manifestations of memory in visual search. Spatial Vision 14(1), 59–75 (2000)
23. Fenrich, P.: What can you do to virtually teach hands-on skills? Issues in Informing Science and Information Technology 2, 47–354 (2005)
24. Baguma, R., Lubega, J.T.: Web design requirements for improved web accessibility for the blind. In: Fong, J., Kwan, R., Wang, F.L. (eds.) ICHL 2008. LNCS, vol. 5169, pp. 392–403. Springer, Heidelberg (2008)
25. Asakawa, C.: What's the web like if you can't see it? In: Proceedings of the International Cross-Disciplinary Workshop on Web Accessibility, W4A 2006 (2005), doi:10.1145/1061811.1061813
26. Huang, C.J.: Usability of e-government web sites for PWDs. In: Proceedings of the 36th Annual Hawaii International Conference on System Sciences, Track 5, vol. 5 (2003)
27. Shi, Y.: The accessibility of queensland visitor information center's websites. Tourism Management 27, 829–841 (2006)
28. Takagi, H., Asakawa, C., Fukuda, K., Maeda, J.: Accessibility designer: visualizing usability for the blind. In: Proceedings of the ACM SIGACCESS Conference on Computers and Accessibility, ASSETS 2004, pp. 177–184 (2004)
29. Chiang, M.F., Cole, R.G., Gupta, S., Kaiser, G.E., Starren, J.B.: Computer and world wide web accessibility by visually disabled patients: problems and solutions. Survey of Ophthalmology 50(4) (2005)

30. Royal National Institute for the Blind, Communicating with blind and partially sighted people. Peterborough (2004)

31. Burgstahler, S., Corrigan, B., McCarter, J.: Making distance learning courses accessible to students and instructors with disabilities: A case study. Internet and Higher Education 7, 233–246 (2004)

32. Velazquez, R.: Wearable assistive devices for the blind. In: Lay-Ekuakille, A., Mukhopadhyay, S.C. (eds.) Wearable and Autonomous Biomedical Devices and Systems for Smart Environment Interface for Navigation of Deaf-Blind in Ubiquitous Barrier-Free Space, Proceedings of the 10th International Conference on Human- Computer: Issues and Characterization. LNEE, vol. 75, ch. 17, pp. 331–349 (2010)

33. Hirose, M., Amemiya, T.: Wearable finger-Braille interface for navigation of deaf-blind in ubiquitous barrier-free space. In: Hirose, M., Amemiya, T. (eds.) Proceedings of the 10th International Conference on Human-Computer Interaction (HCI 2003), Crete, Greece (June 2003)

34. Amemiya, T., Yamashita, J., Hirota, K., Hirose, M.: Virtual leading blocks for the deaf-blind: a real-time way-finder by verbal-nonverbal hybrid interface and high density RFID tag space. In: Proceedings of IEEE Virtual Reality, Chicago, Il, USA, pp. 165–172 (2004)

35. Dipietro, L., Sabatini, A.M., Dario, P.: A survey of glove-based systems and their applications. IEEE Transactions on Systems, Man, and Cybernetics-Part C: Applications and Reviews 38(4), 461–482 (2008)

36. Srikulwong, M., O'Neill, E.: A comparison of two wearable tactile interfaces with a complementary display in two orientations. In: Proceedings of 5th International Workshop on Haptic and Audio Interaction Design, HAID 2010, pp. 139–148 (2010)

37. Douglas, G., Long, R.: An observation of adults with visual impairments carrying out copy-typing tasks. Behaviour and IT 22(3), 141–153 (2003)

38. Dobransky, K., Hargittai, E.: The disability divide in internet access and use. Information, Communication and Society 9(3), 313–334 (2006)

39. Hackett, S., Parmanto, B.: Usability of access for Web site accessibility. Journal of Visual Impairment and Blindness 100(3), 173–181 (2006)

40. Bayer, N.L., Pappas, L.: Case history of blind testers of enterprise software. Technical Communication 53(1), 32–35 (2003)

41. Leuthold, S., Bargas-Avila, J.A., Opwis, K.: Beyond web content accessibility guidelines: Design of enhanced text user interfaces for blind internet users. International Journal of Human-Computer Studies 66, 257–270 (2008)

42. Lahav, O., Mioduser, D.: Haptic-feedback support for cognitive mapping of unknown spaces by people who are blind. International Journal of Human-Computer Studies 66(1), 23–35 (2008)

43. Mioduser, D.: From real virtuality in Lascaux to virtual reality today: cognitive processes with cognitive technologies. In: Trabasso, T., Sabatini, J., Massaro, D., Calfee, R.C. (eds.) From Orthography to Pedagogy: Essays in Honor of Richard L. Venezky. Lawrence Erlbaum Associates, Inc., New Jersey (2005)

44. Jacobson, W.H.: The art and science of teaching orientation and mobility to persons with visual impairments. American Foundation for the Blind AFB, New York (1993)

45. Fletcher, J.F.: Spatial representation in blind children: development compared to sighted children. Journal of Visual Impairment and Blindness 74(10), 318–385 (1980)

46. Wies, E.F., Gardner, J.A., O'Modhrain, M.S., Hasser, C.J., Bulatov, V.L.: Web-Based Touch Display for Accessible Science Education. In: Brewster, S., Murray-Smith, R. (eds.) Haptic HCI 2000. LNCS, vol. 2058, pp. 52–60. Springer, Heidelberg (2001)
47. Espinosa, M.A., Ochaita, E.: Using tactile maps to improve the practical spatial knowledge of adults who are blind. Journal of Visual Impairment and Blindness 92(5), 338–345 (1998)
48. Easton, R.D., Bentzen, B.L.: The effect of extended acoustic training on spatial updating in adults who are congenitally blind. Journal of Visual Impairment and Blindness 93(7), 405–415 (1999)
49. Schultheis, M.T., Rizzo, A.A.: The application of virtual reality technology for rehabilitation. Rehabilitation Psychology 46(3), 296–311 (2001)
50. Standen, P.J., Brown, D.J., Cromby, J.J.: The effective use of virtual environments in the education and rehabilitation of students with intellectual disabilities. British Journal of Education Technology 32(3), 289–299 (2001)
51. Jansson, G., Fanger, J., Konig, H., Billberger, K.: Visually impaired persons' use of the Phantom for information about texture and 3D form of virtual objects. In: Proceedings of the PHANTOM Users Group Workshop, vol. 3 (December 1998)
52. Tee, Z.H., Ang, L.M., Seng, K.P., Kong, J.H., Lo, R., Khor, M.Y.: Web-based caregiver monitoring system for assisting visually impaired people. In: Proceedings of the International Multiconference of Engineers and Computer Scientists, IMECS 2009, vol. I (2009)
53. Helal, A., Moore, S.E., Ramachandran, B.: Drishti: An integrated navigation system for visually impaired and disabled. In: Proceedings of the 5th International Symposium on Wearable Computers, pp. 149–156 (2001)
54. Wasserburger, W., Neuschmid, J., Schrenk, M.: Web-based city maps for blind and visually impaired. In: Proceedings REAL CORP 2011 Tagungsband, Essen, May 18-20 (2011), http://www.corp.at
55. Wentz, B., Lazar, J.: Usability evaluation of email applications by blind users. Journal of Usability Studies 6(2), 75–89 (2011)
56. Borodin, Y., Bigham, J., Raman, R., Ramakrishnan, I.: What's new? Making web page updates accessible. In: Proceedings of 10th International ACM SIGACCESS Conference on Computers and Accessibility, ASSETS 2008, pp. 145–152 (2008)
57. National Federation of the Blind, Assuring opportunities: A 21st Century strategy to increase employment of blind Americans, http://www.nfb.org (retrieved May 5, 2011)
58. Tampokme: the audio multi-players one-key mosquito eater, http://www.ceciaa.com/
59. Egg Hunt, LWorks free game, http://www.l-works.net/egghunt.php
60. SimCarriere, http://www.simcarriere.com/
61. K-Chess Advance, ARK Angles, http://www.arkangles.com/kchess/advance.html
62. Brewer, J., Jacobs, I.: How people with disabilities use the Web?, http://www.w3.org (retrieved on May 15, 2011)
63. Chisholm, W., Vanderheiden, G., Jacobs, I. (eds.): Web content accessibility guidelines 1.0, http://www.w3.org (retrieved on May 15, 2011)
64. Meijer, P.B.L.: The vOICe Java applet-seeing with sound, http://www.seeingwithsound.com/javoice.htm (retrieved on May 15, 2011)

Author Index

Subject Index